With very best wishes for your new appointment and for the future.

Jane & Graham

December 1986

THE BLACK BATTLEFLEET

ADMIRAL G. A. BALLARD

THE BLACK BATTLEFLEET

Nautical Publishing Company Limited, Lymington
and the Society for Nautical Research, Greenwich

DESIGNED AND PRODUCED BY EDITA S.A. LAUSANNE

First published in 1980 by

Nautical Publishing Company Limited, Lymington, Hampshire and The Society for Nautical Research, Greenwich

in association with
Nautical Publishing Company Limited
Nautical House, Lymington, Hampshire, England
and
George G. Harrap & Co Ltd, London

ISBN 0 245 53030 4

Printed in Italy

CONTENTS

FOREWORD

by

Admiral of the Fleet the Earl Mountbatten of Burma, KG, PC, GCB, OM, GCSI, CIE, GCVO, DSO, FRS

Admiral Ballard's writings are historically important and unique. They cover the development of warships at a critical period and deal with the role that was played by the British Admiralty and Royal Navy. His work first appeared as a series of articles in the Society for Nautical Research's journal the *Mariner's Mirror* and as Patron of the Society I am delighted that this learned and active body has been able to instigate this wonderful re-construction of that part of his work which covers the major ships of the period.

In particular I am glad that the book includes HMS Royal Alfred in which my father, the late Admiral of the Fleet the First Marquess of Milford Haven, served as a Midshipman on the North America and West Indies Station from 1869 to 1874. In his recollections, written privately for the family, my father describes the 'RA', as she was known, as follows: —

'*Royal Alfred* was an old line-of-battle ship, never completely finished. Her upper deck was removed and a few, very thick plates of cast iron secured to her sides amidships, constituting her an ironclad. As armament she was given ten 9-inch RML guns, five a side on the main deck, and eight 7-inch, four on the main deck, forward and aft, two on the forecastle and two on the quarterdeck. They were shaped like the old soda water bottles and were entirely worked by hand on huge recoil mountings with fat rope breechings. The ship retained her full rig and sailed very well. A single screw, which could be hoisted up, was driven by an engine of 800 HP, and the single funnel shut down like a telescope.'

My father then goes on to describe the Gun Room where he and his contemporaries lived onboard: —

'The Gun Room was on the lower deck, that is half under water, with three small scuttles through a very thick ship's side and consequently so dark that lamps had to be burnt all day. These soon got out of order or smashed so we were reduced to "Pusser's dips", small tallow candles supplied for the ship's lanterns; these were stuck into empty "Squaro" bottles, square black Gin bottles, of which our senior members kept up a continual supply.'

Despite these privations my father enormously enjoyed his 4 ½ years onboard the *Royal Alfred*, which during that time had steamed some 40,610 miles, and he and the Admiral's Secretary organised an annual 'RA' Dinner in London in 1887 which they kept going until the war put an end to it in 1914.

The editors of this book, Dr Nicholas Rodger of the Public Records Office and Mr George Osbon of the National Maritime Museum, are both members of the Society for Nautical Research and have done their work with great skill. Dr Rodger was mainly responsible for editing the text and Mr Osbon used his expert knowledge of the available illustrations to make this a book which is not only a valuable historical document but one which will capture the imagination of all its readers.

Mountbatten of Burma

A. F.

EDITORS' PREFACE

The contents of this book originally appeared in a series of articles contributed by Admiral Ballard to the *Mariner's Mirror*, the journal of the Society for Nautical Research, of which he was one of the first members. From the first issue of the journal in 1911 until his death in 1948, Admiral Ballard wrote notes and articles on every period and subject of naval history from ancient Egypt to the Great War. Of all the numerous products of his fertile brain and fluent pen, the largest and undoubtedly the most valuable contribution was made by the several series of articles he wrote on the warships of the mid-Victorian Royal Navy, begun in 1929 and continued until his death. The first of these series, 'The Black Battlefleet of 1870', forms the core of this book. It was followed by a further series on the unarmoured warships of that period, and finally by descriptions of a number of later battleships of the 'transitional' era, of which we have selected three to complete this book.

When Admiral Ballard first presented 'The Black Battlefleet' to the readers of a journal then almost entirely concerned with maritime history of earlier periods, he felt it necessary to explain, and even excuse, an account of warships built scarcely fifty years before, some of them still afloat, and most of them remembered by at least a few of his readers. A century after the Black Battlefleet, there can be no doubt of the value of his work. Even in 1929, there were not many men left who were intimately familiar with the Navy of steam and sail, with wooden hulls, muzzle-loaders and the first ironclads at sea. But Ballard had more to offer than the good fortune of a long life and a retentive memory. He was a careful and accurate observer, and an evocative writer in an idiom which was beginning to seem pleasantly old-fashioned and idiosyncratic even in his lifetime. Though he could be discursive, he was a precise and acute historian, who worked for many years in the archives of the Admiralty (not then transferred to the Public Record Office and the National Maritime Museum, and accessible only with special permission) accumulating the technical information and detailed histories with which his work is replete. He was an excellent draughtsman, whose plans and diagrams are not the least valuable of his products. In this volume we have reproduced as many of them as possible, including the coloured ships' draughts, which have had to be re-drawn after the monochrome plates of the *Mariner's Mirror*, the originals being now lost. The ships' plans were redrawn by Miss Helen Downton with technical assistance from Mr Alan Viner of the National Maritime Museum and other illustrations were redrawn by Sister Mary Regis CRW and Mr Barrie Hole. To these he added many old photographs taken by himself and others. Above all, he had the historical sense which so many old men writing of their youth lack. He was not a mere memoirist, relying on fallible memory; he was an historian, supplementing thorough research with the fruits of a vivid recollection. He had the sense to appreciate that the commonplace activities of his professional life were already becoming forgotten curiosities worthy of description. To this we owe his exact and expressive accounts of life in the mess-decks of a broadside battleship, of catting and fishing anchor, hoisting out boomboats or sending down topmasts; things so often mentioned in other sources, and so seldom described in detail.

It is this which gives Ballard's work its unique value; the combination of the careful history of an era of rapid change, with the graphic evocation of what it was like to serve in the Royal Navy in the first ironclads, when the methods as well as the traditions of the sailing Navy were still a way of life. A modern historian might hope to equal Admiral Ballard's research, perhaps even his style, but he could never supply the personal acquaintance which Ballard brought to his subject. The original articles did not reach a large public, and are now accessible only in the larger libraries, but their merits have long deserved a wider readership.

In preparing the original articles for publication, we have tried to tamper with the author's text as little as possible. Nevertheless, certain alterations have been necessary to adapt the articles to their new rôle as chapters in a book. The introductory matter proper to separate articles is to a great extent superfluous in a connected narrative, and we have not hesitated to remove repetitions. These were especially prominent in the accounts of the passage of the Dardanelles in 1878, and of the bombardment of Alexandria in 1882, where substantially the same story was repeated for each ship involved. In these and similar cases we have amalgamated the author's accounts into a single passage. Occasionally Admiral Ballard slipped into more general historical digressions, not directly related to his theme, nor so firmly founded on knowledge of the subject. These we have cut down or removed altogether. With the exception of Chapters XVIII and XIX, which were originally one, the chapters appear in the same order and under the same titles as the original articles of which they are formed.

In the articles, some paragraphs consist largely of statistical information, especially on engines and spars. In the interests of economy of space we have reduced these to tables which will be found in an appendix. We have drawn on other sources, not available to the author, to complete these tables with much fresh information, in order to provide as complete a source of reference as possible. These tables in fact include the only additions of substance which we have made. Some fresh matter has also been inserted in the narrative of the final fates of some ships. This was most necessary for ships which were still afloat when Ballard wrote. In these cases we have added a paragraph of our own, distinguished by asterisks. Throughout the narrative, we have silently corrected a very few errors of fact, and altered chronological references ('sixty years ago') and remarks addressed particularly to the original readers of the *Mariner's Mirror*, which were not appropriate to the present age and context. Finally we have added a few explanatory footnotes at the end of each chapter.

In preparing this volume, we have been aided by many people. In particular the author's son, Brigadier J.A.W. Ballard, C.B.E., D.S.O., and grandson, Mr Roderick Ballard, B.Sc. of Exeter University, have gone to great pains to help us, and the latter has contributed a biographical note of his grandfather which follows this preface. Professor C. C. Lloyd and Mr Brian Dolley, successively Editors of the *Mariner's Mirror*, have worked tirelessly to further the project, as have the President of the Society for Nautical Research, Rear-Admiral E. F. Gueritz, C.B., O.B.E., D.S.C., and the members of the Council of the Society, especially Rear-Admiral P. W. Brock, C.B., D.S.O., himself an acknowledged authority on the period. The National Maritime Museum and the Public Record Office, to whose staff we respectively belong, have furnished valuable assistance. In particular we must thank the Director of the National Maritime Museum, Mr Basil Greenhill, C.M.G., F.S.A., and the Trustees, for making it possible to reproduce so many photographs from the Museum's incomparable collection. To Mr Richard Creagh-Osborne of the Nautical Publishing Company, and Mr Tim Chilvers of Edita S.A., the editors owe a debt which will be apparant to all who read this book.

G.A. Osbon
N.A.M. Rodger

BIOGRAPHICAL NOTE

Lieutenant Ballard, the only naval officer on the staff of General Prendergast during the storming of Mandalay in 1885-86. (NMM. B.8808)

My grandfather was the son of General J.A. Ballard, and was born in India, where he spent most of his childhood. He joined the training ship *Britannia* as a cadet in 1875, and showed early evidence of his abilities. Besides sea service in all types of ship and all quarters of the world, he fought on land in the Sudan, in China during the Boxer Rising, and in the Burmese War. The abilities he displayed had brought him to the rank of Captain by 1903.

These early years showed the development of the gifts as a writer which he demonstrated throughout his life. He twice won the Gold Medal of the Royal United Services Institution for essays written in 1897 and 1899. His essay on the best type of battleship for the Navy was considered by the critics to be a major contribution to the debate on warship design then taking place. Perusal of the family scrapbook reveals that before the First World War he frequently found himself giving lectures which demonstrate his sense of history and capacity for judgement without predjudice or bias. His experience of military life and ability to identify with the army clearly helped to give him a rounded view of strategic problems. It is therefore not surprising to hear of him giving a lecture on Naval and Military Co-operation to members of both Services in 1903.

Shortly before and during the Great War he received important appointments. In 1911 he became Director of the Operations Division on the Naval War Staff, and in 1914 was promoted Rear-Admiral. He commanded the East Coast patrols, and later became Dockyard Admiral at Malta. There he negotiated a pay rise for the dockyard workers, against considerable opposition at home, which earned him the reputation in the island of a man who cared for those over whom he had authority. One gets the impression that he was not easily persuaded to deviate from a course once his mind was made up, even though he suffered by his determination. Certainly his popularity at home waned considerably after the pay incident, but the standard of living of many Maltese families grew accordingly.

The Admiral had a distinguished career in the Navy, but it is what he wrote rather than what he did which has earned him a place in the history of naval affairs. The contents of this book testify to that. He also published three books during his retirement, all of which were very well received. In his *America and the Atlantic,* perhaps the best known of three, he argued the case for an increase in America's naval strength as a logical outcome of historical forces operating on the country.

My grandfather maintained his active interest in naval and military affairs until his death in 1948 at the age of 86. We of a different generation can still find much pleasure in reading of naval affairs of his day, and it is gratifying that this volume will allow his work to become more widely known.

Roderick Ballard

As Flag Officer in command of East Coast Patrols, Rear-Admiral G. A. Ballard, aboard the cruiser *St. George*, with his officers and staff in 1914. (NMM. B. 8818)

ADMIRAL BALLARD'S SERVICE

January 1875	Joined training ship *Britannia* as a naval cadet.
December 1876	Passed out of *Britannia.*
January 1877	Joined the ironclad *Resistance* of the Channel Squadron, taking passage to Lisbon to join his ship.
May 1877	Turned over to ironclad *Achilles* with the rest of the ship's company of the *Resistance:* Mediterranean.
June 1877	Promoted Midshipman.
February 1878	Passage of the Dardanelles.
January 1879	Malta for refit; Ballard in hospital with malaria.
September 1880	Appointed to the corvette *Tourmaline* of the Flying Squadron; 1880-2 sailed around the world.
June 1881	Promoted Acting Sub-Lieutenant.
July 1882	Returned to England after 21 months' absence; posted to the Royal Naval College at Greenwich for a six-month course, followed by courses at the gunnery training-ship *Excellent* and the torpedo training-ship *Vernon* at Portsmouth.
August 1883	Appointed to the torpedo boat depôt-ship *Hecla.*
October 1883	Passed for Lieutenant and promoted Sub-Lieutenant.
December 1883	Sailed in the *Hecla* for the Mediterranean. She took Marines to the Red Sea, where a large party of officers and men from the *Hecla,* including Ballard, were landed to fight on shore.
March 1884	Promoted Lieutenant in the field. Shortly afterwards ordered to join the battleship *Temeraire* at Malta. Meanwhile he served for a month in the corvette *Euryalus.* Later, while the *Temeraire* was refitting, he served briefly in the training barque *Cruiser.*
May 1885	Appointed to the gunboat *Woodlark* in the East Indies. Leaving the *Temeraire* at Port Said, he joined the *Woodlark* at Rangoon.
1885-6	During the Burmese War, the officers and men of the *Woodlark* commissioned the paddle steamer *Irrawaddy* for service up that river. Later, Ballard became Naval A.d.C. to General Prendergast, and took part in the storming of Mandalay.
June 1887	Appointed to the Coast Defence Ship *Gorgon* at Devonport.
October 1887	Re-appointed to the *Temeraire* in the Mediterranean.
August 1891	Appointed to the battleship *Audacious,* guardship at Hull.
December 1891	Appointed First Lieutenant of the gunboat *Linnet* on the China Station.
July 1895	Appointed to the battleship *Sans Pareil,* guardship at Sheerness.
October 1895	Appointed to command the destroyer *Janus* at Sheerness.
May 1896	Appointed to command the torpedo-gunboat *Renard* at Holyhead.
December 1897	Promoted Commander.
May 1898	Appointed Commander of the cruiser *Isis,* Channel Squadron.
February 1902	Appointed to the Naval Intelligence Division.
December 1903	Promoted Captain.
January 1906	Appointed Captain of the cruiser *Euryalus,* in reserve at Portsmouth, and lent for the War Course at Portsmouth.
May 1906	Transferred to command the cruiser *Royal Arthur,* also in reserve at Portsmouth.
June 1906	Appointed to command the cruiser *Terrible.*
August 1907	Appointed to command the cruiser *Hampshire,* Channel Squadron.
December 1909	Appointed to command the battleship *Commonwealth,* Home Fleet.
December 1910	Appointed to command the battleship *Britannia,* Home Fleet.
December 1911	Appointed Director of the Operations Division of the Naval War Staff.
May 1914	Promoted Commodore lst Class and appointed Admiral of Patrols.
August 1914	Promoted Rear-Admiral, flying his flag in the old cruiser *St. George* at Immingham.
September 1916	Senior Naval Officer and Admiral Superintendant at Malta.
February 1919	Promoted Vice-Admiral.
September 1919	Hauled down his flag at Malta.
June 1921	Retired.
March 1924	Promoted Admiral on the Retired List.

CHAPTER ONE

THE BLACK BATTLEFLEET

Some notes on the mid-Victorian transformation in battleship design

The British Naval squadron headed by the *Black Prince* saluting the French flag in honour of the Emperor's Fête Day at Cherbourg in September 1865. (NMM. 58/2582)

N.B. Footnotes are collected on page 258.

Ships of war proper — as distinguished from armed merchantmen — are structures devised for the sole purpose of upholding State policy in so far as that may be effected by using, or threatening to use, force on the water. In the fulfilment of that purpose they often played a prominent part in shaping the destinies of great nations in the past, and the varying changes in their designs as modified from period to period reflect the ideas under which the constructive science of the day laboured to find the surest forms of producing water-borne power.

In the history of those changes as a special branch of knowledge — reaching back with some gaps to early classic times — no chapter has a more arresting importance than that during which Queen Victoria occupied the British Throne. For although the occasions calling for any great use of sea strength in support of the policy of any prominent Power during that reign were comparatively few, it chanced to coincide with the arrival of the iron and steel age, whereby certain of the fundamental practices connected for hundreds of years with the art of the shipbuilder were swept out of existence in the course of a single generation. Before that age the design of the specially built ocean-going fighting machine of any given date had been the product of a slow but very consistent process of evolution following certain definite lines in its main ideas. Then this process suddenly struck out in completely fresh directions. When Victoria was crowned the battleship was a wooden structure in frame and shell, propelled by a natural external agency, and armed solely with a large number of light hand-served weapons carried within the main hull and pointed through its sides, just as she had been since the days of the Tudors. But quite ten years before the death of Victoria the battleship had become an all-metal structure, propelled by an artificially generated internal agency and armed not only with heavy power-worked ordnance mounted outside and above the hull altogether, but with weapons for under-water attack.

Cruise of the Reserve Fleet to Sicily led by the *Agincourt* and *Black Prince*, under the command of Rear-Admiral Sir Astley Cooper Key. Because the First Lord of the Admiralty was on board, he struck his flag in favour of the Admiralty flag. (NMM. 58/3122)

Rapidly, however, though these drastic departures from ancient usage were carried out in comparison with any previous evolution in naval design, the transformation was not completed at one step but through a series of stages; beginning in the early 'fifties as regards the change in motive power, the early 'sixties as regards the change in structural material, and the middle 'sixties as regards redisposition of armament, it may be said to have finished in 1884 with the launch of the last square-rigged ships of the Navy above sloop size, for before that year the principle of an internally mounted armament had been abandoned, and the last timber hulls were about fifteen years old.

The reign of Victoria began in 1837 and ended in 1901. Its middle years therefore were 1869-70, and where the British seagoing battlefleet is concerned — as distinct from coast defence craft and smaller classes — these same years happened to mark with some accuracy the half-way stage of the great transformation in all three of its main lines of movement. For instance, as regards structural material, some of the ships were wooden hulled and some iron. And as regards motive power all were provided with the equipment of both internal and external propulsion, for every vessel carried steam engines in conjunction with an ample sail plan. And lastly in the matter of disposition and nature of armament the old principle of internally mounted hand-worked guns was found side by side with the new principle of upper deck weapons worked by power. Twenty years earlier there were no iron hulls and no engines. Twenty years later there were no wooden hulls and no sails.

As extensive advances in design marked almost every new ship laid down during this transitional epoch the old system of rating by the number of guns carried was necessarily abandoned; for under the new conditions the most powerful vessels carried the fewest. But the Admiralty went much further than that, and discarded also the more general classification hitherto connoted by the terms ship-of-the-line and frigate, without having decided what was to take their place. From the long delay they seem to have found the problem perplexing; and successive Boards gave up the attempt to group them under distinguishing titles of any kind until after fifteen years of deliberation they adopted the designations 'battleship' and 'cruiser'. During the interval the new and really important vessels of the Navy remained un-grouped under any established system of nomenclature whatever; as may be seen from the Navy Lists of the period, in which their appellations are so

H.M.S. *Prince Consort* at Simons Town about 1864 with her original double topsail rig, and wearing the red ensign. (Photograph from the Cape Archives).

variable that not merely ships of the same general type but even those built from the same plans and armed exactly alike are often entered under different designations.

In spite of the variety of individual designs however, a perfectly clear cut general classification of the fleet of 1870 is easily attained on the basis of war functions, and every vessel above 1000 tons displacement can be legitimately described under the modern system as either a battleship or a cruiser; except the few coast defence ships and the two freak sloops. All the armoured units were battleships in the tactical sense, and all the unarmoured were cruisers. For the sake of convenience those terms may be used here.

If the five ships that were on the stocks are included, the British ocean-going battlefleet of 1870 consisted of exactly thirty vessels 'fit to lie in a line', under the then existing conditions of naval warfare; that is to say, armed and protected sufficiently to be more than a match for any unarmoured ship afloat under a foreign flag. A certain number of the old two-deckers had not yet gone to the breakers' yards; but after the experiences of a vessel of this type at Lissa it was quite certain that their only possible rôle in war had become that of heavy cruisers. No longer could they claim membership in the honourable company of the line, which for two hundred years had been formed entirely of their class. And as this was the middle of the great era of transformation the thirty successors which had stepped into their front rank places were not nearly so uniform a collection of fighters in equipment or weight; among which twenty different scales of armament found a place, and eighteen different hull models. Turret ships, central battery ships and broadside ships were all included. Four designs of bow found a place and three of stern. Eight different thicknesses of armour were spread over the sides of the fleet in just twice as many different plans in the details of individual vessels. Trunk engines, direct acting engines and return connecting-rod engines were each represented in the machinery; with single screws, twin screws, disconnecting screws and hoisting screws. It was only in the comparatively unimportant matter of upper gear that any general uniformity existed in this force, for here alone was retained a long established standard of the past. Twenty-three of its units were ship-rigged in accordance with the Admiralty regulation plan, only differing from each other in scale. Three of the remainder had five masts and four were barques; but even in these the gear on the yarded masts corresponded rope for rope in fitting and lead exactly with that on the fore and main of their consorts.

Of the wooden hulled vessels the four *Caledonias,* the *Royal Alfred,* the *Zealous* and the *Repulse* were all originally laid down as 91-gun ships of the old type; but were converted on the stocks and completed as single decked armoured vessels, in which process the *Caledonias* and *Royal Alfred* were lengthened. But the *Lord Warden, Lord Clyde,* and *Pallas* were designed as armoured ships from the first. This group of eleven timber-built but iron-armoured high sea fighting ships occupies a position unique in the annals of British naval architecture.

The six largest vessels in the whole fleet were among the first to be built, and in their case advantage was taken of the rigidity of iron hulls to introduce longitudinal dimensions far exceeding anything previously imagined in the Navy, with the idea of carrying on a single deck offensive power equivalent in weight to that which had been previously carried on two or even three; thereby keeping the centre of gravity low in plating the sides. The fine lines also facilitated speed. Later a partial reaction set in through the belief that speed was less important than handiness for a battleship; and when it was found practicable by

H.M.S. *Lord Clyde* after her launch at Pembroke Dock. (NMM. 58/2320)

improved designs to carry a sufficiently formidable armament on a single deck in a shorter model, provided that a great beam was retained, a reduction in length followed in all new ships. This reversal of development was not carried to the extent of a complete return to the standard lengths of the old unarmoured ships-of-the-line, and even the *Caledonia* Class was lengthened in the process of conversion to armoured type while still on the stocks. But none of the ships laid down after the *Northumberland* exceeded 300 feet in length except the *Captain, Hercules* and *Monarch* which were 320, 325 and 330 feet.

Below water the midship cross sections of the fleet did not differ materially from those of the old unarmoured ships-of-the-line; but above water the sides had of course less height, and moreover less tumble home, so that the maximum beam was carried up nearly to the deck level. This gave the battleships of that period their magnificently spacious top floor, the like of which was never seen in the Navy before or after; as except in the two turret ships and the four *Invincibles* they were quite unencumbered by turrets or superstructures. In comparison the upper deck of the largest ship-of-the-line appeared cramped and small; and it was like entering a ball room from an attic to step on board at the gangway of one belonging to even the smaller classes, such as the *Resistance, Hector* or *Zealous,* on coming from the three-decker guardships. Especially was that so in the flush decked ships which formed the majority; and of those the largest was the *Achilles.* That vessel had a greater upper deck surface than any other ship ever built for the Navy among those free from break of level or athwartship structures fore-and-aft, with very little reduction of beam towards either bow or stern; and it was quite easy to march a battery of field guns and limbers right round it without halt or obstruction from the fore mast to the Captain's after cabin skylight. In making sail, or hoisting in boomboats, or catting anchor, the men on the topsail halliards, or yard tackles or cat fall as the case might be, had a clear run from the order to hoist till the order 'high enough'. And run they did if competing against other ships, so that well attended check stoppers were a necessary precaution.

In the *Invincibles* all this was quite different on account of an upper central battery of which the transverse armoured bulkheads lay right across the deck. As the ships were short the deck space forward and aft of the battery was small, and it was still further cramped by a structural tumble home in the sides to allow fore-and-aft fire to the guns at the four battery corners.

But the decks of that day, whether spacious or cramped, were always — except in turret or coast defence ships — so completely enclosed by the seven-foot bulwarks universal in the Navy of the 'sixties from battleships to sloops that they were as completely shut in as a prison yard. In the small classes, where nearly all hands were stationed on deck in action, a high screen of that kind had its advantages, whether for war or weather, in days when musketry fire still came into play during naval fighting. To that extent also they were not without justification even in large ships of the kind which fought under sail with crowded decks. But in the armoured battleships which fought under steam the deck was deserted at General Quarters except for the helm party, for whom a special protection against bullets could easily be extemporised. Such few officers as were not below stood on the bridge or poop with the signalman, and to these the bulwarks afforded no shelter. Nor in these particular vessels — with upper decks at least fifteen feet clear of the water — did any necessity exist for a massive seven-feet upward extension of the ship's side as a weather screen. Nevertheless this had been an established feature in ships of the Navy for so long that through sheer force of convention it survived its usefulness for many years in spite of its weight and cost, although the low single plate bulwarks of the *Monarch* — forced on her designers to permit firing from the turrets — were obviously worth copying in other new vessels. In the merchant service bulwarks were never too high to see over easily.

All the ships were alike internally in one respect; that is to say they had only three decks. Those with iron hulls were divided into watertight compartments and wing passages below the main deck. Ships' companies berthed and slung their hammocks on this deck, except in some of the *Caledonia* Class, where the absence of wing passages gave more room on the orlop. In the broadside armed vessels the mess tables were hung between the guns as far aft as they extended; but on account of the fewer number and greater concentration of the guns in the so-called 'central battery ships' the battery was not used as a berthing space in many of those. In the vessels where this concentration was carried to an extreme, the men were confined to the main deck forward of the battery. The same applied to the *Monarch* and *Captain* where the midship parts of the main deck were occupied by turret engines and passages. Ample messing space existed however on these mess decks on account of the absence of guns; and in those vessels, as in all the others, the accommodation for the men compared very favourably with that in many ships of later periods.

With the officers it was otherwise except in the *Invincible* Class. All the officers' living spaces were aft — except in some cases where the Warrants were on the fore main deck. The Captain's cabin occupied the stern on the main deck. On either side of the half deck forward of the Captain's cabin bulkhead stood the ship's office, the gunroom, the engineers' mess, and three or four cabins for heads of

Church service on board of a British armoured ship during the 1880s. (NMM. 58/6589)

departments. All these were well lighted and ventilated by large square ports. But the wardroom, and all the cabins for officers below Commander's rank, were right down below on the after part of the lower or orlop deck. In the matter of berthing for officers the battlefleet of 1870 was in fact the worst battlefleet in British records. The wardrooms were small, badly lit apartments placed amidships, with a row of cabins opening directly out of them to port and starboard, and devoid of daylight save that which came through small hatchways on the two decks above. In position they corresponded to the after cockpit of the *Victory,* the floor being well below waterline level. The cabins lining them were too dark to dress in without a candle; as their only light, save that which came through their doors at second hand from the dingy wardroom, was provided by plug scuttles only six inches in diameter, pushed by an iron bar and screw nut into place at the end of a long pigeon hole through a side a yard thick. As these scuttles were not more than four feet above water they were never unplugged at sea, and not always at Spithead or the outer Nore.

The compartment next abaft a lower deck wardroom contained some more cabins and the pantry. Abaft that again came officers' store rooms and the tiller compartment, through which ascended the propeller well in vessels with lifting screws. In those ships a yoke was necessarily substituted for a tiller. Where the hull was iron the foremost side of the wardroom was usually formed by one of the transverse watertight bulkheads of the main structure, but with wooden hulls it was only a light boarding. The space immediately forward of this dividing line in either case was generally known as the steerage, and used as the sleeping quarters for the gunroom officers. Here their chests were ranged right across the ship, and their hammocks slung over them. And here they washed on the open deck in tin baths of the skimming dish pattern, which they had to provide for themselves, as no such thing as a bathroom had yet been invented in the Navy for either officers or men. In the steerage a ladder-way to a hatch on the half deck provided the only daylight, and in some ships the only exit.

In the iron hulled ships another watertight bulkhead enclosed this space on its forward side, and the series of compartments known as the 'flats' extended thence right along the orlop deck to the bows. In the centre of the first of these working forward from the steerage was the large open engine room hatch, which was not surrounded by a bulkhead enclosure as in more modern ships but merely a coaming. It was plumbed by two equal sized hatches one above the other on the main and upper decks, which gave the engine room plenty of light; but as they were not enclosed by 'tween deck bulkheads the rising heat from the engines spread about the ship below instead of being carried off on deck. In most of the ships the engineer officers' chests and hammocks were in this compartment. Next came two long compartments lying over the boiler rooms, of which the centre line was occupied by funnel and ventilator casings and the rest of the space by shell lockers and the ship's company's bag racks. Forward of these the orlop deck lay above the fore holds, tank rooms, magazines, shell rooms and lower store rooms, and the orlop flats contained the stokers' mess deck, the sail room, the hawser tiers, the cells, and a series of upper store rooms. As the wing passages extended for the greater part of the ship's length along the orlop deck level in the iron ships, no ports existed on that deck forward of the small wardroom cabin scuttles; and as the only hatchways giving direct light from the upper deck were those over the engine room, the flats forward of that compartment had nothing to light them but candle lanterns. In the wooden ships however — which had no wing passages — a line of small scuttles ran right along this deck, giving light and ventilation in harbour.

In its motive equipment the battlefleet of 1870 combined the last stage of the old order with the first of the new. When Queen Victoria came to the throne the oceanic application of force on any large scale was still dependent on the wind as it had been from time immemorial; and all the world-wide sea conflicts antecedent to her reign from the days of the Tudors had been the clashes of atmospherically propelled armaments. Before twenty years had passed however, a new motive agency had been devised which permitted such vastly greater certainty in manoeuvring as to introduce a revolutionary change in the science of sea warfare; and thenceforward it was inevitable that battlefleets would never again actually engage under the old system. But the use of this new artificially generated power was attended by drawbacks from which wind movement was free. Wind was a motive force available on the spot, inexhaustible in quantity and immune from any possibility of enemy interception. It conferred therefore a virtually unlimited radius of action on fleets designed to harness its energy. This consideration necessarily carried weight in the days when the new source of movement as yet demanded so extravagant a combustion of coal that none of the battleships thus far built could reach even Gibraltar at full speed, whereas their sailing predecessors could easily fetch the Antipodes without calling anywhere. Strenuous controversy revolved round this question, for while nobody suggested that battleships should be expected to fight under sail, there were some who contended that this was sufficient reason for stripping them to the last spar; whereas just as many others argued that our maritime Empire required that they should be able, if necessary, to protect British interests in the most distant parts; which, in those days of few coaling stations and no Suez or Panama Canals, demanded the power to move — even if slowly — though fires were drawn and funnels lowered. This view certainly seemed to be supported by our past history and prevailed at Whitehall, especially as most of the contemporary foreign battle ships were masted. Every vessel in the fleet was square rigged and heavily sparred accordingly, and hence it came about that its detached units made unbroken oceanic passages up to 8000 miles; remaining eight or ten weeks under canvas which they certainly never could have done under steam.

But steam remained supreme when it came to fighting, and as the fleet was built to fight, this retention of the old means of movement was of secondary importance to the introduction of the new. In this earliest stage the apparatus for using the latter differed so profoundly from modern designs that an engineer lieutenant of the present day would have been as unfamiliar with the machinery of the 1870 battlefleet as an executive lieutenant would have been with its main topsails and muzzle-loaders. The mid-Victorian period was the era of single propellers, telescopic funnels, rectangular fire tube boilers working at 30 lb. pressure or less, and horizontal reciprocating engines with only two cylinders and no compound use of steam, of which the maximum running was rarely above 60 revolutions to the minute. Except in the armoured corvette *Pallas,* neither vertical nor compound engines had so far found acceptance at the Admiralty, nor yet cylindrical boilers, though these various improvements were just beginning to attract the attention of the large steamship companies. Twin screws on the other hand were almost unknown in the merchant service but making their first appearance in battleships. Horizontal engines had some bad drawbacks; but it was considered that these were more than counterbalanced in a fighting ship by the advantage they possessed of keeping the machinery below the waterline, and thus protected against direct fire.

Judged by the standards of their day the battleships of 1870 were very fine steamers when driven hard, quite as fast in their measured mile speeds of 12 to 15 knots as the best contemporary mail liners and faster than any contemporary cruisers, with two or three exceptions. Speeds recorded in Stokes Bay, however, were only supreme tests under exceptional conditions for purposes of comparison, and as such always subject to a discount of about 20 per cent in the full power trials of a ship in seagoing commission, which at that period only took place once a year. For a case in point the *Monarch* and *Achilles* on one occasion carried out their annual trial in company. Both had been timed at well over 14 knots between the beacons in their acceptance runs, but neither could make good 11 in this particular test when driven against wind and sea with topgallant masts on deck and yards pointed, though keeping neck and neck for the whole twenty-four hours. Yet both had better steaming lines than the great majority of their consorts. Under like circumstances very few of the shorter ships were able to exceed 9 knots. The usual passage or cruising speed of an armoured squadron in 1870 was in fact no more than 5 or 6 in ordinary weather. This approximated to the economical mean of any average group of the battleships of the day; and apart from the saving of fuel it avoided prolonged discomfort to those on board, for at any effort to travel more than 8 or 9 the ponderous alternating thump of a two cylinder engine with the laborious grind of a propeller like a windmill combined to shake a ship till every rope and spar aloft was on the racket.

Low as they seem to modern ideas however, these standards of speed were a great advance from a tactical standpoint on anything attainable under canvas. It was not in speed but endurance that the new method of propulsion compared unfavourably in its early stages with the old. Viewed strategically in relation to worldwide sea power its defects were serious. Fires were required under all boilers to move at even 10 knots, and when that was the case no ship in the fleet could in practice rely on her coal lasting for more than 120 hours steaming, though on paper some were credited with 150 hours from the misleading calculations based on measured mile performances. With speed kept below 6 knots the coal consumption was reduced by two-thirds, and it was then possible for some of the ships to cover distances up to 2500 miles under steam, which was just sufficient to cross the Atlantic at its narrowest from England with a small reserve in hand for possible weather emergencies[1]. To pass to the outer oceans beyond the Atlantic it was necessary in the 'sixties to double one of the Capes, and those regions were therefore inaccessible to any vessel travelling under steam from the British Isles unless she could stow coal somewhere in addition to her bunkers. Merchant steamers solved the difficulty by using a cargo hold, but battleships had no space available below the main deck and their seagoing qualities were adversely affected if coal in heavy quantities was carried above that level. Cases occurred in which it did happen to be so carried; but it was always recognised that such an arrangement would have been totally impracticable under conditions of oceanic war, battleships so burdened being quite unfit to meet an enemy and doubtfully fit to meet extremes of wind and sea.

In so far as its steaming capabilities were concerned, therefore, the battlefleet of 1870 was little more than a home guard. Extreme innovationists contended that nothing else was necessary, since all the armoured battleships then existing were under European flags, the decisive engagements of any prospective maritime war were bound to occur in European waters. This argument might have prevailed if all the foreigners had been solely dependent on steam for movement, and subject to its limitations. It happened, however, that many were capable of travelling under sail also, with scarcely any limit as to distance; and this consideration, in conjunction with several others of a less practical nature, decided the Admiralty Boards of the 'sixties to retain a similar power in British ships in spite of the protests of an ultra-progressive section of naval

Coaling ship during the mid 1880s, a scene typical aboard almost any of the broadside or central battery ships. (NMM. 58/6644)

'Man the yards, the Queen's coming', an incident during the Review of the Fleet in 1889. (NMM. 58/6625)

reformers. At this distance of time from the controversy most people will agree that under the circumstances Whitehall acted rightly. Ten years afterwards, when compound engines and other improvements had trebled the steaming radius of our battlefleet, the still continued retention of heavy square rigs in many new vessels had certainly no such justification.

Four of the vessels were barques and three were five-masters. The *Captain* was ship-rigged but with tripod lowermasts. All the remaining twenty-two were ship-rigged in accordance with the established plan of long standing, under which in varying scale according to size all vessels from corvettes upwards were fitted out for sea in the mid-Victorian Royal Dockyards, with no variation of any kind except in the running-in bowsprits of the battleships.[2] In this respect, therefore, a far greater degree of uniformity existed than in any other. Hulls differed, machinery differed, armaments differed, but everything aloft followed the same model so closely that not one ship of the twenty-three could be identified if her hull was obscured and only her spars were visible, except perhaps the two armoured corvettes by their smaller gear. Every item of the running rigging from the main tacks to the royal halliards, from the spanker brails to the jib downhaul, followed precisely the same lead in each; and every item in the standing rigging was turned in and fitted exactly alike rope for rope, though a close scrutiny by an expert eye might perhaps detect five topmast and nine lower shrouds a side in a few of the ships instead of three and seven as in the others. As regards the gear on the fore and main the barques were fitted precisely the same as the others; and in every vessel, whatever her rig, the spars on the fore and main were of equal dimensions, so that one spare set did for both.

By 1870 iron lowermasts of 110 to 121 feet total length and 36 to 40 inches diameter, with fixed iron caps, had become universal in armoured seagoing ships, whether iron or timber hulled. Steel yards and topmasts on the other hand had been discarded after a short trial, as being difficult to repair on board if damaged in action; and by 1870 all spars except the lowermasts were wood, the lower and topsail yards being built and hooped and the others single trees including the bowsprits. Ten battleships were masted with a total hoist of 165 to 175 feet from main truck to deck, fore and main yards 96 to 105 feet long and fore and main topsail yards 70 to 74. Seventeen others had a rather lesser scale with a hoist of 155 to 160, fore and main yards 88 to 91, and fore and main topsail yards 62 to 68. The remaining three were small battleships masted on the scale of big corvettes. The tops were large and nearly rectangular at the sides, the topmast crosstrees rounded on the fore side to the shape of a strung bow, with a grating between the horns on each side. As no pole extended above the royal rigging the trucks lay right on top of the royal funnels. Running-in bowsprits were square in section and iron capped; lying nearly horizontally on the topgallant forecastle and projecting between the knightheads. Lack of steeve caused frequent loss of jib-booms when pitching in a seaway. A trysailmast with gaff was attached to each lowermast.

All lower rigging was iron wire, topmast rigging shroud laid hemp but wire in the newest ships. Topgallant rigging was always hemp. No chain was used except for the slings and trusses of the lower yards, the crosstree necklaces, the solitary bobstay of a running-in bowsprit, and sometimes the martingale stay. Yard fittings such as parrels, quarter strops, jackstays, head earring strops, lift and brace block strops were wire; but yard tackle pendants, foot-ropes, stirrups and flemish horses were invariably hemp. All blocks, hearts and deadeyes were wood, and only the hanging and tye blocks of the topsail halliards and the jib and fore topmast staysail halliards were iron stropped. The sole gin blocks were those below the after end of the lower trestletrees as a lead for the chain trusses. All the running rigging was hawser-laid, three-stranded tarred hemp with the government thread denoting Portsmouth, or Chatham make. The fore and main tacks and sheets were tapered rope for ease in handling. No left-handed rope was used anywhere aloft.

Except in the four barques the square sail plan was comprised of the units of canvas — with slight modifications in cut — of the standardised 'ship rig' established in the British Navy half a century previously; that is to say, a fore and a main course, three topsails, three topgallantsails and three royals. Two square suits were supplied. But the fore-and-aft canvas differed from the early Victorian plan in the substitution of large trysails for staysails between masts, of which only the main topmast staysail was retained. The trysails had one reef on the foot and were not laced to the gaff like a spanker, but hauled out on the head and laced to the trysail mast on the luff. The gaffs were slung at the peak and throat with single short wire pendants instead of running halliards. Jibs and flying jibs travelled on lacings but fore topmast staysails on hanks. Clews of courses and topsails had an eye and cringle for shackling to the sheet blocks, but those of topgallantsails and royals were fitted with a toggle to take an eye in the sheet. All topsails had four reefs fitted with lines and earrings, and two reef tackles a side. In reefing the earrings were passed round the yard and the reef line secured by beckets all along the jackstay. As topsails were much wider on the foot than on the head, the spread of the sail across the third or fourth reefs was greater than the length of the yard between lifts; so that when treble or close reefed the temporary head of the sail extended outside the lifts and braces at each yard arm. With a close reefed topsail the yard was right down on the lower cap and the sheets slightly eased. Courses were pierced for two reefs, but that was merely a survival of obsolete practice, as only the upper was ever fitted with lines, etc. Maintopmast studdingsails had been abolished in the 'sixties but the others remained. No spare suit of studdingsails was carried however.

As rough weather seaboats the armoured vessels of 1870 were better than their unarmoured predecessors when lying-to, but rather worse when off the wind. Greater length enabled them to ride more easily in meeting a head sea, and a better grip on the water reduced drift and leeway; but greater weight caused sluggishness in running under sail and a tendency to excessive yaw. As gun platforms the eight largest were admirably steady under all conditions, but some of the others were heavy rollers, especially the *Lord Clyde, Resistance* and *Prince Consort* Classes when bunkers were full. None was ever unsafe at sea in the worst of weather except the *Captain,* which had only seven feet freeboard and capsized through sheer inability to stand up to her sails in a squall. Of the ships under 8000 tons the four *Invincibles* were the steadiest in a heavy sea.

Though very few vessels in the fleet had either balanced rudders or steam steering gear, all except the six largest turned well; but treble or quadruple wheels with helm parties of six men at least were necessary for steering at anything approaching full speed, or running before a sea. All of them were conned from light bridges far aft, or, in pooped ships, from the break of the poop. The wheels were under the break of the poop, or, in flush decked ships, close abaft the mizen mast. No view beyond the deck was obtainable from the wheel itself on account of the height of the bulwarks, as the wheels were down on the deck and not on the bridge. Chart houses were practically unknown, and except for an engine room telegraph and voice-pipe, the bridge, or break of the poop, was totally devoid of appurtenances.

The weapons of the fleet, like the vessels that carried them, represented a midway stage in a complete transformation of the ancient order. It was the age of the wrought iron rifled muzzleloading gun, which came between the cast iron smoothbore and the breechloader. Breechloaders of a kind had been tried in the shape of the original Armstrongs — with a breechblock inserted athwart the bore after the style of a Krupp — but proving unsafe had been discarded in favour of a return to the muzzle-loading principle for all battleship armaments. But so rapid was the increase in size when once wrought iron ordnance was adopted that within ten years five new calibres were introduced, each greater than the one before, and the weights of individual pieces increased sixfold. All five calibres were represented in the fleet, with one that had preceded all; so that among them the ships mounted 6-, 7-, 8-, 9-, 10- and 12-inch weapons. Practically every vessel carried two calibres — a few even three — but not on the modern principle of a clearly differentiated primary and secondary armament; for every gun was mounted for engaging an armoured enemy. In each ship, however, one particular calibre predominated to which the others were accessory even if heavier. For example, the 7-inch batteries of the *Agincourt, Caledonia* and *Resistance* Classes constituted their main fighting equipment, though all carried two to four larger guns; just as the 100-pounder carronades of the *Victory* were accessory to a numerous broadside throwing much lighter projectiles. In some cases on the other hand it was the heavier calibre that formed the main offensive power of a ship, to which the smaller guns played a subordinate part; as in the 10-inch battery of the *Hercules* or the 12-inch turret guns of the *Captain.*

Even with three calibres, however, the armament of a battleship of 1870 was much simpler than that of her successors a few years later, as she required no defence against either destroyers or aircraft, and carried no minor ordnance but two or three boat guns and a field piece. Torpedoes not having been invented, she was absolutely safe against any enemy except a vessel of her own description.

In the fundamental matter of disposition or arrangement of guns, the principle of an externally carried centre line armament was just receiving first recognition in seagoing battleships by its adoption for the *Captain* and *Monarch.* All the other contemporary battleships carried about 90 per cent of their total weight of metal mounted internally to their main hull except the *Invincible* Class, and all including the *Invincible* Class carried the guns at their sides instead of amidships. This was of course a perpetuation of the old broadside system only modified in detail to permit the use of much heavier weapons; but the rapid increase in size of guns enforced a drastic reduction in the number carried with each new ship, so that the ten or twelve on the main deck broadside of a *Caledonia* or *Northumberland* became only five in the *Bellerophon* and four in the *Hercules.* Decreased thus, the armament only extended over a fore-and-aft space of a half or even a third of the vessel's total length; which gave rise to the term 'central battery ships'. In the *Invincible* Class the concentration was carried to a further extreme than in any other battleships built to date for the British Navy; for only three of their full broadside of five 9-inch guns were mounted on the main deck and the other pair were in an enclosed armoured battery above them, so that, in effect, the ships were armoured two-deckers.

The substitution of artificial for manual power in working guns was making its first appearance in the steam turning engines of the two turret ships. But in every other respect the entire armament of the fleet was hand-served, even as regards the loading and elevation of the 12-inch turret guns. The 6-inch guns were mounted on wooden trucks (and officially designated 64-pounders), but all calibres above 6-inch on iron mountings of the carriage and slide pattern. Every size up to 10-inch had rope breechings. The 6-inch were trained by hand-spikes, the 7-inch by tackles, the 8-, 9- and 10-inch by winches and the 12-inch by steam. All sizes were laid for elevation by hand lever. In loading, all projectiles of 8-inch size and above were hoisted to the muzzle by shell whips, but the smaller sizes were lifted by hand. Those of 10- and 12-inch calibre were rammed home by the whole gun's crew manning bellropes on the rammer staff, while 9-inch and under were rammed home by the loading numbers only. All guns were radial vented and fired by lanyard and friction tube. The universal propellant was very smoky black R.L.G. powder, which was made up for loading in three different sizes of charge for each calibre, known as the 'battering', 'full', and 'reduced'. The small arm in use was the short stocked 'navy' pattern Snider rifle, with cutlass type of sword bayonet.

In the matter of external appearances the mid-Victorian armoured sea force might very well have been known as the Black Battlefleet of our history. Chequered sides had

The British squadron, with the *Agincourt* leading the port column, entering the Dardanelles in the early morning of February 14th, 1878. (NMM. 58/4841)

An artist's impression of the vessels involved in the Cartagena Rebellion of 1873, with the British squadron in the distance under the command of Admiral Yelverton, flying his flag in the *Lord Warden.* (NMM. 58/4064)

gone[3], and the conspicuous white superstructures that came in with the mastless ships of a dozen or fifteen years later had not yet appeared. In the interval, though no regulations as to styles of painting existed, the service paint allowance imposed a certain degree of uniformity which became conventional. Under this, all battleship hulls were unrelieved black, save for a narrow white boot-top or band running round the ship outside the hammock nettings — which was sometimes double — and copper or oxide bottoms. The white masts and black yards of an earlier age had disappeared, and the spars were all painted the same in each individual ship, though between ship and ship the colour varied from a buff or stone to a light brick red. Funnels were black or buff. Boomboats were black but the smaller boats were usually white.

Such is a very brief generalised account of the British line-of-battle at the half-way stage of the great Victorian transformation in fighting ship design. Its period did not coincide with that of great naval conflicts, and only three of its component units ever came under enemy gunfire. Its interest, therefore, belongs much more to the domain of naval architecture than of naval war. But it received and maintained the tradition of British maritime supremacy established at Trafalgar, and its very obvious existence retained for Britain unquestioning recognition as a Great Power in the world of politics during an extremely unsettled era in Continental Europe when war succeeded war. It served its purpose, therefore, by preventing undeclared enemies from drawing the sword if not by defeating declared enemies in arms.

The *Black Prince*, on the extreme right, with other British warships at Larnaca, Cyprus, dressed overall and saluting the Duke of Edinburgh on his birthday in 1878. (NMM. 58/4920)

The *Defence* and *Resistance* saluting the Royal Yacht, *Victoria and Albert* with Princess Alexandra of Denmark being taken to England for her marriage to the Prince of Wales. (NMM. 58/2139)

CHAPTER TWO

THE 'MINOTAUR' AND 'AGINCOURT'

H.M.S. *Agincourt* aground on Pearl Rock in Gibraltar Bay in 1871, with the *Hercules* manoeuvring into position astern in order to effect a refloating. (NMM. 58/3643)

With few exceptions each ship or class in that fleet had some peculiarity or distinction shared by no other; either as a feature of design or equipment, or as the first exhibitor of some new development or the last of some old. Thus for instance the *Minotaur* and *Agincourt,* with the *Northumberland* — their sister in external dimensions — were the longest and largest single screw fighting ships ever built, as well as the longest and largest to carry their heaviest guns mounted between decks. They were also the only fighting ships ever rigged with five masts. As the *Minotaur* and the *Agincourt* were sister ships in every detail except their engines, all that may be said of the appearance, hull, rig and armament of one of them applies equally to the other.

The *Minotaur* was an iron-hulled vessel with a rounded ram, a semicircular and nearly vertical stern over a low counter, a dead straight sheer plan from knightheads to taffrail (like all mid-Victorian battleships) and very little tumble home in her side. Below water her lines were very fine for a fighting model of her day. She was a most excellent sea-boat and steady gun platform, but unhandy under steam and practically unmanageable under sail. On first completing for sea she carried a hydraulic steering engine, which proved too unreliable for use, and as steam gear was not installed till eight years later, she remained virtually a hand-steered ship for two commissions, with a fourfold wheel on the quarter deck and another on the orlop.

Internally she was constructed with the three complete decks which existed in all the battleships of 1870, that is to say, an upper, a main and a lower. She also carried an open topgallant forecastle like the majority, and a poop and stern walk, which were practically restricted to flagships among the armoured vessels of the day. As in all the iron-hulled ships, her whole interior below the main deck was divided into watertight compartments by transverse and longitudinal bulkheads, with wing passages extending over the greater part of her length. The ship's company berthed on the main deck between the guns. In wooden contemporary frigates and some of the wooden-hulled battleships they were berthed on the orlop deck, and the move up in iron vessels was an improvement from a sanitary point of view, as giving more light and air. But it was not popular with the men, who got all the air and light they had any use for in the fore or main top, and preferred that their messes should be what they called 'snug'. Moreover, mess tables slung between guns had to be cleared away not only at weekly General Quarters but also at battery divisional drills, which disturbed the watch below. The construction of the iron ships however made the change necessary, and the medical returns proved its benefits.

With the officers it was otherwise, and the wardrooms remained down on the ill-lighted and worse-ventilated lower deck, for no reason set forth in any Admiralty record, or even possible to guess at, unless as the invincible prejudice of Whitehall against innovations which are not absolutely unavoidable. In some of the shorter vessels, where the captain's quarters were on the half deck aft, the restricted space available did perhaps complicate the possibilities of a change; but no such difficulty arose where the *Minotaur* was concerned, where the half deck at the stern was an unenclosed void lighted by four large ports. Every commanding officer of the ship and every inspecting Admiral recommended with monotonous emphasis that the empty space should be made the wardroom, and every such recommendation was turned down by the Board without comment.

As the *Minotaur* and *Agincourt* were peculiar also in having no transverse battery bulkheads, it was possible in them, as in none of their consorts, to stand not only right at the very stern on the main deck, but to look forward its whole length to the hawse-pipes. From this absence of the usual screening, the inner life of a ship on all sides was more exposed here than in the others; and from the half deck at eight bells, after the six boatswain's mates had piped to dinner, was presented a full view of 700 drill-exhausted men stowing away pea soup and 'Fanny Adams' to a loud babel of forecastle language resounding from every iron beam overhead. From the same spot at the same moment might also usually be heard the drill-cracked voices of half a dozen lieutenants down the wardroom skylight clamouring for shandygaff, and a little further off, from the direction of the gunroom doors, a confused racket of knives, forks and midshipmen's heated argumentation.

As first designed the *Minotaur* was meant to carry a

H.M.S. *Minotaur*, an artist's impression of her taken from the *Illustrated London News*, as she appeared at her launch, with neither a poop nor topgallant forecastle, and only her lowermost armour plating secured. (NMM. 58/2229)

H.M.S. *Minotaur*, her gundeck as it appeared about 1870. Showing two of her 9-inch weapons, and beyond them some of her smaller 7-inch guns. Mess tables can be seen set up between the 7-inch guns and fading into the distance illustrating the point made by Admiral Ballard when he stressed the unobstructed view which could be obtained only in her and her sister. (NMM. 7112)

Sail plan of a 5-master in 1870 drawn to scale. 30,600 square feet without stunsails.

mixed armament of 68-pounder smoothbores and 100-pounder Armstrong breechloaders; but before launching that was changed to twenty-six muzzle-loading rifled guns, of which four 9-inch and eighteen 7-inch were carried on the main deck broadsides with two 7-inch chase guns, two under the forecastle and two under the poop. Eight brass 24-pounder howitzers on wooden trucks stood on the poop for saluting purposes and had only blank ammunition.

In 1875, when the ship paid off for a long refit, her armament was again changed to seventeen 9-inch M.L.R. guns, of which two were under the forecastle, one under the poop and fourteen on the main deck broadsides. She was also equipped at the same time with the first search-light in the British fleet. Later still she received the earliest mark of Whitehead torpedo, and a pair of light breech-loaders on the quarter deck for instructional purposes when breechloaders were adopted in the early 'eighties.

Her armour consisted of a wrought-iron casing which enveloped her sides as completely as the shell of a crab, without leaving a soft spot anywhere from 5 feet below the waterline to the level of the upper deck, though she had no conning tower or internal bulkheads. It was 5 ½ inches thick on the broadside tapering to 4 ½ inches at the bow and stern, and being fairly adequate to resist the impact of contemporary projectiles conferred a protection greatly enhancing her fighting value; though a great deal of its weight would have been better devoted to an increase of armament as in the *Northumberland*. Relatively to modern plating equally thick, its toughness was only about one-fourth, and as any defence against a modern 6-inch gun the ship might just as well have been plastered with pie-crust.

In the matter of external appearance the most noticeable feature of the *Minotaur* was undoubtedly her long and imposing row of masts; but in spite of its spectacular effect

H.M.S. *Minotaur*, as completed in 1867 with her double topsail rig which was so very quickly discarded. (NMM. HAW.1)

she and the other two vessels making a similar display aloft were the dullest performers under canvas of the whole masted fleet of their day, and no ships ever carried so much dress to so little purpose. In the original Admiralty order her five masts were designated fore, second, main, fourth and mizen; and it was laid down that all five should have pole topmasts, but only the first four should carry yards. The square canvas on those four was to consist of a course, a 'cap topsail', an upper topsail and a topgallant sail all of specified dimensions. On that plan she was fitted accordingly for her experimental cruise. But the rig was reported against, and changed early in her first regular commission by the removal of all yards from the fourth mast and the substitution of single topsails, fidded topgallant masts and royals for the original spar and sail plan on the fore, second and main. After the change the masts were re-named fore, second, main, mizen and jigger.

The new spars were on the second class scale carried by seventeen vessels in the 1870 battlefleet; but among them was a sub-division into two groups arising from the fact that eleven — including the *Minotaur* — had been first completed for sea with lower masts intended for the double topsail rig. On those masts the tops were brought 5 feet further down from the cap in order to lessen the interference of the topmast rigging with bracing up the cap-topsail yards; which alteration from the ordinary service standard of proportions reduced the drop of the courses and the height of the trucks above the deck in the same degree. And when double topsails and pole topmasts were abolished, the tops were left where they were to save expense; so that to a seaman's eye the fore or main mast of the *Minotaur* or *Prince Consort* looked very different to that of the *Bellerophon* or *Lord Warden,* though any spar on one had practically the same dimensions as its opposite number on another ship.

All three topsails, topgallant sails and royals were the same cut. At first three courses were supplied of which the fore had 6 feet less drop than other two, but as the second mast stood between the funnels the second course got badly smoke burnt and was abolished. Trysails were carried on the fore, main and mizen, and a spanker on the jigger mast. It was customary to carry a spare topsail and topgallant yard crossed on the mizen and a spare topgallant and royal on the jigger.

Except with a strong breeze well abaft the beam, the *Minotaur* refused to answer her helm properly under sail alone, and the best speed she ever reached without steam was 9 ½ knots under plain sail and port stunsails, with a wind of force 5 to 6 on the quarter, a clean bottom and disconnected propeller. Her weight and stiffness brought a heavy strain on her gear in a hard blow. On the wild night on which the *Captain* was thrown on her beam ends and went down, the squadron had treble or close reefed topsails set, and in the same hour in which the *Captain*'s lights disappeared the *Minotaur*'s fore topsail blew clean out of its roping, her second topsail carried away its sheets and was only furled by all the second topmen after more than an hour's struggle, and her main topsail also carried away its sheets, and having defied all efforts to furl it, thrashed itself to ribbons and had to be cut away. The modern bluejacket can form no conception of what his predecessors had to go through in working aloft on such a night as that.

The *Minotaur*'s complement of 705 officers and men was not founded on her Quarter Bill as would naturally be supposed, and its origin was a good example of official methods seventy years ago. In the days when the fleet was graded in 'rates', based upon a simple and more or less standardised series of truck gun armaments, each rate had a rigidly fixed complement assessed on the Quarter Bill of a ship of the grade. But when provision had to be made for the classification of an entirely novel type of vessel and armament by the building of the *Warrior*, as the forerunner of an armoured fleet, a problem arose which perplexed Whitehall more than enough. Properly the rate of a vessel under the existing rule was fixed by the number of her guns, but as that would have classed the most powerful ship in the world down among the frigates if applied to the *Warrior*, even a mid-Victorian Board of Admiralty realised that it would look odd. So the number of men she would require was taken into consideration and found more suitable; as the Gunnery branch had calculated her Quarter Bill at 740, and though that did not tally exactly with any of the standard complements, it was not far from that of a third-rate ship-of-the-line which was 705. The *Warrior* was classed as a third-rate accordingly though more than a fighting match for all the first-rates in the fleet.

But Red Tape had not finished yet. This vessel might no doubt require 740 men for fighting purposes, but as a third-rate it was obviously impossible that her complement could be either more or less than 705, and at that it was fixed by order of the Board. And when a precedent had once been established in this fashion all the rest was easy when the case of new ships of approximately equal size came up for consideration. Thus the *Minotaur, Agincourt, Northumberland, Achilles* and *Black Prince* were all classed as third-rates with the same complement as the *Warrior*, though the *Minotaur*'s Quarter Bill required about 30 men less and the *Northumberland*'s about 60 more. Third-rates they remained till the system itself was eventually abolished, but even then their original complements were left to stand to the very end of their seagoing careers, though three of them were entirely re-armed. And in actual prac-

H.M.S. ***Minotaur***, ***ca.*** 1883, after her alternate gunports had been enlarged, her main armament modified, light quick-firing guns added along her topsides, and torpedo net booms housed above her gunports amidships. (NMM. AAF/3)

tice the Admiralty made their own decision a dead letter from the start by permitting all these ships to carry large numbers of supernumeraries when in full commission. It was very seldom in fact that the *Minotaur* had less than 800 officers and men on her books on any given day. As regards internal organisation, the ordinary Service Watch and Station Bill for a three-masted rig did not of course suit the five-masters, and a special one had to be made out. The Quarter Bill of the *Warrior* as a broadside armed vessel was the first ever drawn up on the principle of a full gun's crew to each gun in place of a full crew to every pair as with truck guns; and as this was necessary where slide mountings were carried, it applied to all the armoured fleet.

So far everything that has been said about the *Minotaur* may be taken as equally applicable to the *Agincourt,* but in the matter of machinery they differed considerably. The *Minotaur* was driven by Penn's trunk engines, much the best known variety of marine engine of the period, and a form which might have lasted a great deal longer than it did had it been capable of use in any compound fashion. By eliminating piston rods it occupied the minimum of floor space without loss of stroke, and for a horizontal lie offered the further advantage of large surfaces for the sliding parts, which reduced wear. In such favour were these engines at the time that they were installed in eight of the ten largest ships of the fleet of 1870, and might have been placed in all if the supply had equalled the demand. In the *Minotaur* the twin cylinders lay on the port side of the shaft, and weights were balanced by placing the condensers in the space opposite. The propeller was fitted to disconnect by clutch, and left handed in order that the alternating push and pull of the connecting rods should react upwards on the trunks and so save downward wear of the bearing surfaces still more.

Ten rectangular boilers of the return fire tube pattern — ranged in two rows fore and aft with the stokehold between them — supplied steam at 25 lbs. pressure when new (afterwards reduced to 15 lbs.) with furnace uptakes leading to two telescopic funnels of very small size to modern ideas. Having only one pair of small ventilators and no fans, the stokehold was very hot, but the engine room was light and airy owing to the large hatchways opening through all decks above it.

On the four full-power runs of her acceptance trials at load draught she was sent along at a mean speed of 14 ¼ knots. This was no higher than the speed of some older ships in the fleet, but it was justifiably considered a success, as she was a bigger vessel with the same sized engines, and the fact that she was no slower was entirely due to better lines. But like all the rest of them, her weak point was a very small radius of action.

Outside their engine and boiler rooms, neither the *Minotaur* nor the *Agincourt* had any steam machinery except a steam capstan, and during the latter part of their careers a steam steering engine. But two-thirds of the fleet of 1870 were without even those. The main engines of the *Agincourt* were made by the Maudslay firm, partly because Penn could take no more orders at the time, and partly because it was desirable for many reasons that big Admiralty contracts should be spread. As Penn had a monopoly of trunk engines, Maudslay had to use some alternative kind and chose the return connecting-rod type, which in some respects was the trunk engine's antithesis.

With one, the connecting rods went straight from the trunk to the crankshaft, with the other two connecting rods reached from the piston rod right across the engine room to guides on the opposite side, and thence back to the crankshaft. But the cylinders were smaller and lighter than in trunk engines, and those of the *Agincourt* with a diameter of 101 inches gave just as good results in horse power and speed, with slightly better results in economy of fuel when new. Their weak point, which was a chronic source of minor trouble all through her sea career, lay in the excessive wearing down of the glands in the cylinder covers through which the very heavy unsupported piston rods travelled to and fro. The ten boilers were very similar in every respect to those of her sister ship; but as she carried a few more tons of coal and used slightly less for a given speed, her radius of action was about 80 miles greater.

Except that each carried the addition of a 14-oared Admiral's barge, the boat equipment of these two vessels was precisely the same as that of all the larger battleships of the period. And except that their chain cables were 2 ½ inch instead of 2 ⅜, the whole of their ground tackle and apparatus for working it, as well as their outfit of hawsers, etc., was the same as that of the *Warrior.*

The *Minotaur* was launched from the slips of the Millwall Shipbuilding Company in the days when shipyards still lined the lower Thames. She was ordered in 1861 as the *Elephant,* though what caused the change of name at a very early stage will never now be known, as it was the outcome of a verbal and unrecorded Board Room discussion. An enormous crowd on both banks witnessed her launch in December 1863, and unlike several of her contemporaries she showed no hesitation about taking the water in excellent style. Owing to frequent changes in final points of design two years were passed in completing her for sea, and then another eighteen months in experimental trials and alterations in armament and rig, during which many different forms of new heavy gun mountings were tried out as patterns for general service use. It was not till April 1867 therefore that she began her eighteen years of

H.M.S. *Minotaur*, as Flagship of the Channel Fleet, at anchor in Bantry Bay about 1884. (NMM. PX.21206)

regularly commissioned service as leader of the old Channel Squadron by hoisting the flag of Rear-Admiral Warden.

As leader of the Channel Squadron she was more constantly under the immediate observation of the British public of her day than any other ship, and the subject of more frequent reference in the Press. To a bank-holiday crowd her five masts were a guarantee of invincibility, for it was stoutly maintained that a ship with five masts must be bigger than a ship with only three and therefore more formidable. And if another five-master was present she was almost sure to be lying further out[1] and therefore looming smaller to the eye. From a diplomatic standpoint the moral impression these ships created in ports abroad was not without its value, and foreign governments took them very seriously for many years. In other words the existence of the *Minotaur* was always important, even though her career was rather specially uneventful in the larger sense during a generally uneventful naval period. By being a day late for the bombardment of Alexandria she missed the only British fleet action fought in half a century. Her one collision, in an era when unhandy ships and slow reversing engines made fleet collisions regrettably frequent, only cost her a bowsprit and fore topgallant mast, though she came within an ace of sinking the *Bellerophon*.

From 1867 till 1873 she remained continually in commission and flew the flags of Admirals Warden, Symonds, Yelverton, Wellesley and Hornby. In 1873 she was paid off for long re-fit and re-armament, and in 1875 returned to her former station for twelve more years of unbroken

H.M.S. *Minotaur* at Portsmouth *ca.* 1875 after her refit and rearming. (NMM. PX.21220)

sea service, during which time she carried Admirals Seymour, Hay, Hood, the Duke of Edinburgh, De Horsey, Fellowes and Hewett. Her last appearance on an historic occasion was on June 23rd, 1887, when she lay at the head of the southern line at Queen Victoria's Jubilee Review at Spithead, flying the flag of Sir William Hewett, V.C., K.C.B. But by that time she had completed twenty-two years afloat, and was already under sentence of retirement as a long outclassed fighting ship. At the end of the year she was finally paid off and relegated to the non-effective list.

* * *

She waited six years to know her fate, and in 1893 was appropriated for subsidiary duties in the training service; so, after eighteen years of existance as the premier keel of the narrow seas, she spent twenty-nine as a floating laundry for scrubbing hammocks. In 1918, as *Ganges II,* she flew the flag of Rear-Admiral Hickley, Flag Officer Harwich and Admiral commanding the training service. Towards the end of 1919 she was relieved by the cruiser *Blake.* She was considered by the Marine Society, who were looking for a replacement for the burnt-out *Warspite,* but the *Minotaur* was too large, and so, in 1922, she was sold for breaking up. Next year she was towed to Swansea, sixty-two years after the laying of her keel; and surely some thousands of ghosts in whiskers and the uniform of every rank and rating of the mid-Victorian Navy must have watched the process, including one who stood for some years as second in succession to the Throne.

* * *

If the *Minotaur* in her day was the pride of Spithead the *Agincourt* was the pride of the Sound, as she was always a Plymouth ship. Laird's famous yard at Birkenhead obtained the contract for her construction, and her keel was laid in a dry dock in 1861 under the name *Captain,* changed very soon after. A special plant of moveable cranes had to be installed for transporting and lowering the scantling and weights unprecedentedly heavy for that yard. She was floated out in 1865, and as she was even longer than the *Minotaur* in completing for sea (through no fault of Laird's) it was not till the end of 1868 that she was ready for service.

Thenceforward it was her lot almost throughout her

H.M.S. *Agincourt,* in dry dock about 1870 showing her elaborate bow decorations. (NMM. A.2256)

H.M.S. *Agincourt*, heavy evolutions aloft in progress whilst at anchor in Bantry Bay in 1884, with her crew in the process of sending down both her royal and topgallant yards. (NMM. PX.2120)

H.M.S. *Agincourt*, at anchor in Bantry Bay about 1884 as the second Flagship of the Channel Fleet drying her sails. (NMM. PX.212207)

active career to dance attendance on her elder sister. For fourteen years of her seagoing service her station was on the *Minotaur*'s beam as leader of the second division. But apart from that her experiences were the more varied and on the whole the more adventurous. As soon as she was ready for sea a most unusual job for a battleship awaited her, which was to act as an ocean tug with 500 tons of coal stowed in bags on her gun deck in addition to full bunkers. In the month of June 1869 the Bermuda floating dock was lying in the Thames ready to go out, and as no tugs capable of making an ocean trip existed in those days the *Agincourt* and *Northumberland* — both brand new vessels — were commissioned to tow it to Madeira, and there transfer it to the *Warrior* and *Black Prince* for the second stage of the voyage. They took it over from river tugs at the Nore accordingly on June 23rd and proceeded in a tandem tow with the *Agincourt* in the lead, Madeira being duly reached without incident after an eleven-day passage, where they left it in charge of the other two ships and returned home.

The *Agincourt* was then commissioned as flagship of the second in command of the Channel Squadron and remained as such till 1873. During this phase of her career she was nearly lost by grounding hard on the Pearl Rock in 1871, when the squadron were steaming out of Gibraltar Bay westward bound and she was leading the inshore column. It was of course quite contrary to the unwritten law and practice of the Navy for the junior column to be in that position when a fleet was passing near the land, and it transpired at the subsequent Court Martial that if she kept proper station she was bound to take the ground, through the flagship making insufficient allowance for the set of the tide. Fortunately the weather remained calm or she would never have come off again in any condition to float. For three days and nights the whole fleet worked against time in transferring all her moveable weights, including guns, to lighters brought out in haste from Gibraltar; but when nothing more was left to shift she still stuck fast in spite of all the efforts of every available tug to move her, till the *Hercules* was ordered to see what she could do. Lord Gilford's successful seamanship on that occasion was the talk of the fleet for long after, and all vessels present were manning the rigging and cheering when it was seen that the *Agincourt* was at last on the move. It was found that after all she was not too seriously damaged to return home for docking under her own steam. Great public interest was shown regarding the Court Martial, in which the verdict went against the officers concerned. Both of the Admirals were superseded.

H.M.S. *Agincourt*, in 1887 with light quick-firing guns behind shields, and searchlights added along her upper sides, and torpedo net booms stowed above her amidship gunports. (NMM. AAE.1)

H.M.S. *Agincourt*, photographed in 1889, almost at the end of her active career. Reduced to a three-masted barque, and most of her main armament put ashore. Steaming covers to protect the sails to the lower yards of both her main and mizen. (NMM. AAE.2)

When the *Minotaur* paid off in 1873 for re-fit and re-armament, Admiral Hornby, who then commanded the squadron, transferred his flag to the *Agincourt,* whose own place as second flagship was taken by the *Northumberland.* And when the *Minotaur* resumed her place in 1875 the *Agincourt* herself was paid off for similar treatment. Being re-armed and once more ready for sea in 1877, she was commissioned as the flagship of Commerell, a new second-in-command in the Mediterranean; so she did vary her station at least once, and that happened to be at a critical time, when the Empire was on the verge of breaking off relations with the Russia of the Tsars. The Russo-Turkish war was at its height, and in those days the policy of England was uncompromisingly directed towards keeping the Russians out of Constantinople and the treaty for closing the Dardanelles to armed ships in force. The Tsar wanted both Constantinople and the Dardanelles for himself, and as his armies had invaded and occupied all European Turkey outside Constantinople itself (but up to the neck of the Gallipoli peninsula) public opinion in England became so restive that in February 1878 the British Government telegraphed orders to Hornby, who was lying with the Mediterranean battlefleet at Besika Bay, to go through and show a squadron at the entrance to the Golden Horn. He took six vessels accordingly, with the *Agincourt* as second flagship, and entered the Dardanelles on a dark winter morning in a gale and snowstorm. The ships were at General Quarters in view of the possibility that the Russians might have obtained control or possession of the forts on the Gallipoli side; though long before Chanak was reached the blizzard was so dense that the ships and shore were obscured from each other, and the squadron had become so completely broken up through the impossibility of keeping station, that each vessel had to grope her way through independently; none knowing anything about what was happening to the others, as in those days before wireless ships that could not see each other had no means whatever of keeping touch if outside syren signalling distance. In the end all got through and joined up next day, but only very sound seamanship saved a fiasco, and even as it was Hornby's flagship took the ground for several hours.

The incident sobered Russian policy at once and led to the long-protracted Berlin Congress, during the sitting of which the bulk of Hornby's command remained in the Sea of Marmora; but the *Agincourt* was sent back to join the Channel Squadron in her old position as rear flagship. At that she remained for another eleven years, and as a second-in-command only held this appointment for twelve months, no fewer than ten Rear-Admirals succeeded each other on board her in that time; so that from first to last she carried fifteen Flag Officers, which was three more than the record of the *Minotaur.* Like the latter she missed the bombardment of Alexandria, but her men took part in the shore operations. At the Jubilee Review of 1887 she lay at the head of the northern line for the last time as opposite number to her elder sister, whose signals she had repeated for fourteen years; and in 1889 she was paid off and followed her on to the non-effective list.

* * *

When relegated to harbour service she was first employed as an overflow ship for the training establishment at Portland, and then converted into a Medway coal hulk. With the passing of the coal-burner, even this service was no longer required of her, and on Trafalgar Day 1960, whilst Her Majesty the Queen was launching Britain's first atomic submarine, a very tired old lady was gently coaxed up-river to Ward's shipbreaking yard at Grays, and into her final berth.

H.M.S. *Agincourt*, her final degradation as coal hulk C.109. Originally the cranes on her starboard side were fitted with grabs for unloading the coal from the colliers, while those on the port were used for dumping the filled bags of coal on to the decks of the waiting warships alongside for coaling. (NMM. AAE.3)

H.M.S. *Agincourt*, as the coal hulk C.109, photographed in September 1960 when she was stripped down awaiting disposal. Stripped of most of her armour, her timber backing is clearly visible, as well as her filled-in former gunports. (GAO.4127/42)

CHAPTER THREE

THE 'NORTHUMBERLAND' AND 'ACHILLES'

H.M.S. *Northumberland*, stuck on the slipway, while tugs endeavour to float her (NMM. 58/2684)

Although in external appearance the *Northumberland* seemed at a distance to be the *Minotaur*'s double, she was in reality considerably different, being one of the twelve vessels in the battlefleet of 1870 that had no full sister. On a closer inspection it could be seen that her gun ports were not only larger but separated by noticeable gaps from the ventilating ports forward and aft of the battery, instead of presenting one unbroken line all alike in size and spacing. Moreover, the roof of a large conning tower on the quarter deck was just visible over the bulwarks. In exterior dimensions, rig and engines, she was practically identical to the *Minotaur,* but her offensive and defensive equipment differed considerably, and her internal plan differed in detail. Without any sacrifice of seaworthiness she was a much better fighting ship than the other two five-masters in every essential except a trifling inferiority in speed; the higher war value being due to the restriction of the armouring to parts directly vital to combatant efficiency, instead of its extension to the whole hull above water as in the others. By reducing its area, enough buoyancy was gained to permit an increase of 120 tons in weight of armament, which meant 560 lb. more weight of broadside fire beside the addition of a 4 ½-inch armoured conning tower. Internally she differed from the others in having her main deck battery enclosed by transverse armoured bulkheads, and in carrying stern guns on the main as well as the upper deck.

The *Northumberland*'s distribution of weights produced a trim whereby she floated rather deeper aft than the others, though her mean draught was about the same, and her officially calculated displacement even slightly less. The reports of her trials at this time make it evident that she rode as well if not better than the others in a heavy sea, and yawed less in running. But her trim was slightly less favourable for speed. Under steam at full power the loss was only about half a knot; but the coal consumption was always rather higher for any given number of revolutions, and the loss of speed under sail in a strong breeze was at least a knot and a half. Except in this one point however she was much the most formidable of the five-masters as first designed and completed for service, being a higher embodiment of the offensive principle as regards the weights carried on the same sized hull.

The armament of the *Northumberland* when first completed for sea consisted of four 9-inch, twenty-two 8-inch and two 7-inch, all rifled muzzle-loaders. Two of her 8-inch were under the forecastle and two under the poop; the remaining twenty-four guns being on the main deck, with the two 7-inch at the stern and the others on the broadside. All were constructed on the same general principle of wrought iron coils shrunk on over a steel inner tube. The 9-inch gun weighed 12 tons, and its adoption for naval service a few years earlier had been preceded by a great discussion as to whether it would ever be practicable to handle such a tremendous weapon on board a ship. It was sighted to 7000 yards, the sights being on the gun itself. The mounting consisting of a carriage of wrought-iron plating travelling in and out on an iron slide made of a pair of double T girders joined front and rear, with a slight

H.M.S. *Northumberland*, as originally rigged with five masts, and with false gunports painted over the gaps at either end of her battery in order to simulate the appearance of the other two five-masters. (NMM. HAX.3)

H.M.S. *Northumberland*, a heavy roll during manoeuvres about 1890. (NMM. 58/6751)

H.M.S. *Northumberland*, looking forward on her gun deck after her 1875 change of armament, showing both her 9-inch, and smaller 8-inch weapons. As compared with the similar picture of the *Minotaur*, the armoured bulkhead in the distance restricts the extent of the view. (NMM A.2258)

Below right:
'Night Quarters' aboard a broadside armoured ship in 1873. The scene is illuminated by the light of battle lanterns, while the Captain of the Gun directs the efforts of his training numbers. (NMM. 58/4017)

Below left:
A 7-inch M.L. gun on naval carriage, and slide Elswick plate compressor. N.B. The breeching is not shown.

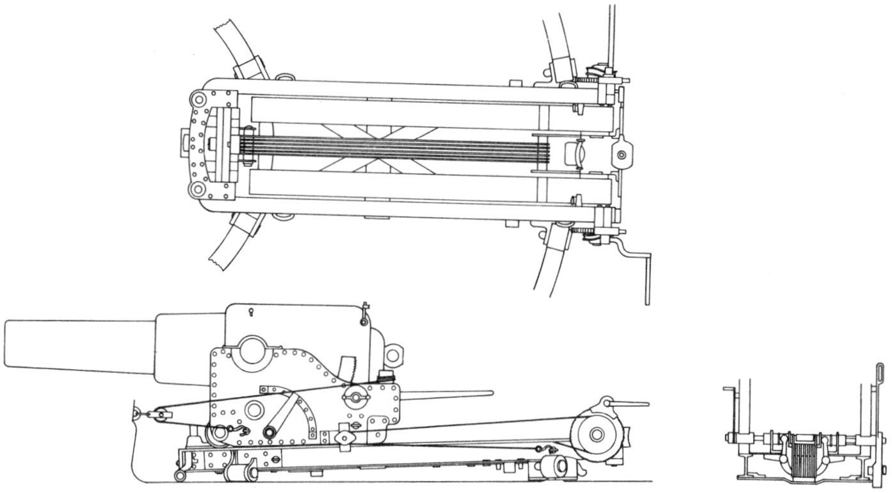

downward slope towards the former. The front end was connected by two bars to the pivot under the port, and the whole slide trained right or left on transverse brass rollers moving on flat brass concentric racers on the deck.

When fired, the gun and carriage recoiled far back on the slide into the loading position and stayed there; the recoil being checked partly by a stout hemp breeching and partly by a series of vertically hung compressor plates under the carriage, sandwiched between a corresponding number of long plates lying lengthways in the centre of the slide, against which they acted by friction when all were jambed together by a rocking lever apparatus on the side of the carriage. After the bore had been carefully sponged the gun was loaded (charge first of course) by the crew manning the shell whip and the loading numbers ramming home, with the rammer staff worked through the loading scuttle in the lowered iron port. The port was then triced up, the loading numbers sprang clear, and the gun was run out by slacking up the compressor lever and letting the gun's own weight take it down the slide, which had to be done with care if the ship was rolling at all heavily. When out, it was laid for elevation by hand levers inserted in drums on the carriage, gearing into arcs on the side of the breech. After these had been clamped and the levers withdrawn the charge was pricked through the vent, the friction tube put in and the long firing-lanyard hooked, the other end of which was held by the captain of the gun, standing right in rear of the slide to look along his sights. Here he gave the necessary orders for training to the hands on the winches — which moved the slide right or left by geared connection with a cogged arc on the deck — and pulled as his sights came on.

As all this complicated process was effected by the coordination of purely manual efforts, it afforded scope for a very high standard of team work, which with a good gun's crew reached such a pitch that they could fire a round a minute. Smartness of drill counted for much more with an inspecting officer than accuracy of shooting, and being the accepted criterion of efficiency produced results to correspond. Hence it came about that the military correspondent of *The Times,* who attended all the army manoeuvres at home and abroad, once asserted in print that no gun drill in the world equalled what was to be seen in the British Navy except that of the Prussian field artillery. Certainly the battery deck of a smart ship during general quarters in those days did present a remarkable display of high pressure muscular activity organised on an impressive scale.

On the other hand the shooting was incomparably below modern standards. No recognised system of range-finding existed, and as the target was merely a cask with a staff and rag of bunting, which was hardly ever actually hit, the scoring was practically guesswork. But even so, the practice of foreign navies was usually worse. At the battle of Lissa much ammunition missed enemy vessels scarcely a ship's length distant, and the few foreign officers who witnessed the bombardment of Alexandria expressed considerable surprise at the number of hits scored by our ships on the forts at ranges of as much as 1500 yards.

The 8-inch guns that formed more than three-quarters of the total weight of the *Northumberland*'s armament were merely a reduced model of the 9-inch except in a few details. The 7-inch were much the same on a still smaller scale, but trained by tackles instead of winches. Though only two were mounted in the *Northumberland,* they provided about 60 per cent. of the whole armament of the 1870 battlefleet as a war force.

During her long refit beginning in 1875 she received four carriages for the 16-inch Whitehead torpedo, which was then being introduced as a new weapon in several navies that had purchased its secrets from the inventor. It had a pointed nose, a nominal speed of 7 knots and a nominal range of 1000 yards, but its movements were apt to be unreliable at best. The carriage was a light open steel framework with an impulse rod, on low rollers and pointed through a main deck port.

Like all the battleships of 1870 the *Northumberland* was also for a time equipped with the Harvey towing torpedo; a flat case of iron and wood containing powder with a detonating trigger, which floated vertically with its upper edge just awash, on being dropped from the ship's side and towed like a paravane from the main yard. With this contrivance it was necessary to pass within thirty yards of an enemy to score a hit, and as it very frequently fouled and theoretically blew up its own ship in dropping or picking up, it was abolished eventually as having rather less than no practical value.

In the matter of complement this ship was classed as a third rate under the obsolete and totally unsuitable system still in force at the time of her launch, which allowed her 705 ranks and ratings all told, though her Quarter Bill required about a hundred more. It was never altered, even when she was partly re-armed in 1875 by the substitution of two 9-inch guns under the forecastle and one under the poop for the 8-inch originally in those positions, and the transfer of the after 8-inch pair down to the main deck at the stern in place of the original 7-inch. After these changes her total armament comprised seven 9-inch and twenty 8-inch. In 1887 she received a pair of light breechloaders, more for instructional purposes than anything else. But though her officially decreed complement was never really adequate to develop her full offensive powers, the deficiency was always made good by supernumeraries.

As a five-master her masts were differently named to those in the *Minotaur,* being known as fore, main, mizen, jigger and after jigger; and she was unique in the mid-Victorian navy as a ship with stunsails on two masts. Otherwise her sail plan was identical with the *Minotaur*'s. After paying off in 1875 for a long refit, the second and fourth masts were removed leaving her barque rigged; but all yards in the new rig as well as fore and main topmasts and topgallant masts were increased to the dimensions of those in the *Achilles.* As the old lower masts remained, the change produced a sail plan differing from all ordinary service proportions, for her topsails equalled the largest in the battlefleet, while in depth her courses were only equal to the smallest except those of the corvettes. Under canvas she was the slowest mover in the whole masted navy with either rig, probably because her trim was more by the stern than that of the other five-masters, slow though the latter were themselves. Her best record under the most favourable conditions was a bare 7 knots.

Her main engines and boilers were in all essentials duplicates of the *Minotaur*'s, but her load draught speed was a quarter of a knot less and her coal consumption a little more. It was her distinction to be the first ship in the British Navy to be fitted with steam steering-gear, though of a pattern very different to anything known during the last eighty years. In this machine the connection between the wooden rotating drum of the engine and the tiller was maintained by hide tiller ropes, of the same description as those used for hand steering-gear in all other vessels.

In the matter of anchors, cables, boats and general sea equipment her establishment was the same as in all contemporary ships over 9000 tons.

The *Northumberland* was built by Messrs Mare and Co., the predecessors of the Millwall Shipbuilding Company; her keel being laid on the bank of the Thames near that of the *Minotaur* in the same year 1861. But whereas the latter was only two years in hand before launching, the former was five; partly on account of frequent alterations in details of design, and partly because she was carried to a much further stage of completion on the stocks. This imposed so great an additional weight on the cradle and launching ways as nearly to lead to irretrievable disaster; and few ships have begun their career in more inauspicious fashion, especially before so distinguished a gallery of spectators, including many Cabinet Ministers and large parties from both Houses of Parliament. For an hour after the dog-shores were knocked away she refused to budge, and meanwhile the tide had started to ebb. That was bad enough, but far worse followed when at last she slid half-way down and there stuck with her after part hanging not waterborne and threatening to break her back. For exactly a month she remained suspended thus in spite of ceaseless efforts to make her move, in which the Admiralty co-operated with the builders by placing the resources of Sheerness dockyard at their disposal. These were applied to the manufacture of large timber pontoons for lashing under her quarters by chain cables at low tide; and to shoring her up while the cradle was strengthened under her weight. A first effort to get her afloat at the next spring tides a fortnight later had no success, though a whole flotilla of tugs were hauling at her stern. But at the next spring tides again an exceptionally high river gave the necessary lift to the pontoons, and she finally left the bank with a rush that threatened to carry away all the checking appliances, but was eventually brought up by good luck. As all the lower Thames region was interested and excited over the whole affair the crowds along the waterside were even huger than on the proper launching day, and according to *The Times* the enthusiasm was indescribable when she took the water at last.

Even then however her vicissitudes were not at an end. The builders were only waiting to have her handed over to go into liquidation, and as the final instalment of their debt from the Admiralty was not due till the next naval estimates had been passed, the official liquidators claimed her as an asset of the company and kept her lying in the Victoria Docks till the debt was discharged. There she remained for eight months before the Admiralty were permitted to take possession; but after that her completion for sea was carried out expeditiously and she was ready for service in 1868.

Unlike the other five-masters the *Northumberland* was not always a flagship, though she had admiral's accommodation under a long poop. All three of them were too large and too expensive in the matter of coal and upkeep to be really suitable for distant stations — apart from their exceptionally poor sailing powers — and as only two flagships were needed in home-waters she only hoisted a flag when one of the others was out of commission. In this way she held at various times all three grades of a ship's precedence; that is to say, a senior flagship, a junior flagship and a private command. In the last-mentioned capacity she was the largest vessel ever in permanent commission to fly a pennant

The upper deck of the *Northumberland* from the poop, taken after she had been reduced to three masts. (NMM. 58/6216)

till the *Inflexible*'s day thirteen years later; and it was under a pennant that she began her service in 1868 in the Channel Squadron with Captain Roderick Dew in command; a very well known character in the mid-Victorian navy of whom many yarns were spun for long years after. It was Captain Dew who first painted the *Northumberland*'s yards black in order that she might not be submitted to the indignity of ever being mistaken for the other five-masters, though both happened to be flagships. The same officer was credited with the different naming of the masts to that in vogue in the others, and was in the habit of setting a lower stunsail triced up on the 'main yard' as a flying kite when his 'mainsail' — or second course as it was called in the others — had been abolished by Admiralty order.

In her first year of service the *Northumberland* was detailed in company with the *Agincourt* to tow the Bermuda floating dock as far as Madeira. The dock was commissioned as one of H.M. Ships for the voyage, partly in order that it might enjoy the usual rights and privileges as such in the event of using foreign harbours en route, and partly in order to supply it with a working crew of service ranks and ratings under a Staff Commander of the old navigating branch. The towing gear consisted of one 26-inch, one 19 ½-inch and one 13 ½-inch cable, all of course hemp with Elliott's eyes at both ends for shackling to lengths of chain. Of these the first and the last connected the *Northumberland* to the dock, and the second connected the *Agincourt* to the *Northumberland*. Astern of the dock the paddle frigate *Terrible* acted as a rudder with a 12-inch cable from her bows. The voyage was without particular incident in its first stage, and at Madeira the dock was duly handed over to their successors.

H.M.S. *Northumberland*, a close up of her bow showing her decorations. From the style of painting, it suggests the picture was taken about 1875. (NMM. A.2257)

Parting of the *Northumberland*'s cable on Christmas Day 1872 whilst in Funchal Roads, causing her to fall across the bows of the *Hercules*, and sustain some underwater damage. (NMM. 58/3991)

From 1868 till 1873 the *Northumberland* remained in the Channel Squadron under a succession of captains. During this period references to trouble with her cables appear somewhat frequently in Admiralty correspondence, and when they had parted more than once in coming to an anchor, Whitehall asserted that it was due to snubbing with too much way on; not an unreasonable assumption, as a vessel of her type carried her way about four times as far as the wooden liners in which all the captains then in the navy had learnt their business. It was left to an 'Act of God' — to use the language of marine insurance — to prove that her captains were not at fault. The wrought iron contract-made chain supplied to the fleet in those days was on the whole the best in the world — most foreign navies obtained their cables in England — but very occasionally a contract below standard quality passed the testing house and found its way into the service; and it was with inferior metal of this kind that the *Northumberland* was equipped for sea. Possibly its exceptional size for that period was rather beyond the real capacity of the contractors to produce in first class workmanship throughout, though the other two five-masters had little trouble with theirs. But whatever the cause, the consequences were serious and might have been disastrous. The squadron were lying in Funchal Roads, an exposed anchorage at best, when a violent gale in the small hours of the morning ushered in Christmas Day of 1872, and once again the cable of the *Northumberland* failed under the strain. When it parted she fell across the bows of the *Hercules,* and as the cables of the latter were of the best, the one in use withstood the weight of both ships, with the result that the ram of the latter functioned precisely as rams were intended to function, but on a wrong target. For a brief space the *Northumberland* remained impaled till released by the veering of the *Hercules'* cable and the holding of her own on a second anchor, but by that time her bottom had been ripped well open and her wing compartment flooded. The nearest dock large enough to take her was the then nearly new Somerset dock at Malta — one of the smallest there now — and to Malta she was escorted accordingly. At that yard the ripped plating was set up in a frame as a memento of the first under water repair to an iron hull in the history of the yard; and it was still there forty odd years later during the Great War, when relics of that sort might have been collected by the score, and all the dry docks in the island were engaged in repairing damages of a similar kind for four years.

In 1873 when the *Agincourt* was promoted from junior to senior flagship for a spell, the *Northumberland* took her place as second in command, and carried in succession the flags of Rear-Admirals Hancock and Lord John Hay. In 1875 when the *Minotaur* re-joined the squadron after a long refit, the *Northumberland* was paid off for similar treatment after seven years continuous sea service, and underwent the previously mentioned changes in armament and

H.M.S. *Northumberland* as she appeared in 1890 with her rig reduced to a barque, and light armament added along her topsides. The gap in the line of gun ports clearly shows the extent of her battery and those used for ventilation at either extremity. (NMM. AGD.1)

rig. For four years she remained in dockyard hands and reserve, and was then, in 1879, again re-commissioned as a captain's ship in the Channel Squadron. As such she completed two more full commissions and bore a share in the minor operations which fell to that squadron in the Egyptian war of 1882. In 1885 she was again paid off and refitted, and in 1887 relieved the *Minotaur* as senior flagship when the latter's long sea career came to an end. As such the *Northumberland* flew the flags of Admirals Hewett and Baird — the former officer being in command when he died — but in 1890 she was transferred to the First Reserve Squadron in place of the *Hercules,* as flagship of Admiral Tryon who was afterwards lost in the *Victoria.* There she only remained for a year however before being relegated to harbour service.

* * *

After removal from the seagoing fleet in 1904 her first employment was as a training ship for stokers in one of the Chatham basins, as the *Acheron.* Then in 1909 she was stripped down and became the coal hulk *C8* at Invergordon, renumbered *C68* in 1926. She was sold to Ward for breaking up in 1927, but resold for further use at Dakar as the *Stedmound,* and finally sold for breaking up in 1935.

H.M.S. *Achilles,* as seen when under construction in dry dock at Chatham in 1862. Workmen can be seen employed on her partially completed lower deck, and others erecting the watertight bulkhead between her two boiler rooms, and with the horizontal stays for her wing bunkers already in position. Unfortunately, in order to impress the general public as to the size of the vessel, if the artist is to be believed, the average height of the men shown in the picture cannot be more than about three feet, when it is considered that the internal depth of the vessel at this point below her lower deck was about twenty-seven feet. (NMM. 58/2059)

Her name had been on the effective list for twenty-four years, of which seventeen were spent in commission.

* * *

The *Achilles* was another vessel with no sister ship, and in her case individuality of design extended to outer shape as well as internal plan. In some respects she was rather noteworthy in the annals of naval architecture. Though not by any means the first iron hulled fighting vessel ever built, she was the first to be built in a government dockyard, and the laying of her keel at Chatham in 1861 initiated the gradual abandonment of timber shipbuilding in the State-owned yards of the world. She was the largest broadside-armed ship without a poop ever belonging to any navy. With her original four-masted square rig she had the largest area of canvas — 44,000 square feet exclusive of stunsails — ever carried by a ship of war; and when reduced to three masts she was the largest ship-rigged vessel that ever flew British colours, either naval or mercantile. In this rig she was a very taking vessel to the eye. She sat the water on a remarkably even keel, and it was said of her that the placing and proportions of her spars and funnels were symmetrically perfect in their relation to the general outline and shape of her hull. She was a splendid seaboat and steady gun platform.

The general purpose underlying the design of the *Achilles* was the production of a vessel intended as an improvement on the *Warrior,* the first armoured British seagoing ship, and with her sister the *Black Prince* the largest fighting ship in the world so far built. She was given the same main dimensions as the *Warrior,* therefore, but her bow and stern were finished in a new fashion, her load draught and displacement were rather greater, and her protection made more complete by the addition of a waterline belt to the 4 ½-inch armour covering the battery which was the only plating in the design of the *Warrior.* To carry the extra weight, however, the *Achilles* had to sacrifice 100 tons of coal. In the first draft of her plans the bow and stern were the same as the *Warrior*'s, but both were changed before her keel was laid; the new bow being without either the above-water knee of the *Warrior* or the under-water ram of the *Defence* and consisting of a nearly vertical but slightly outward-curving stempost; while the new stern was the first of the semi-circular type with a straight tower-like rise from a counter just above the water. She was the first ship to be designed for a four-bladed non-hoisting propeller.

Though her internal plans were never exactly duplicated in measurements in any other vessel they embodied the normal features of the iron built ships of the early 'sixties,

H.M.S. *Achilles*, photographed about 1864, as she originally appeared rigged with four masts. (NMM. A.527/22)

with their three decks, watertight subdivision and main deck armament. She had no poop, but an open topgallant forecastle rather longer than usual. The ship's company's messes ran the whole length of the battery, and the battery itself was divided off from the officers' part of the ship aft and the cable deck forward by transverse armoured bulkheads, as in most of her armoured consorts. The officers' quarters were arranged in accordance with the usual custom of the period, whereby the wardroom was down on the orlop level, as well as most of the cabins, but the gunroom and engineers' mess were on the half deck. As she had no poop, the captain's quarters were on the main deck level right at the stern.

Though several of the vessels comprising the British battlefleet of 1870 underwent one change of armament at some period of their career, only the *Achilles* was re-armed twice. She began her service carrying Armstrong breechloaders on her upper deck and smoothbores below. Of the former she mounted four 7-inch; one being at the bow, one at the stern and two on the quarter deck. Of smoothbores she carried sixteen 100-pounders and six 68-pounders all on the main deck; the 100-pounders being the last and largest smoothbore in the British Navy and the 68-pounders the last and largest cast gun. Both were on wood mountings. In 1868, after four years at sea, her whole twenty-six weapons were changed for an equal number of the then new wrought iron rifled muzzle-loaders on iron carriages, with which the entire battlefleet was being armed in various calibres. Four 7-inch of this modern type of ordnance replaced her upper deck breechloaders, while four 8-inch and eighteen 7-inch replaced her twenty-two main deck smoothbores. At that period such a change was very easily carried out when the new guns approximated in dimensions to the old. As no part of the mountings was a fixture, scarcely any alteration in the structure or fittings of the ship was required, beyond adapting the shot racks to receive elongated in place of spherical projectiles, and making some minor adjustment in the pivots for the slides; and all the guns and mountings could be hoisted in or out by the ship's own appliances. But when the *Achilles* was re-armed for a second time six years later, a good deal of structural alteration was necessary, as the guns were now nearly double the size of their predecessors, though reduced in number from twenty-six to sixteen. Twelve 9-inch, mounted in alternate ports, were substituted for her former twenty-two main deck weapons, and the ports had to be enlarged to receive them, by cutting away the armour all round the edges. Two 9-inch also replaced the 7-inch at the bow and stern, requiring new turntables, and of her second armament only the two quarter deck 7-inch remained. Being always classed as a third rate — like the five-masters — her complement remained fixed at 705 throughout all these changes, which was about fifty too few for her first and second armaments, and nearly 100 too many for her third.

In the matter of protection she was the most advanced ship of her day when launched, being the first to be armoured on the waterline as well as the battery. Except for a slight waterline tapering at the ends, her plates had a uniform thickness of 4 ½ inches, which was the standard for all ships laid down before 1864. Having no conning tower she was handled entirely from a light narrow bridge resting

H.M.S. *Achilles*, as she briefly appeared in 1865 after her bow mast and bowsprit had been removed, and before her fore mast had been moved. (NMM. 8702)

on her seven feet bulwarks just before the mizen mast, with a small box-like charthouse having just room for two people at the table. Except for the engine-room telegraph and voice pipe, and another voice pipe to the battery, this bridge had no control appliances, and all steering orders had to be conveyed by hail or gesture to the fourfold hand wheel about 50 feet further aft. She carried no steam steering-gear or bridge-wheel till already seventeen years old, though a hydraulic steering apparatus which proved too unsatisfactory for use was installed in 1868.

The main machinery of the *Achilles* was practically the same throughout as that later put into the *Minotaur*. On her first trials she lost two propeller blades through using too coarse a pitch; a mishap affording valuable guidance for subsequent vessels. Except for a steam capstan, she had no machinery of any kind fitted outside the main engine and boiler rooms.

As first completed for sea this ship was a four-master; the 'bow mast' and mizen mast being a pair in dimensions, and the fore mast and main mast another. With this rig it was officially reported that she had held her own in a sailing trial with the wooden liner *Edgar* when the latter's propeller was not hoisted. But after nine months experience the four-masted rig was found to be unsatisfactory with the wind before the beam, and the bow mast and bowsprit were removed to try the effect. She now carried too much weather helm, and to reach a properly balanced sail plan the fore mast was shifted 25 feet forward and the bowsprit restored. With that ship-rig she was one of the six most heavily masted vessels in the fleet of 1870, whose six spar plans were all alike except in length of bowsprit and minor differences in the lowermasts.

Except for the three iron lowermasts all spars were wood. The lower rigging, bowsprit shrouds, jackstays and jib martingale were iron wire; the slings and trusses of the lower yards, and the bobstay, chain. Every other item of standing and running gear alike was hemp. The sail area exclusive of stunsails was about 33,400 square feet, to which the stunsails added about 15,000 more. Under canvas the *Achilles* was very stiff. She was once caught aback by an exceptionally violent squall, which whipped the main and mizen topgallant masts out of her, and split the topsails before a rope could be let go, and yet she scarcely heeled ten degrees.

A fore or main topsail in the ships with this sail plan contained about 4000 square feet, with leeches of 60 feet and a spread of 95 from clew to clew. 'Pocket handkerchiefs' of that size reached about the limit which it was humanly possible for thirty men to hand in a hard gale; that being the most that could get on the yard. For if once the sheets or halliards were started, the gigantic bag of canvas, inflated by tons of wind pressure, flapped and thundered with such uproarious violence — however much the hands on deck might strive to steady it by brace, buntline and reeftackle — that the yardarms were jumping like the point of a fishing rod, and every topman on the footropes had to hang on like grim death and labour like a Samson to gather in his bit in reefing or furling. As a rule they were good men at their job, but old logs can bear witness that the struggle sometimes lasted for hours, and on rare occasions even ended in their defeat. When that happened they had no alternative but to use the knives which were a tool of their trade.

But the commanding officer who saw the last remnants of a topsail cut adrift to whirl away to leeward and vanish in the chaos of the elements was very seldom in any mood to condone the failure to save it however impossible. For in the mid-Victorian masted navy, competent smartness aloft was always regarded as the ultimate test of a ship's general condition of discipline and readiness in all things; because, although affording no proof of war efficiency, it did provide immediate and plainly visible evidence of high pitched and well organised effort on the part of officers and men alike. It was accepted, in fact, as a sufficient guarantee of a proper spirit of emulation under capable direction. All ranks and ratings therefore strove to show the way to the rest of the squadron in smartness aloft. The captains of the tops were the best of the petty officers, the upper yardmen the pick of the able seamen. It was the ambition of a lieutenant to excel in this line as officer of the watch, rather than in accurate station keeping; and the ambition of every youngster in the gunroom to attain to the dignity of a top midshipman, though knowing well enough that it was a ruthlessly persecuted existence much associated with penal edicts in the leave book.

To foster and maintain the atmosphere of perpetual rivalry thus created, incessant competitive drills were carried out aloft, both at sea and in harbour, which lent a certain zest to the otherwise rather monotonous routine life of a large squadron. This was something entirely different to the annual rivalry of modern battle practices, or regattas and sports. It was not a yearly but a daily opportunity of being first ship in a fashion that all could see. Every morning accordingly when in harbour, except Saturdays and Sundays, an evolution took place in the way of sending up topgallant masts or crossing upper yards before hoisting the colours, and every evening after quarters some upper gear was sent down. In addition to this, sail drill or heavy evolutions such as shifting topsail yards were carried out on Monday forenoons. At sea it was usual to have a watch evolution twice a day and a hands' drill in the evening; besides the ordinary making, shortening and trimming of sail to meet the conditions of wind and weather.

As an example of the drills and exercises carried out aloft, the log of an ordinary ten-day cruise of the Channel and Reserve Squadrons in company may be quoted. In the cruise selected the ships present were the *Minotaur, Agincourt, Northumberland, Achilles, Audacious, Valiant, Warrior, Hercules, Lord Warden, Hector, Defence* and *Penelope.*[1] On a Thursday evening they weighed from Berehaven, and on getting outside sent down royal yards and pointed other yards to the wind. On Friday morning they exercised general quarters, and in the evening crossed royal yards, made plain sail, shortened and furled all sail. On Saturday, struck topgallant masts. On Sunday there were no drills, but the yards were braced round twice to shifts of wind. On Monday the watch was exercised morning and afternoon in shifting main topsail. In the evening they sent up topgallant masts, crossed upper yards, made all plain sail, reefed topsails, shook out reefs, furled sails, sent down upper yards, struck topgallant masts. On Tuesday, the watch exercised at fire quarters, which included getting up the tackles for hoisting out boomboats. On Wednesday they exercised each watch in shifting jibboom, and in the evening sent up topgallant masts, crossed upper yards, made plain sail, furled again. Thursday being small-arm Company day, no watch drills took place aloft; but in the evening hands made plain sail, set stunsails one side, shifted topsails and furled sails. On Friday morning they prepared for action aloft, which meant unbending all square sails, sending topgallant masts, upper yards and stunsail booms on deck, and running bowsprits in. They then exercised general quarters, and after general quarters bent all sails again, sent up all spars that were down, and ran out the bowsprits. On Saturday they moored in Vigo Bay, hoisted out boomboats, and in the evening sent down upper yards.

'Send down Topmasts', in preparation in clearing for action, a scene familiar to all those serving in the vessels of the Black Battlefleet, and could have been repeated on every masted vessel. (NMM. 58/6522)

It might perhaps be imagined that the strain of this daily striving for precedence all through the year (except when refitting) would wear stale in time, but it was kept from that by constant accessions of fresh effort whenever a ship was recommissioned or relieved.

Every new ship's company had their reputation for smartness to establish, against which every old one had a reputation to maintain. Unfortunately in both cases a tragic price had sometimes to be paid in a form leading to half-masted colours and three volleys with blank. The men often grew incredibly reckless under the stimulus of competition, and needed constant watching for their own safety; but no vigilance could prevent accidents of some kind. When the signal to cross upper yards (or whatever it might be) was whisked down from the main truck of the flagship as the sign to begin, and the immediately shouted order 'mastheads' in every ship present sent all the barefooted upper yardmen in the fleet flying up the steep slope of the topmast rigging, a missed grab at the bellrope under

Coaling the *Achilles* at Malta July 1877. (NMM. 58/667)

handling of a ship. Both in stature and temperament he somewhat resembled the celebrated Lord Cochrane; combining as he did a six foot physique and undoubted qualities of leadership with a continual tendency for assertive opposition to higher authority. Presumably it was this which prevented him from ever hoisting his flag, though as a captain he was selected for the finest commands, and his reports on the qualities of the *Achilles* were most useful in connection with the designs of later vessels. It was on his recommendation that her four-masted rig was altered.

After four years in the Channel Squadron she was paid

the crosstrees meant a backward fall of about 120 feet. Or in the rush with the topsail sheets that followed the order 'let fall', the heavy flap of a half set sail catching unawares the boom tricer who was bearing it clear of the top rim, sometimes meant a loss of balance and a pitch forward to hit the deck head first. In these ways as many as half a dozen or more fatalities often occurred in a single commission[2], nor were the top midshipmen always exempt. But occasional accidents were regarded as inevitable and did nothing to lessen the spirit of competition. A signal from the Admiral 'evolution well executed' was enough to elate a ship for days — the more so as all the fleet saw it — and a signal 'repeat the evolution' brought a proportionate sense of public humiliation. Such were some of the ideals and traditions under which the ranks and ratings of the mid-Victorian black battlefleet drew their rations and earned their pay; of which either the quality of the former or quantity of the latter would provoke immediate and indignant protest 'through the proper channel' if offered to their successors of the present day.

In this fleet the *Achilles* first took her place as a new ship in 1864 under Captain Vansittart; having been built in dry dock at Chatham by a selected party of Admiralty shipwrights, who had under gone special training in order to start the innovation of iron shipbuilding in H.M. Dockyards. Readers of Dickens will probably recall the famous novelist's account of his visit to Chatham Yard to see this prodigiously formidable new vessel that was in hand. Like all the battleships completed for sea at that period, she began her career in the Channel Squadron, of which the flagship was still a wooden two-decker. Vansittart was another of the noted personalities in the service at that time, who had done first class work against Chinese pirates as a brig Commander in his younger days and was claimed by his numerous admirers as having no superior in the

Rear-Admiral Sir William Hewett, V.C., K.C.B., under whom Admiral Ballard had the honour to serve as the most junior midshipman of the *Achilles* in 1877. Sir William won his V.C. at Sebastopol during the Crimean War, and a Knight Commander of the Bath while still only with the rank of Captain. Although apt to be quick tempered, he was very popular with both his officers and men. (NMM. 58/6158)

The watch on deck of the *Achilles* in September 1877, in front of her stern 9-inch muzzle loading chase gun. From left to right, Midshipman Armstrong, Lieutenant Hyland, and Midshipmen Ballard (seated) and Prance. (NMM. 58/660)

H.M.S. *Achilles* at sea in 1884 with the Channel Fleet. The *Neptune* can just be made out over her forecastle. (NMM. PX.212209)

off for her first re-armament in 1868, and re-commissioned next year as guardship of the First Reserve in the Portland district. This form of service was common among the larger vessels of the battlefleet during that period of history, as providing an economical method of keeping expensive ships in permanent readiness for war or political emergency at the shortest possible notice. In the First Reserve they carried a half complement of active service ratings, which could at any moment be made up to full numbers in twenty-four hours by drawing on the district coastguard. A ship so manned had the finest veteran material on her Quarter Bill, quite equal to the present day fleet reserve, much more expeditiously embarked without any need of proclamations to call them out, and engaged in useful public duties when not required afloat. The system had only one serious drawback and that was its lack of sea practice for captains and officers, as the ships only left harbour for quarterly gun practices and summer cruise.

For five years the *Achilles* remained First Reserve guardship at Portland, and in 1874 was paid off for a second change of armament. Being again ready for service she was commissioned in 1875 as First Reserve guardship of the Liverpool district, but in 1877 exchanged stations and ships' companies with the *Resistance* in the Channel Squadron once more, and was now commanded by Captain Sir William Hewett, who at that time was one of the only two recipients in all the services of the Crown of a bar to the Victoria Cross. Within a few months however she was despatched to the Mediterranean to join the fleet under Admiral Hornby, watching the progress of the Russo-Turkish war from Besika Bay, and was selected by him as one of the six vessels he took through the Dardanelles in 1878. In 1880 she returned to the Channel Squadron, with which she continued to serve till her seagoing career came to an end in 1885, twenty-one years after her first hoisting of the pennant. For her last commission she was barque rigged, and thus her sail plan was twice changed as well as her armament.

* * *

After being paid off and dismantled in 1885 she lay in the Hamoaze for sixteen years. She commissioned at Portsmouth in April 1901 for harbour duty as depot ship at Malta, under the flag of the Admiral Superintendant. In the following year she was renamed *Hibernia,* taking the name of the old three-decker which she replaced. In March 1904 she changed her name again, to *Egmont,* to leave her adopted name free for the new battleship then under construction. After spending thirteen years at Malta as base ship (the only part of her entire service spent as a flagship), she was brought home shortly before the Great War to be fitted as a torpedo school ship. In June 1918 her name was again altered, this time to *Egremont,* and in June 1919 to *Pembroke.* She was finally sold to the Granton Shipbreaking Co. on January 26th, 1923.

CHAPTER FOUR

THE 'WARRIOR' AND 'BLACK PRINCE'

H.M.S. *Warrior* on the stocks in December 1860, almost ready for launching. Amidships the timber backing can be seen, and a start has been made on securing the lowermost strakes of her armour plating. (NMM. 58/1888)

It may perhaps be assumed that the majority of readers are fully aware that the *Warrior* was the first armoured battleship, but it may not always be completely realised what a remarkably novel chapter she opened in the chronicles of naval architecture. Some 170 successors have flown the white ensign since that day, but of not one of all these vessels, nor of any vessel before her since the reign of Henry VIII, could it be said, as it was said of the *Warrior,* that at the date of her launch she was in herself a match for all her existing line-of-battle consorts put together. When the *Warrior* and the *Gloire* — her French opposite number — left the slipways, all war values had to be reckoned on a perfectly clean slate. Old long-established standards simply ceased to count, and the topgallant masts of the three-decker were due to disappear over the horizon for the last time.

Foremost among the problems produced by the then novel proposal to pile hundreds of tons of iron slabs along a ship's sides above the waterline, arose the question of stability, to which the only two answers were each attended by grave objections. Stability could be maintained by a great increase in beam or a drastic reduction in height of hull above water, but in no other way. A great increase in beam however meant a great decrease in speed. And a drastic reduction in height of side (or freeboard) meant the abandonment of the age-long practice of carrying a battleship's armament on two or three decks, with a corresponding loss of broadside fire, unless the ship could be vastly lengthened. No forests in the world yielded timber at once long enough in the bole, and rigid enough to stand the rackng stresses of a head sea beyond certain limits as to keel length in a wooden vessel, which limits had been reached and even passed with unfortunate results in new wooden ships actually afloat. These ships sagged in perfectly calm water under the mere weight of their machinery, though built of the utmost strength that timber would permit.

If designs were to be appreciably lengthened, therefore, it could only be done by adopting an iron frame and shell, such as the progress of the science of iron working had at last succeeded in producing some twenty years earlier, and to which in a non-fighting ship there was no serious drawback. But for sufficiently good reasons every navy in the world absolutely rejected iron as a structural material in a ship of war. Some ten or twelve years before the *Warrior* was planned six iron frigates had been built by contract to Admiralty order as an experiment, and the result of subjecting their sides to a test of firing had proved sensational. It showed that even the very best existing wrought and rolled iron was so brittle and full of carbon that the impact of a shot smashed it into something very like flying razor blades. Men going into action behind it might just as well

Figurehead of the *Warrior* taken before 1873, showing her original short bowsprit. (NMM. 7488)

have been stationed to be fired at through plate glass. These frigates, therefore, were straightway disarmed and converted into transports; and for some few years more timber continued to hold its ancient sway in all fighting ship construction.[1]

Meanwhile, however, the iron masters had not only increased the toughness of their iron, but constructed mills capable of producing armour plate able to defeat the heaviest shot and shell. A covering of this was tried therefore on the so-called 'floating batteries' specially built to attack the Russian sea forts towards the end of the Crimean war, and answered the highest expectations. Thereupon the French naval architect, Dupuy de Lôme, decided to apply it to a seagoing ship in spite of its very top heavy nature. But French industry was unable to produce an iron hull, so he could only make the *Gloire* a few feet longer than the old wooden two-deckers.

The British Admiralty could not possibly allow that innovation to pass unheeded, and when Scott Russell and Isaac Watts laid before them an iron-hulled and iron-armoured design utterly different from anything ever seen by anybody they went ahead and built the ship it represented. When once an iron keel and frame were considered permissible, no limitations interfered with lengthening the single gun deck to any extent; and hence the measurement of the new ship between perpendiculars exceeded that of the largest three-deckers by 120 feet. The magnitude of that extension may perhaps best be grasped by glancing at the fact that it was greater than the accumulated increases of keel length of the whole succession of first-class fighting models designed during the previous 300 years, which was little more than 100 feet in all. Thus, what had been for three centuries an average annual creep in growth of about 4 inches, became at one stroke 480 times accelerated. Nothing in the least degree comparable to such a revolutionary step in British fleet architecture is to be found in the whole records of the Constructor's and Surveyor's department at Whitehall. And this was the *Warrior*.

The *Warrior*'s armour covered her sides for a length of 208 feet of her central hull, and completely enclosed a main deck space with ports for twenty-eight guns by internal transverse as well as external longitudinal plating with a thickness of 4 ½ inches; but her ends were left unprotected. Against the holing of about 170 feet of her length, therefore, she had no better security than her predecessors; but against the effects of holing she had a much higher insurance by the application to a battleship for the first time of the system of internal watertight subdivision. The cavernous holds of a wooden liner[2] were only partitioned by light boarding and removable battens in so far as they were partitioned at all; but the holds of the *Warrior* were divided up into a series of self-buoyant chambers or compartments.

Thus the *Warrior* introduced four great departures simultaneously from all traditions and customs in British Admiralty shipbuilding; that is to say, an armour protection, a length of 6 ½ beams, an iron structure in frame and shell, and a new safeguard of flotation. Such a wholesale abandonment of long-established practice, in a service previously ruled by uncompromising veneration and adherence to precedent, was something quite unheard of, and deserves proportionate recognition. But even in giving credit where it was officially due, it must in truth be admitted that the general public apprehension awakened by the launching of the *Gloire* had a good deal to do with Admiralty activity. The British Parliament and Press were in vociferous mood.

When built the *Warrior* had no poop, but a light open poop was added when she was twelve years old, though it was practically only a shelter deck, as it contained no cabins or living quarters. She was one of the very few ships in the 1870 battlefleet that never carried a topgallant forecastle. In the arrangements for berthing the ship's company she was the first of the large vessels with a single gun deck in which the men's messes were on that deck instead of the deck below, the move up being necessitated by the introduction of watertight wing compartments. But the wardroom and the majority of the officers' cabins were still on the orlop level, as in a frigate. From the point of view of war efficiency her only important drawback, as compared to the wooden ships-of-the-line which she utterly ousted from their long-retained pride of place, was unhandiness of movement; but, in the absence of such aids to manoeuvring as twin screws and balanced rudders, that defect was inseparable from her much greater length. Until the *Warrior* was fitted with steam steering gear twenty years after launching, she was steered by a fourfold hand wheel abaft the mizen mast and conned from a light bridge resting on the quarter deck bulwarks.

H.M.S. *Warrior* in dry dock during her 1872 refit when she was given a poop as well as a steam capstan. (NMM. PX.21221)

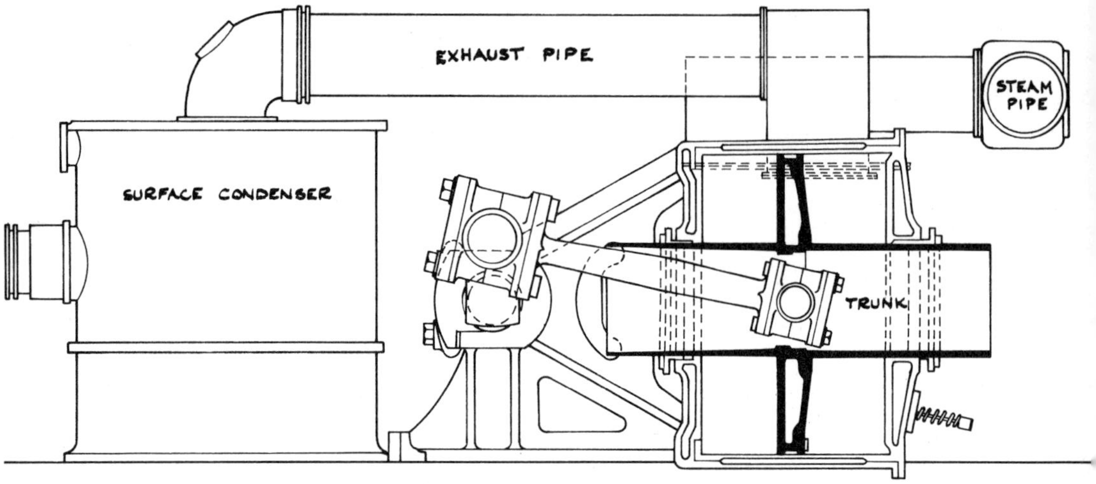

Section through the cylinder of a trunk engine of the type put in H.M.S. *Warrior.*

Visitors inspecting the engine room of the *Warrior* as portrayed in 1861. (NMM. 58/1948)

This ship had other and avoidable defects, however, which were subsequently recognised and eliminated in the designs of vessels built afterwards. Chief of these was the distribution of weights at full load, which brought her down nearly a foot by the bows. This was a bad handicap to fast movement in a head wind, and a frequent subject of official report by her captains. When the squadron was steaming against a gale, the *Warrior* (and *Black Prince*) would be snouting up green Atlantic, while the *Achilles,* of the same length and lines was rising easily over the oncoming seas. To some extent this was made better by the balancing weight of the poop added later; but her trim was never good, and a contributory and quite unnecessary factor in this state of affairs was the long, heavy iron knee at her bow, which answered no useful purpose and detracted alike from her seagoing and her fighting efficiency. Unlike the knee in a wooden vessel it answered no useful purpose as a holding for the bowsprit gammoning, and owed its existence solely to the conventions of the period as to the proper profile of a ship's bow above water. Not only did it burden her in a head sea but, of course, deprived her of any power of effective ramming, for which her massive stempost would otherwise have been suitable; and in those days ramming was certainly a method of attack open to possibilities, however much they may have been over-estimated. Indeed, the only word that could be said in favour of the *Warrior*'s type of bow was that she could never ram and sink a consort as two other British battleships did in later years; which, for an unhandy vessel whose seagoing movements were always made as one of a squadron in close company, was at least a peace-time recommendation, and probably saved her from sending the *Royal Oak* to the bottom in the one collision of her long career. A peace-time recommendation, however, which entails a war-time disadvantage has obviously no proper place in the design of a ship built entirely for war purposes. Her wide handsome frigate stern was another obsolete convention, imposing unnecessary weight without any compensations of practical value.

Though originally intended to carry 68-pounder smoothbores only, the actual armament of the *Warrior* on first going to sea was a novel combination of 68-pounders and rifled Armstrong breechloaders. One 100-pounder Armstrong was mounted on each bow on the upper deck and one on a turntable at the stern, with four 40-pounder Armstrongs on truck carriages on the quarter deck. Below she carried eight 100-pounder Armstrongs and twenty-six 68-pounder smoothbores, all except six inside the main deck armoured enclosure. In 1867 these forty-one guns were replaced by thirty-two rifled muzzle-loaders, which was the greatest number of rifled weapons above 6-inch calibre ever carried by any vessel in the British navy. Of these, eight 7-inch were disposed on the upper deck, a pair in the bows, a pair in the afterpart of the waist, a pair on the quarter deck and a pair at the stern. Twenty more 7-inch with four 8-inch were placed on the main deck, all behind armour. Four 20-pounder Armstrong were supplied as a saluting battery. She never during her career carried Whitehead torpedoes.

The *Warrior*'s engines were much the most powerful ever made for any navy up to the time she was launched. Penn's firm were the makers, and the design was the model for the machinery subsequently put in the *Minotaur,* except that the propeller was quite different and the boiler pressure only 20 lb. Her measured mile performance of 14.3 knots left all previous records for a fighting ship nowhere, and remained unsurpassed by any battleship for some years; while her coal stowage of 850 tons (reduced to 700 tons by 1868) was unequalled for a decade in any navy in the world except by her own sister ship. Her two-bladed 10-ton 23 ½-foot Griffith's propeller was the largest hoisting screw ever made, except her sister ship's. It was raised by the old-fashioned shears instead of the later pattern strongback. The tackle consisted of two separate 8-inch hemp pendants rove double — one on either side — each shackled to a fourfold purchase lying fore-and-aft, with the forward blocks hooked to bolts in the deck on opposite sides abreast the mizen mast, and the hauling parts led right along to the waists. The gear could be rigged with the ship moving, as the reaving lines were kept in place always, but it was necessary to heave to for the actual process of hoisting or lowering, the engines being locked with the slot in the tail shaft exactly vertical to allow the tongue in the cheese coupling to lift out. Hoisting was a 'clear lower deck' job, each watch its own side, stamp and go from the order 'hoist away'. About 600 men were on the falls, and it may readily be supposed that of all deck evolutions, pulling up the 'ruddy old twiddler' was the most popular among the stokers. When high enough it was held in place by large pawls and the gear unrove. In cases where a dock was not available such propellers could be hoisted right up on deck for emergency examination or repair. A spare topmast was rigged as a derrick supported from the mizen mast, which was specially stayed.

Outside her main engine and boiler compartments the *Warrior* carried no machinery except a steam pump in the next compartment forward, which, when she was twelve years old was geared to work a capstan on the fore main deck; and as till then her anchors and cables were the heaviest in maritime history before or since to be entirely dependent on manual working, her cable deck appliances are worth some description in detail. But for those who are unacquainted with the technicalities of early Victorian seamanship, it is necessary to glance back for a moment to the equipment of a still earlier period to understand the origin of the design of her own.

In the first place, it has to be remembered that the present all but universal practice of carrying the lead of the hawsepipes up to the forecastle did not come in till years after the *Warrior* was built. All the battleships of the early 'sixties worked their cables on the main deck. Secondly, although every vessel in the British Navy carried chain by that time, the system under which the old hemp cables of an earlier day were hove in still survived in its distinguishing feature of the use of a 'messenger'. As the massive hemp cables were too large to take round the barrel of a capstan for heaving in, they had to be attached by some method to a smaller one that the capstan could receive. With this object a messenger was carried; that is to say, a relatively small-sized cable taken three times round the maindeck barrel of the capstan — which stood right aft under the quarter deck — and thence led forward on both ends (one on each side of the deck) to the manger in the bows, where each end was passed round a roller and lashed point to point to the other. This made it an endless traveller between the capstan and the manger, constantly rotating — forward on one side, aft on the other — as the capstan revolved; to which, while thus in movement, the much bigger main cable was gripped as it came in through the hawsepipe, by continually winding lengths of small rope, known as 'nippers', round both, and as continually unwinding them as they arrived aft to let the cable pay down the navel pipe. Two specially smart hands, who faced each other just abaft the manger, kept passing the nippers with the cable and messenger between them, while

Sketch of general stationing for weighing with a chain messenger (not to scale), side view and plan. Hands furthest forward pass the nippers, while other hands holding the nipper ends walk aft with them as the cable comes in. Just abaft the main mast, hands remove nippers to allow the cable to pay down the navel pipe, while others carry removed nippers forward again for re-passing as required. The quarter deck capstan has ninety hands on the bars, and ten to twenty on the swifter.

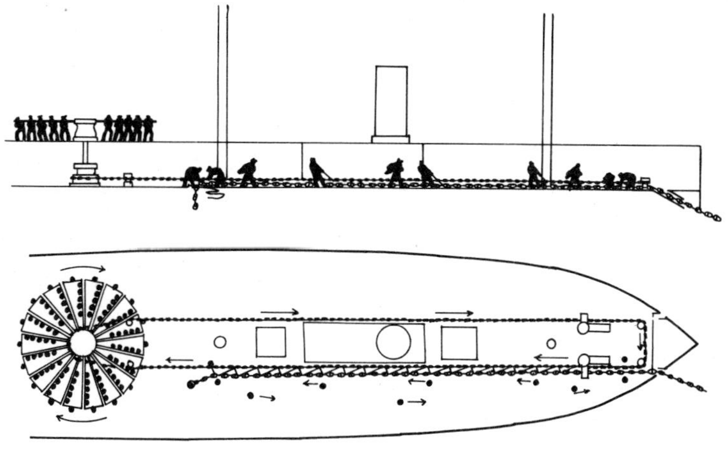

two others kept removing the nippers at the other end. Passing nippers was a job for experts only, requiring dexterity and strength in the very exhausting labour of keeping pace with the ceaseless movement of the messenger and transferring that movement to a bower cable under heavy strain. And it was not without an element of risk.

For more than forty years after chain cables had replaced hemp in the British Navy this system of heaving in remained in use because no method of bringing a chain cable to a capstan had been devised. But the old hemp messenger was replaced by a chain one with studless links, which caught on teeth forming a sprocket wheel on the lower part of the capstan barrel, and it was a chain messenger that was used in the *Warrior.* Her chain lockers lay abreast the main mast and her capstan had the usual two barrels on the same spindle, one on the quarter deck and one below, though as a rule only the former was manned. With five men on each of its eighteen bars and twenty more on the swifter, such a capstan accommodated more than a hundred hands, and when in use presented the appearance of a densely packed human whirlpool, all stamping in time to the music of the full band. Even in the ships with the earlier forms of steam gearing, it was usual to rig and work the hand capstan in conjunction with the steam connection; and thus when the Black Battlefleet was getting under weigh the operation was accompanied by as much sound in the air as the festivals of Nebuchadnezzar's golden image or a modern Aldershot Tattoo. The instant the flagship hauled down the signal 'shorten in', every commander ordered 'heave round', and every band struck up one of the recognised capstan tunes that all naval bandsmen were expected in those days to know by heart, though all are forgotten for ever now; perhaps because most of them differed from the chanties of the merchant service in having neither words nor name. All these were played of course to the swinging cadence of the quick step; to which music the anchors were also catted and fished. But if weighing homeward bound after a three or four years' commission abroad, the time was usually taken at the double, and even then the band was apt to get behind the feet.

The ground tackle of the *Warrior* consisted of four Admiralty pattern wooden-stocked anchors (two bowers and two sheets), all of 95 cwt, one 28 cwt iron-stocked stream anchor and two iron-stocked 19 cwt kedges. The bowers were stowed between the cathead and the billboard in the immemorial style; and catted and fished by large hand purchases similar to those in eighteenth-century models, except that the fish davits were fixtures, and the upper parts of the catfall were rove through a treble block on the iron cathead instead of sheaves in the cathead itself. The sheet anchors were stowed outboard each side in the fore part of the waist, on hinged iron crutches to throw them clear on letting go. For stowing again they had first to be catted and then transported aft by purchases on the fore and main yards. The stream anchor was kept unstocked and lashed up and down the main mast, and the kedges slung on each quarter. The chain cables were 600 fathoms of 2 ⅜-inch and 100 fathoms of 1 ⅝-inch. In the tiers were a 19-inch and a 13-inch hemp cable. The messenger was 1 ⅝-inch unstudded chain.

The *Warrior* was the first of the six battleships masted on the maximum spar plan for armoured vessels except as regards head gear. Her bowsprit was fixed and she carried a flying jib. With the exception of the *Black Prince* she was the only armoured battleship ever fitted with wooden lowermasts and caps. To lighten these two vessels at the bow their first bowsprits were only 25 feet long and stepped on the upper deck; but when their trim was improved by the poops afterwards added, their bowsprits were lengthened to 49 feet in accordance with the intention of their original building plans.

Under canvas the *Warrior* was one of the four fastest armoured ships. Her officially recorded speed of 13 knots in a trial under all plain sail and weather stunsails was only surpassed by the *Royal Oak* and equalled by the *Royal Alfred* and *Monarch;* and in a ship of more than 9000 tons dead weight it must certainly be acknowledged as remarkably good, even allowing for the advantage of a hoisting propeller. That performance was an early indication of the supreme value of fine lines, standing as it did on a par with the best records of the contemporary three-deckers, though the *Warrior* was nearly 3000 tons heavier than the largest, and spread slightly less sail area. She had the advantage, however, of being stiffer and able to hold on to it longer; and, in order to maintain a margin of safety for her masts and yards under the greater strain, they were all rather stouter and heavier in proportion to their length than the spars in any wooden vessel. This became an established rule in all the masted battleships that followed. On the other hand, she was quite unreliable in stays on account of her slow turning powers, and could never be worked to windward in narrow water like the shorter vessels in the fleet. Her sluggishness in answering helm under canvas was especially marked in paying off from the wind, and once caused a serious collision. She was next astern to the *Royal Oak* on a dark night, blowing hard, with the wind on the port beam and the squadron sailing westward off the Scillies with fires banked, when the *Royal Oak* lost some of her way temporarily through the parting of some gear, and the *Warrior* over-ran her, but with helm hard up to try and pass to leeward. She was too slow in turning, however,

H.M.S. *Warrior*, a group of her officers taken about 1863. The young officer leaning on the binnacle is Lieutenant Fisher, who in later years was responsible for the *Dreadnought* and other momentous changes in the service. (NMM. 7117)

and struck the *Royal Oak* on the lee quarter, crashing along her side as far as the waist and shearing off everything, including boats and starboard main and mizen chains.

As no royal dockyard was equipped with iron shipbuilding plant when the *Warrior* was designed, the Admiralty had recourse to the private shipbuilding industry on the lower Thames, where iron merchant steamers of the very largest size had been launched in considerable numbers, and placed the contract in the hands of Messrs Mare, of Blackwall, the forerunners of the famous Thames Ironworks Company. She was ordered in 1859 and very expeditiously built, being ready for launching before the close of the following year. Never before or since have such crowds of visitors come to stare and speculate on the rising framework of a floating war machine; of whom some of the old school declared that the Admiralty had gone mad, though the general majority expressed the opinion that the Board was showing unusual signs of understanding its business. One parliamentary wit asserted that when he went down to Blackwall he found that all the deafening clangour of the shipwrights and rivetters was drowned in the uproar of controversy among the onlookers round the scaffoldings. The mid-winter launch was not exactly a spectacular success, as the grease froze hard on the ways the night before, and six strong tugs hauling together at her stern for an hour were required to fetch her afloat by main force.

Being ready for sea a few months later, the *Warrior* was taken round to Portsmouth, where she hoisted the pennant in August 1861, for service in the Channel Squadron under Captain Cochrane. Throughout her first commission of three years she remained the most powerful ship in the world, except her sister vessel the *Black Prince,* which joined her in 1862. During that period they were as much the reigning favourites with the *Illustrated London News* and the public of Southsea beach and Plymouth Hoe as the five-masters became later. Lying at Spithead or in the Sound, their long black hulls formed a conspicuous and entirely novel contrast to the short chequer sides of their obsolescent consorts, which still shared the anchorages though in fast decreasing numbers. 'Ravens among daws' was one frequent comment, 'black snakes among rabbits' another.

And, if they were popular on one side of the Channel, they were regarded with decisive respect in certain very important quarters on the other. There were those in France who detested England as a sanctuary for political refugees, and were prepared to use the huge French army to redress the grievance. But Napoleon III had a broader vision of amphibious strategy than the army chiefs of France — or of England either for that matter — and saw clearly enough that, with such obstructions to be negotiated as the *Warrior* and *Black Prince,* the prospect of the Imperial drum majors ever swinging their batons in the Old Kent Road, was not bright enough to justify a rupture of the correct relations towards Great Britain, which he was careful to preserve to the end of his reign in spite of all pressure to the contrary.

It was in her first commission that the *Warrior* escorted the Royal Yacht, bringing Princess Alexandra of Denmark on her way to become Princess of Wales and afterwards Queen of the Realm. In 1864 the *Achilles* relieved her and she was paid off for re-armament, remaining out of commission till 1867 when she rejoined the Channel Squadron for five years of further continuous service. During this period of her career, in company with the *Black Prince,* she towed the Bermuda dock on the second stage of its voyage out in 1869, taking it over from the *Northumberland* and *Agincourt* at Madeira. It was a single-line tow with the *Black Prince* in the lead. Each ship carried several hundred tons of coal above her full bunker capacity, most of it stowed on the main deck. The passage occupied twenty-five days, and all went well till the dock was handed over, in open sea and a fresh breeze, to the group of small ships that were to tow it into the harbour, of which it took complete charge and narrowly escaped being wrecked on the reefs.

Two years later the *Warrior* happened to be the *Agincourt*'s next astern when the latter piled on the Pearl Rock. Only very smart handling saved her from adding a violent collision to the complications of an already sufficiently regrettable situation, and three more officers to the list of those who went temporarily short of their swords. It transpired afterwards that the people on her bridge were getting rather anxious about the course the flagship was steering just before she grounded, and were fortunately on the look-out for an emergency.

In 1872 the *Warrior* was paid off for a second overhaul — receiving a poop and steam capstan — and recommissioned in 1875 as guardship of the Portland Coastguard district. After six years' duty in the Portland district she was transferred as coastguard ship to the Clyde, where she passed three years more drilling the local R.N.R. and swinging to her anchors at the Tail of the Bank, except for the annual squadron cruise, the annual refit at Portsmouth — during which in 1879 she received a steam wheel — and the quarterly firing practices.

* * *

In 1884 she was finally paid off as a sea-going ship, remaining in reserve for some sixteen years, till she was taken in hand for conversion into a hulk for the Vernon Torpedo School. In March 1904 she took up duties as an auxiliary workshop under the name *Vernon III.* When the establishment moved ashore the *Warrior* became surplus to requirements. However, a survey of the hull showed her to be perfectly sound, and she was converted into a floating pontoon for the Llanion Oil Depot at Pembroke. Apart from a break during the last war when she was employed as a depot ship for M.T.B.'s, she still serves as a pontoon, and even today she makes regular visits to Milford for dry docking. There for the present she lies, the ancestress and now the sole survivor of the whole British armoured fleet, having outlived all her posterity, but still waterborne and employed for a practical purpose.[3]

H.M.S. *Warrior* as she appeared about 1875. In addition to the alterations to her forecastle and lengthened bowsprit, she has new funnels and a charthouse just forward of her mizen. The *Thunderer* can be seen in the distance beyond her forecastle. (NMM HAH/3)

H.M.S. *Warrior* photographed in 1961 in her final rôle as a floating pontoon for oil tankers berthing at the Llanion Oil Base at Pembroke Dock, South Wales. Officially known as C.77, part of her original name can be read from her name board just above her counter. The well into which both her screw and its banjo frame were hoisted can be clearly seen, but the screw and frame do not seem to have survived. Her poop is now used as a watchman's quarters, but she must be unique amongst all the Victorian ironclads to step a television mast and aerial. (GAO. 4128/5)

Figurehead of the *Black Prince* photographed about 1875. The black painted canvas screen was rigged to give some degree of protection to the men using the heads. (NMM. A.2259)

Except in two matters of detail the *Black Prince* as completed for sea and first commissioned was an exact duplicate of the *Warrior*. Her propeller had a slightly finer pitch, and she always had a steam capstan, the first in the British Navy. After the re-armament of both vessels she carried two 7-inch guns less than the *Warrior* on her upper deck and two less below; and when she received a poop it housed admiral's accommodation. In every other respect these ships were precisely similar in design and equipment; but the *Black Prince* was never quite the equal of her elder sister in speed either under steam or sail. The differences were naturally small but experience proved them to be constant and not accidental. On the measured mile the *Black Prince*'s best record at load draught was 13.9, and the most she ever logged under sail was 11 under whole courses, doubled reefed topsails and topgallant sails, with a wind force 7 to 8 on the quarter. Close hauled under single reefs, however, she equalled the *Warrior*'s 7.5. Like her sister, she carried first a short bowsprit and then a long one.

In the *Black Prince, Warrior* and *Achilles* (when three-masted) the span between the fore and main was too great to make it possible to use the ordinary yard tackles without straining both yards and boats. Each was supplied therefore with a special purchase so designed that the fore yard did not come into use for hoisting out boats at all. This consisted of two 16-inch double blocks, of which the upper had a very long tail of 10-inch hemp reaching right up to

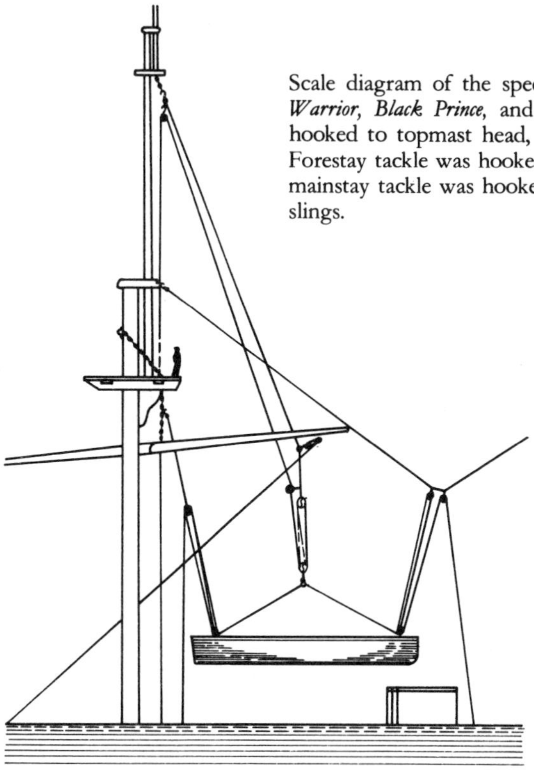

Scale diagram of the special boomboat tackles used in the *Warrior, Black Prince,* and *Achilles.* The main purchase was hooked to topmast head, with an outhaul on the yardarm. Forestay tackle was hooked to fore and main caps, while the mainstay tackle was hooked to a spare link in the main yard slings.

H.M.S. *Black Prince*, as she appeared about 1885, with searchlight mounted on her poop, and light quick firing guns mounted along her topsides. (NMM. 6151)

the main topmast crosstrees, where it hooked to a chain necklace round the masthead, the object of that being to relieve the main yard of most of the weight of the boat. The fall led up from the lower of these through a leading block on the neck of the upper, thence to another leading block under the crosstrees and back down on deck. On the long tail travelled a heart to which was stropped one block of an outhaul tackle, the other being hooked or toggled to a strop round the main yardarm. The yard, in fact, served merely as an outhaul spar, and was braced forward just enough to take the boats clear of the main rigging. When competing against ships using the ordinary permanent yardtackles these long complicated purchases were a handicap, as they had to be got up specially for each occasion; but under the Victorian standards of smartness aloft the whole rig-out was sometimes lifting a boat within four minutes of the hauling down of the signal 'out boomboats'.

The boat allowance of all vessels over 6000 tons displacement in 1870 was practically the same, whatever the complement might be of officers and men, and consisted of three boomboats, three cutters, three gigs or whalers and a dinghy. All boats had oars and sails, and one had removable engines as well. Flagships carried in addition a 14-oared barge or an extra gig, and sometimes both. The boomboats were two 18-oared launches and one 14-oared pinnace, all diagonal built, and all rigged with two standing lugs and a staysail. One of the launches could be converted into a twin-screw steamboat at short notice, for which purpose two propeller shafts were permanent fixtures under her quarters. If required for service under steam the propellers were screwed on before hoisting out. As soon as she was in the water the central thwarts were unshipped and the machinery lowered into her by a yardtackle, bolted down on its bearings, and connected to the shafts, the engines being a permanent attachment to each side of the boiler so that the whole went out in one piece. Two coal boxes and a watertank were added, and the launch was ready for raising steam. Her best speed was no better than six knots, and she had the theoretical defect of negative buoyancy if swamped; but she was very useful when the ship was lying at Spithead or the Nore, and so robust a seaboat, that no instance of her getting swamped seems to be on record. With engine and propellers unshipped she was just as handy under oars or sail as the other boats. In practice she was far more in use without her machinery than with it. For armament the launches of 1870 carried a 12-pounder Armstrong on a wooden carriage and slide, and the pinnaces a 9-pounder similarly mounted.

The cutters were two 12-oared and one 10- or 8-oared, all clinker built and all rigged with a dipping lug and mizen. For armament they carried a rocket tube shipped on one gunwale, for use with the Hale war rocket. The single-banked boats were a 6-oared captain's galley — classed as a gig in store lists — and either a pair of 5-oared whalers or a 5-oared whaler and a 4-oared gig. All these were rigged with a single standing lug. A dinghy and copper punt completed the list. All three cutters and two of the single-banked boats were carried at davits for sea — always turned out — the third single-banked boat and the dinghy and punt being hoisted in and stowed on the boomboats. In vessels with iron davits each of the five hanging boats had her own, two pairs on each side of the ship and one pair at the stern; but in vessels with the old straight wooden davits each pair took two boats, and only one pair of davits a side was carried. The only exception to this was in the five-masters, which carried two pairs of wooden davits a side. No davits ever masked the fire of a gun, and none were ever placed forward of the quarter deck.

In this twentieth century the art of the oarsman has become almost as obsolete in the fleet as that of the flying-jib stower; but in the mid-Victorian Navy it was held in honour. The one steamboat in a ship never carried her engines except as duty steamboat for the day for the whole squadron, or by special permission of the admiral at a remote outer anchorage; and whenever a boat was away, her crew were expected to put their backs into the job and make her travel, always bearing in mind that they were under constant observation from the flagship. All communication from ship to ship was conducted under oars or sail, and a cutter's crew thought nothing of pulling in all the way from Spithead or Plymouth Sound to the dockyard and back again.

Built by Napier and Son of Govan, the *Black Prince* was the largest vessel that had ever been launched on the Clyde when she slid down the ways in February 1861, before an enormous crowd in pouring rain. The event excited great local enthusiasm, and was most successfully carried out. Fifteen months later she was ready for sea and hoisted the pennant at Plymouth for service in the Channel Squadron under Captain Wainwright. At the end of a four years' commission she was selected as flagship of the Admiral Commanding on the coast of Ireland, but as he spent most of his time in administrative work in his official residence at Queenstown the ship was only fitted to accommodate him and his staff for temporary periods. In 1867 she was paid off for re-arming and general overhaul, and in 1868 commissioned as guardship of the First Reserve in the Clyde district, where she did duty for six years till paid off in 1874 for a second general refit. This time the Admiralty decided to give her a poop with full flag-officer's quarters, and a steam wheel. She also had her wooden lowermasts replaced by the iron masts of the *Ocean,* which ship had

H.M.S. *Implacable III*, originally the *Black Prince*, as a boys' training ship at Devonport, photographed shortly before her disposal. (NMM. AAA/2)

H.M.S. *Black Prince*, dressed overall and her yards manned, about 1885. (NMM. 57/1602)

been placed on the sale list; and it was rather paradoxical that the timber sticks of an iron hull should be removed to make way for iron sticks from a hull of wood. When again ready for sea in 1875 she hoisted the flag of Lord John Hay as Second in Command in the Channel Squadron, and afterwards those of Rear-Admirals Phillimore and Dowell. During this stage of her service she was especially noted for an exceptionally handsome scale of ornamentation, even for a period when that was commonly carried much further than it can be now. Her long battery deck presented a most conspicuous effect, as she was one of the very few ships in which the guns were burnished on the bare metal of their entire surface to the lustre of silver instead of being browned and oiled in accordance with the then prevailing custom. She also carried the most genuinely artistic figurehead of the British Navy, representing her celebrated namesake on a 15-feet scale in his black armour under a white and gold mantle.

In 1878 she reverted to the pennant under H.R.H. the Duke of Edinburgh. As befitted her name, therefore, she had a member of the royal family for captain in the closing stage of her service, an honour not shared by any other of the ships which were in the battlefleet of 1870. Under the Duke she was sent to Canada, being by far the largest masted battleship that ever crossed the Atlantic, or went so far from European waters in any direction east or west. She started from Gibraltar and reached Halifax in seventeen days, partly under steam only, partly under sail only, and partly under steam and sail. As she only went out to be present in the ceremonial functions connected with the appointment of a new Governor-General, her stay was very short, and she then left for Spithead under orders to pay off. Her homeward passage of nineteen days took place in one of the worst North Atlantic winters on record, under short reefed canvas almost the whole way and with the loss of her head gear.

* * *

That crossing ended the *Black Prince*'s regular sea service, and on paying off she was relegated to the Fleet Reserve for eighteen years at Plymouth, whence she only emerged temporarily manned on occasions of annual manoeuvres. In 1896 she was removed from the effective list, being then thirty-five years old, and fitted as a harbour training ship at Queenstown, where she was renamed *Emerald* in 1903. This had a political object — to meet the complaint of the Irish M.P.'s that all the R.N. training ships were in English or Scotch ports; but the number of Irish boys joining proved too small to justify the expense of a special ship of their own, and after the experiment had lasted some years she was brought back to Plymouth to relieve the overcrowding in the old ships forming the west country training group. There she continued to serve as *Impregnable III* till removed to make way for a hulk some forty years her junior, and placed on the sale list, whence she went to the Dover Shipbreaking Co. on February 21st, 1923, being then in the language of obituary notices 'in the 64th year of her age'.

CHAPTER FIVE

THE 'BELLEROPHON' AND 'HERCULES'

H.M.S. *Hercules*, floated out from the dock in which she had been built at Chatham in 1862. (NMM. 58/2924)

The six battleships of 1870 formed a group apart, as the only specimens of a style in naval architecture unrepresented at any other period in British history before or since. They differed radically from all predecessors and contemporaries in their great proportion of length to beam, in which respect they resembled the ships of thirty years later. They were also the six largest ships in the British Navy in 1870 in point of displacement tonnage. Next to them in size, though quite unlike in design, came the *Hercules* and *Monarch*; but, as these two were also quite unlike each other, it simplifies matters to take the *Bellerophon* with the *Hercules,* as both being central battery ships.

The *Bellerophon* and *Hercules* — with the very much smaller *Pallas* — were the first of the eighteen armoured vessels of the central battery type in the history of the British Navy, of which only ten were afloat by 1870, including three of very small size. The principle followed as a direct consequence of the extremely rapid advance in the size of ordnance between 1860 and 1865; during which brief period the weight of the maximum naval gun was increased nearly threefold through the abandonment of cast iron construction in favour of forged iron and steel. Where the total burden of armament was intended to remain substantially unaltered as imposed on a ship, a double or treble weight in the individual weapons naturally demanded a corresponding reduction in the number placed on board; and as, in spite of its size, one of the new guns did not require very much more lateral training space than one of the old, the fore-and-aft dimension of a battery containing this reduced number each side was sometimes less than half that of a battery containing the much more numerous smaller calibre. For example, the aggregate weight of the guns and their mountings in the *Hercules'* battery of 84 feet long was roughly the same as that in the *Black Prince*'s battery of 212 feet; which obviously permitted a great saving in the side surface to be armoured.

In fact the original reason for making the first armoured ship very long as compared with a three-decker had disappeared soon after her completion through the unprecedently rapid growth of the gun; and the question then arose as to whether great length in proportion to beam possessed sufficient recommendations in other ways to justify its retention in future ships. It did permit lines conducive to speed, and minimised the resistance offered by a head sea. But it was detrimental to easy handling; and fine extremities which were not self-buoyant (such as those of

H.M.S. *Bellerophon* about 1893 when she was port guardship at Pembroke Dock. Shows her stern gun ports below the Admiral's stern walk, and her torpedo net booms stowed on her broadside. (NMM. AAG. 2)

the *Warrior* and *Achilles*) imposed a burden on the central portion of the hull. These drawbacks tipped the scale in no uncertain fashion against the reproduction of long models for nearly a generation. Power of rapid movement was at a discount where battleships were concerned, and the speed of those already under the white ensign was regarded as ample, being beyond dispute superior to anything attained by their foreign contemporaries. On the other hand, rapid turning powers had come into great demand. The older officers, who had been brought up in short wooden vessels, complained that the *Warrior* required almost the whole width of the deep water space at Spithead to make a sixteen point alteration of course; while the younger were quite convinced that the entire future of naval warfare would be dominated by the ram, and talked continually about 'ramming tactics' and 'end-on fighting'.

To the very few voices still raised in demand for better speeds the reply was made with perfect truth that British battleships were already the fastest in the world; and for quite fifteen years after the launch of the *Warrior* her Stokes Bay record of 14 ¼ knots was considered by the Admiralty as a standard which it was superfluous to exceed and not even imperative to equal. As that was the official view, the new Chief Constructor, Edward Reed (a man of very active and original mind), proceeded in the first design of his period of office to submit a draft representing the ideals of concentrated power and handiness, but with no higher speed than the older vessel and less strategic reach under steam. This launched the *Bellerophon* and initiated the vogue of the short battleship, built with the idea that her ram had such possibilities in its use that her design must be adapted to seize them. Now it cannot be denied that the inspiration underlying this idea was wholly in accordance with the best traditions of the British Navy. It breathed the spirit of the uncompromising tactical offensive, without which no fighting service can make world history. To vessels attacking with the ram, as at Lissa, Nelson's signal 'engage more closely' would have had no meaning; and in the 'sixties of the nineteenth century, before the rapid-firing, hard-hitting, long-range breech-loader was known, ramming tactics had some practical justification. But like many other theories with a sound enough foundation under the conditions of their original conception, the idea became an obsession defeating its own ends and outliving its day. The short handy battleship, of ramming purpose but indifferent speed, became even shorter and slower under the domination of an exaggerated belief in her type, till she was not fast enough to catch up with an enemy of even moderate steaming power, and her ram ceased to have any value. She might at this stage of her evolution have received the official designation of battletub as accurately descriptive. The difficulty with the battletub was not to make her turn easily, but to prevent her from turning to every point of the compass instead of keeping in the wake of her next ahead; and it required anything up to six or seven days to get certain of her kind through the Suez Canal, even with tugs ahead and astern to prevent them from nosing the banks. The *Bellerophon*, however, was only a first step away from long keels, and gave no trouble of that kind. She was in fact exceptionally easy to handle under steam without being difficult to hold to a straight course.

Having the new 9-inch ordnance available — which the Admiralty had decided to adopt after much hesitation — her designer started with the advantage afore-mentioned of a much reduced requirement for battery length in his plans, in proportion to the total weight of broadside armament, equivalent to twenty of the 7-inch guns which were the main weapon in existing ships; and by putting that weight into 9-inch halved the number and found he could mount them five a side, in a battery only 98 feet long. This reduced the side area requiring protection so much that it became possible to increase the thickness of what armour remained from the existing standard of 4 ½ inches to 6 and still save hundreds of tons.

To carry this box of arms he drew a hull which had something of the shape (on a slightly reduced scale) of the central body of the *Warrior,* after cutting off 40 feet forward and aft and closing up the ends of this rather pontoon-like structure with a bluff semi-oval bow and stern; producing a ship with great buoyancy in proportion to length and beam, but no stream lines worthy of the name. Compared to the slim tapering *Warrior* the *Bellerophon* looked as full-breasted and full-buttocked as a canal barge, though her maximum beam was 2 feet less than the *Warrior*'s. But the absence of long fine ends which were not self-floated reduced the burden on the centre body; and another economy in weight was effected by restricting her coal stowage to 640 tons and her boilers to eight in number, which was all she had space for.

The net result of the plan was a vessel £35,000 cheaper than the *Warrior,* with 1600 tons less displacement at full load; yet much more formidable in offensive and defensive power. Taken as a whole, therefore, the *Bellerophon* was a decided success in the application of the theories beginning to find prevalence at the time as to the prime attributes of a front rank instrument of sea war. Like all Reed's ships she was trimmed rather heavily by the stern, being as much as 4 feet deeper aft than forward. That had the drawback of increasing her maximum draught; but was helpful in steaming head to sea, and gave a good immersion to the propeller. Her worst fault was a very heavy and uneasy roll in a beam sea with a certain period, due to a low centre

H.M.S. *Bellerophon* in the process of fitting out for sea, with a sheer hulk alongside. She is riding very high in the water, and may only have part of her armament on board. The small schooner appears to be engaged in coaling her. (NMM. A.9068)

of gravity which was another feature in this Chief Constructor's earlier models. That, of course, was a serious defect in a gun platform, which is what every fighting ship was built to be in the days before torpedoes.

As Reed always held progressive views he introduced novelties not only in the general design of the *Bellerophon* but in its details. Her ram projected farther than that of any vessel previously launched, and a shield of armour covering her true bow rose from the waterline belt high enough to be a protection for two chase guns on the fore part of the main deck. A light false overhanging bow, containing the ship's company's head (so placed experimentally), extended upward outside this shield and curved forward over the ram. In the old ships with a knee bow the head was a primitive affair, consisting merely of a pair of long latrine troughs with multi-holed lids, lying fore-and-aft each side above the knee timbers abreast the inner part of the bowsprit; each with its soil pipe running down to the water in the angle between the knee and the butts of the bow planking. These troughs were concealed from outboard by black painted canvas head screens; but they were absolutely open to the sky, and men using them were fully exposed to rain, spray or snow. If the ship was pitching heavily they had to be temporarily abandoned for a makeshift rigged up in the lee fore chains by hitching ropes along the sheerpole. As these arrangements were ventilated by all the winds of heaven, they had much to recommend them from a sanitary standpoint. When the knee type of bow was discarded some alternative placing of the head became necessary; and as this was a matter of sanitation for several hundred men it required careful consideration. In the earliest ram-bowed ships it had been placed in the fore part of the main deck; but ventilating difficulties had proved that to be objectionable; and in the *Bellerophon* the head was moved up again to the highest part of the ship forward by the erection of the false bow described above, entered from under the topgallant forecastle, decked over and provided with scuttles. Here again, however, the ventilation proved unsatisfactory with a wind aft; and in most of the later vessels of the 'sixties the heads were built to starboard and port in the waist, level with the upper deck and projecting outboard with an external soil pipe to the waterline. Exceptions were the *Lord Warden,* the *Repulse* and the four ships of the *Audacious* class, which had the same arrangements and bow profile as the *Bellerophon;* resembling a cross in a varying degree between a clipper stem and a ram, without definitely partaking of the character of either.

At her other end the construction of the *Bellerophon* followed a precedent established not long before when converting certain wooden ships on the stocks into armoured vessels, whereby the wide overhung stern of convention had been abandoned in favour of an oval underhung stern devoid of quarter-galleries and ornamentation. The *Bellerophon*'s sloped inwards above the counter, and was elliptical in form, but with the major axis of the ellipse in the fore-and-aft line instead of athwartships as in a frigate. This type of stern lightened the ship aft, and improved on the older kinds in every way except that it greatly reduced the accommodation for officers. Under the counter she carried another novelty in the shape of the first balanced rudder in the British Navy, though until she was twenty years old it was only moved by hand.

Inboard her chief innovation was a new arrangement for berthing the men, rendered possible by the reduced space occupied by the guns as compared with that in the older iron-hulled ships in which the line of guns extended over the greater part of the length of the main deck, and the mess tables were slung between them right fore-and-aft from one armoured bulkhead to another. In the *Bellerophon* the battery was less than 100 feet long, leaving sufficient accommodation for 600 men outside it. As obviating the necessity of clearing the mess deck on every occasion of quarters, firing, or divisional gun drill, this was a boon to a ship's company of a century ago, whose existence in the days before canteens, bathrooms, electric light, bakeries and other alleviations of seafaring life was utterly comfortless. The bare bleakness of the mess deck with its long range of plank tables and stools had as little suggestion of physical ease as a prison cell. It was damp and chilly in a cold climate, and damp and hot in the tropics. It was swept by searching draughts if the ports were open, and nearly pitch dark if they were closed, glass scuttles not having been invented. It was dimly lit at night by tallow candles inside lamps at long intervals, and as there were no drying rooms it reeked of wet serge and flannel in rainy weather. In short the living quarters of the mid-Victorian bluejacket, stoker, or marine were as widely dissociated from any ideal of a home in the usual sense as could well be imagined.

Moreover, he was always in a crowd by day or night. His work and his leisure, his eating, drinking, washing and sleeping were all in crowded surroundings. He swallowed his bully beef and hard tack, his pea soup, 'copper rattle', and rum, at a mess table so congested that he had absolutely no elbow room and scarce space to sit. He washed himself twice a week on deck at the same time as he washed his clothes, in the two tubfulls of cold water which formed the allowance for the whole twenty-five men in his mess, in the middle of a splashing mob at other tubs all round; and he slung his hammock at night among hundreds of others so tightly packed that they had no swinging room however much the ship rolled. Even in the heads he had no individual privacy. All this doubtless had compensations of a kind; for human nature is strongly gregarious by inclination, and this crowd were all of his own breed, mainly drawn indeed from his own native district, without any alien element. He knew nothing of the lonely solitude of the shepherd or farm hand, and whatever else he may have lacked he never lacked company of a kind which he thoroughly understood. But it was exaggerated in degree, and always subject to the restrictions of a rigid discipline, and in proportion as this was relieved in its minor inconveniences, by so much was his lot rendered less hard. Ships in which the messes were outside the battery therefore were popular in the fleet with the ratings.

But if the men's accommodation was better than in the older ships the officers' was worse. Reed mounted main deck chase guns in the stern of all his earlier ships, which usurped the space used for the captain's quarters in older vessels. These quarters were, therefore, shifted to the side and reduced in size to those of a corvette. And by the change they in their turn displaced other officers' cabins which were pushed down below; so that in the *Bellerophon* not only were the wardroom, gunroom and engineers' mess on the lower deck but practically all the cabins. Being originally flush-decked aft (like nearly every battleship built during the 'sixties except the five-masters) it became necessary to provide her with a poop for Admiral's accommodation when she was selected as a flagship. She always had a topgallant forecastle.

One of her internal improvements in equipment consisted in the installation of a so-called 'patent capstan'; which, by gripping the cable itself instead of a messenger, obviated the necessity of placing the capstan and chain lockers a long way aft from the hawsepipes. In the *Bellerophon* these were situated well forward between the funnels and the fore mast. A patent capstan stood on the centre fore-and-aft line like a messenger capstan; but it was iron instead of wood, and in place of carrying a sprocket wheel to catch the links of a messenger the lower part of its barrel was formed of a cable holder exactly like those in present-day use. To shorten in cable the slip was put on, the compressor thrown back, and enough slack hauled up from the locker to bring a bight to the capstan; off which it was prevented from jumping by transposable rollers dropped in sockets in the deck close round the barrel. As the links passed the last roller in heaving round they were pulled from the capstan by chain hooks and hauled to the navel pipe for paying down. By the abolition of the messenger one of the oldest major contrivances in the Navy disappeared from use, but the change saved a vast amount of labour and time, and disencumbered the main deck of the long stretches of cable lying each side from navel pipes far aft to the bows. For her first two commissions the *Bellerophon*'s capstan was entirely hand worked, but in her refit after the second the steam pump, situated on the lower deck in the compartment next forward of the boiler rooms, was fitted with a clutch and gearing to rotate the capstan also.

Another novelty in the *Bellerophon*'s equipment was the introduction for the first time in the British Navy of the method of catting the anchors by cat and ground chains, instead of hooking the cat purchase itself to the ring. In Reed's ships the ram was so bluff and the bow so much underhung by it that an anchor bore very heavily on the ship's side if hove right up to the hawsepipe. To avoid this the cat and ground chains were devised, and remained a service fitting till hawsepipe anchors were adopted forty years later. They were still in use in some ships into the beginning of this century, except of course that by that time they were not shackled on the inner end to a threefold hand purchase as in the *Bellerophon,* but tailed to a wire pendant taken to a steam capstan. The *Bellerophon*'s tackle consisted of the usual two bower, two sheet, one stream and two kedge anchors; in her case all iron-stocked. The bowers and sheets all weighed 4 ½ tons and the stream 2. She stowed 575 fathoms of 2 ¼-inch chain, 100 fathoms of 1 ½-inch chain, one 17 ½-inch hemp cable and one 15-inch.

In her armament the *Bellerophon* had the unique distinction among the battleships afloat by 1870 of being before the end of her service completely provided with the long type of breechloader; which was designed during the early 'eighties to supersede all muzzle-loading ordnance and eventually developed into the patterns in use in the British Navy of today. Moreover, she was the only central battery ship ever to undergo any kind of change in her main deck armament. The muzzle-loading combination for which she was designed (and carried for nineteen years) comprised ten 9-inch in the central battery, with a pair of 7-inch at the bow and three 7-inch aft. Her only breechloaders at this stage were four small Armstrong 12-pounders for saluting. Of the three stern 7-inch, two were on the main deck and the third on the upper. The two bow 7-inch were both on the main deck behind the armour screen mentioned above; but it proved impossible to keep their ports open when steaming head to sea. At best the experiment of placing them so low had a doubtful chance of practical success, and Reed only tried it once again (in the *Hercules*) before abandoning it altogether.

H.M.S. *Bellerophon* taken about 1893, with searchlights mounted on her fore bridge, and light armament along her topsides. (NMM. HAY/1)

By the end of the 'seventies the Admiralty had decided after long and exhaustive discussion to follow the lead of the principal continental Powers in adopting breechloaders. The order accordingly went forth whereby the muzzle-loading gun joined other relics of a glorious past. An absolutely different class of ordnance took its place; to carry which all ships laid down thenceforward were specially designed, and some of the old were altered in so far as that was practicable without wholesale reconstruction. In such of the latter as were armed on the centre line principle the change presented comparatively little difficulty, in spite of an increase of 50 per cent. in the new type's ratio of length to weight for all calibres. But it was no simple matter in the broadside main deck armed vessels to devise a mounting that would allow guns 20 feet long to be secured for sea with their muzzles inside the port; and this was considered essential, except where the angle of an embrasure recess was deep enough to allow the muzzle to lie within the plane of the ship's side.

For the experiment of placing heavy breechloaders on a main deck broadside the *Bellerophon* was selected, though she had no embrasures. Reasons for her selection are obscure, as she was already one of the oldest ships on the effective list, with neither compound engines nor twin screws; but it was probably due to the fact that she was about to be taken in hand for a long refit just as a new 8-inch breechloader was being turned out, which approximated in weight to her 9-inch muzzle-loaders though 7 feet longer. Ten of the new guns were substituted for the old in the central battery; while four 6-inch breechloaders, in pairs at specially constructed embrasures under the poop and topgallant forecastle, replaced the five 7-inch chase guns of the original armament; and six 4-inch breechloaders the 12-pounder Armstrongs. These 4-inch were the earliest form of a special anti-torpedo craft defence.

On paper the change made the *Bellerophon* a much more formidable battleship, as the range and penetration of the new armament was about double that of the old. But in practice the experiment had only a qualified success, as the design of mounting was not very satisfactory and the battery space was too cramped for a thoroughly effective use of guns of that length. Thereafter, in consequence, the installation of breechloaders in the older vessels was restricted to upper deck position.

When her armament was being changed in this way she was also provided with two carriages for 16-inch Whiteheads, placed at ports on the main deck, but outside the battery armour. Her side armour was 6-inch — the thickest in the world at the time of her launch — with 5-inch on her athwartship battery bulkheads; and she carried an armoured 'pilot tower' on the quarter deck, but without any wheel inside it till fitted with steam-steering gear at the same time as her change of armament. Her full complement was 550 ranks and ratings.

From the first her designer realised, in connection with the highly important matter of her machinery, that, although the *Bellerophon*'s engines would have about 1500 tons less dead weight to send along than those of the *Warrior,* the shape of her underwater body would more than cancel the advantage.

A vessel of her build does not split the water in a fair cleavage. She shoves a great heap of it in front of her, and drags another astern by suction to fill the void of her passage; and it was one of the service jokes of her early days that the flag-lieutenant used to report to the admiral that the *Bellerophon*'s bow wave was in sight long before he reported the ship in sight herself. Less machinery space was available in her design than that of the *Warrior;* but improvements in boiler construction compensated for that by enabling eight boilers with a pressure of 27 lb. to supply more drive to the pistons than the *Warrior*'s ten with a pressure of 20 lb. only. She was trunk-engined by Penn, with the 112-inch cylinders of which that firm cast so many; but, in view of her exceptionally bluff lines, it was deemed advisable to ease the immediate load on her machinery by reducing the pitch of her screw. This, of course, demanded a more rapid rotation to obtain any given speed of ship than in the vessels with a coarser pitch; and on her full-power trials it was found that with a maximum of 74 revolutions to the minute as compared to the 54 of the *Warrior* each ship travelled 14 ¼ knots at load draught. On a purely paper calculation the older vessel should have been slightly the faster, but the *Bellerophon* proved to be one of the ships with the rare and accidental peculiarity of no propeller slip. Thanks to this and a fine pitch she required very little boiler power at low speeds, and steamed very economically for a vessel of her size (without compound engines) until driven hard.

Under canvas the *Bellerophon* was a dull mover in spite of a disconnecting propeller; because although requiring as much force to send her through the water as the *Warrior,* she spread a smaller sail area. Her best sailing speed ever logged was 10 knots before a moderate gale under treble reefed fore and main topsails and reefed foresail; and her longest 24-hours' run was 182 miles under single reefed topsails, whole courses and topgallant sails. Moreover, it was a curious fact that she was very sluggish in her steering under canvas, though extremely easy to handle as a steamer. As her sail plan was well balanced this could only be explained by assuming that a large body of following water hung closely under the counter through the suction of her very full run. When the propeller was revolving this was thrust back against the rudder and gave the helm ample effect, but if the engines were stopped the rudder was in dead water. As a result she constantly missed stays in trying to tack; and in spite of drawing 4 feet more aft than forward she absolutely refused to wear, which was rather remarkable. Indifferently though she sailed, however, she covered a vastly greater mileage under canvas during her career than the great majority of her armoured contemporaries, through her exceptionally continuous employment on a distant station after her first two commissions.

For nineteen years the *Bellerophon* was ship-rigged, with spars (on fore and main) of the same scale as those of the *Minotaur*; but as she was not one of the vessels originally fitted with double topsails her tops were placed 5 feet higher on the lower masts and her lower yards proportionately raised, thereby increasing the drop of her courses by the same amount. The heels of the topmasts were also necessarily lifted in a like degree, and thus her total height from main truck to deck was 158 feet to the five-master's 152. In 1885 the yards on her mizen were removed. She carried the usual boat complement of a battleship of her period with the addition of an admiral's 14-oared barge when she was employed as a flagship.

The *Bellerophon* was built at Chatham in the historical dry dock which was the cradle of iron ship construction in the Royal dockyards, and was herself only the second Government built iron-hulled vessel. Her keel was laid as soon as the *Achilles* had been floated out; and her own floating took place in April 1865. When ready for sea the following year she hoisted the pennant at Portsmouth under Captain Tatham for service in the Channel Squadron; being then the most powerful ship under the white ensign. With that squadron she remained for five years and then spent a year in the Mediterranean. It was during her Channel Squadron time that she narrowly escaped ending her career by a collision with the *Minotaur,* when forming column on leaving Belfast Lough. Thanks to having a good swing on under her own helm she avoided the flagship's ram and received a blow from the bluff of the latter's bow, whereby a leak was started, but the water was kept under till she reached dock.

At the end of her year in the Mediterranean she was recalled home to be paid off and fitted as a foreign station flagship. The *Royal Alfred*'s second commission on the North American and West Indies station was drawing to a

H.M.S. *Bellerophon* photographed whilst under sail between Nassau and Bermuda in 1891 towards the end of her commission on the North American Station. By this time she had been rearmed, and reduced to a barque. (NMM. HAY/5)

close, and so the *Bellerophon* was chosen for the next flagship and her poop was added. In October 1873 she hoisted the flag of Rear-Admiral Wellesley and took her departure to begin her duties on the station, which were eventually extended to four long commissions covering an aggregate of fourteen years. This, though broken at half-time for refit and re-armament in England, established a still standing record for length of service by any British armoured vessel as flagship on the same foreign station.

Her passage out was made via Madeira to get the benefit of the north-east trade winds. It was marred by a regrettable occurrence, shedding incidentally a light on a bygone side of seafaring life. In the days of long oceanic passages extending over many weeks when wireless telegraphy was unknown, it was quite common for ships meeting in mid-ocean to communicate by boat and exchange newspapers and general information. When therefore the *Bellerophon,* being about a week out from Madeira under sail with fires drawn, sighted the Liverpool steamer *Flamsteed* eastward bound from the West Indies, she made a signal to the latter asking for newspapers. The captain of the *Flamsteed* very obligingly stopped his engines while the *Bellerophon* sent a boat, and as the ships were drifting apart he again very obligingly gave the boat a tow back. But through a misjudgment in handling he approached with too much way on and ran right into the battleship, stoving in his own bows. Working parties were rushed on board to help in stopping the leak and jettisoning cargo, but all in vain. Within a few hours the *Flamsteed* sank in mid-Atlantic, after passengers and crew had been transferred to the scarcely damaged *Bellerophon;* which had then to raise steam and take them to the Cape de Verdes as the nearest port to await shipment for home, while she herself resumed her voyage to Bermuda where she arrived after a good run of eighteen days.

For two successive commissions of more than three years each she served continuously on that station; flying in turn the flags of Admirals Wellesley, Cooper Key, and Inglefield. Then she was ordered home, paid off, re-fitted. In 1885, when all this was completed, she returned to her former duties (which the *Northampton* had carried out in the interval) for seven years more, as flagship to Admirals the Earl of Clanwilliam, Lyons and Watson; coming home again finally in 1892 when twenty-seven years old.

After that she became port guardship at Pembroke Dock till struck off the effective list in 1904 and relegated to harbour service as a stokers' training hulk at Plymouth as *Indus III.* In this employment she lay for many years in the Hamoaze till sold to P. and W. McLellan for breaking up on December 12th, 1922. She arrived at Bo'ness in March 1923, sixty years after the laying of her keel.

Although the *Bellerophon* was the prototype of the British central battery ship, she was one of the twelve vessels in the battlefleet of 1870 that had no sister or even half-sister in her actual dimensions and construction. By their individuality each of these vessels acquired a character in the service distinctively her own; this applied in a marked degree to the 'Old Billy' as she was known not only throughout the Navy, but to the shore populations, white and black, throughout the station where for so many years she was a familiar object in the harbours of Halifax, Bermuda and the West Indies. No ship outside home waters was more sought after by officers of all ranks, indifferent though her officers' accommodation might be as compared with that of her later contemporaries.

The 'sixties of the nineteenth century were a decade of intensive advance in the application of iron to the requirements of sea war. Half a dozen new calibres of wrought iron rifled ordnance made their appearance during those ten years, each one larger than its immediate predecessor, and exactly as many increased thicknesses in armour were successfully rolled in new mills. Thus it came to pass that before the *Bellerophon* was even half built (though she was built quickly) the plans for another central battery ship, to be her superior in all things but speed, were already well in hand. When the *Bellerophon*'s plans had received Board approval they presented a design intended to carry the 9-inch gun which was then the latest and heaviest in existence; but while her framework was still rising at Chatham the Ordnance authorities were anticipating the birth of a 10-inch gun at Woolwich, and the armour makers were offering the Admiralty plates that would defy the *Bellerophon*'s attack at the acceptedly normal fighting range of 1000 yards. Foreign Powers being active, it was quite impossible for the British Admiralty to stand still, and drawings were accordingly prepared for a new vessel suitable to receive this increased offensive and defensive equipment. These were in readiness for work to begin in the dry dock where the *Bellerophon* was being expeditiously hammered together as soon as she was floated out, and they bore the caption *'Hercules'.*

When completed for sea the *Hercules* was the most formidable ship in the world in her turn. She was the first to carry 9-inch armour; and in the British service the first to carry a pointed instead of a rounded ram, as well as recessed embrasures at the angles of her main battery to increase the arc of fire of the end guns. Her bow had the unique peculiarity in a British ship of a gun port cut exactly amidships through the stempost at main deck level. Her stern was of the same general type as the *Bellerophon*'s, but armoured as high as the upper deck. She had a topgallant forecastle from the first, but no poop till she was prepared for flagship duties after her second commission. Though her proportion of beam to length was practically the same as in the *Bellerophon* on a larger scale, her maximum beam did not extend so far towards her extremities, and thus she had better lines. She was also a steadier gun platform and a better sea boat except for an excessive tendency to yaw in running. Thanks to increased experience and improved methods in iron shipbuilding her hull only cost £ 40,000 more than that of the *Bellerophon,* though her displacement was 1100 tons greater, and her machinery actually cost less. She carried a balanced rudder with a threefold hand wheel under the break of the poop, and another threefold hand wheel below on the main deck.

Inboard the general arrangements and sub-division of the *Hercules* were on the same general principle as the *Bell-*

H.M.S. *Hercules* showing her 9-inch chase gun, the picture taken about 1875. (NMM. PX.21213)

Sketch showing proportion of furlers to each yard on the fore and main in ships with the spar plan of the *Hercules*. In furling the sails on arrival in harbour the topgallantsails and royals were furled before anchoring and the topsails and courses after. But if the sails had been loosed to dry they were all furled simultaneously, under the executive orders 'away aloft', 'lay out', 'second reef earring and furl', 'up bunt whips', 'lower the booms', 'down from aloft'; the time occupied between the first order and the last being from five to seven minutes. This was followed by the pipe 'square yards, haul taut ropes'. All square sails were furled on the head, except the topsails which were furled on the second reef on account of long leeches. The sketch shows the men passing in the leeches before gathering up the slab. Details of the rigging are incomplete.

erophon's; but she was noteworthy as the first capital ship in British history where the cables led in on the upper instead of the main deck. The change in her case was not made from any particular belief that it might be advantageous in a general way. Indeed, it was roundly condemned by some of her officers on the grounds that it made the cables more likely to get foul of the ram. But the presence of a centre line main deck gun with a port cut right in the stem rendered the construction of a manger impossible; and in the absence of a manger a wholesale flooding of the main deck through the hawsepipes was inevitable when working cables in a head sea. If the cables came in on the upper deck no manger was necessary, as a similar flooding there mattered nothing; and in any case was less probable on account of the extra height above water. Like the *Bellerophon* she was fitted with cat and ground chains. Her 6-ton bower anchors were of the Rogers' pattern with fixed iron stocks, but her 5 ½-ton sheets were 'Admiralty' wooden stocked. She carried the usual stream and two kedges, all with movable iron stocks; except that her sheets were before instead of abaft the fore rigging, all her seven anchors were stowed like those of the *Warrior* and her cables were also identical to that ship's.

In the matter of offensive power the *Hercules* was one of the four battleships in 1870 with the rather undesirable complication of a three-calibre armament, though designed to receive the new 10-inch gun as her main weapon, which no other vessel then carried. The carriage was lower in proportion to the size of the gun than in the smaller calibres, but the slide so much higher that it had to carry a raised rear platform for the gunlayer to stand on instead of the deck. Recoil was checked by an external bow compressor acting frictionally on the slide girders, and a 10-inch breeching. The slide rollers were circumferentially grooved to straddle a ridge in the arc of the racers. Eight of these guns were mounted in the central battery, of which the four at the battery angles had two ports each, one on the broadside, and one in an embrasure, or as it was called then an 'indented' port. Prize-firing records show that the highest possible rate of aimed fire for an exceptionally smart gun crew was one round in 70 seconds with the 18-ton muzzle-loader.

In addition to her 10-inch ordnance the *Hercules* carried a 9-inch main deck chase gun in the bow with three ports, and a similar gun at the stern; also a pair of 7-inch on the upper deck under the topgallant forecastle and another pair on the upper deck right aft. The main deck bow gun suffered (though in a lesser degree) from the same disadvantage as the *Bellerophon*'s, that it could not be used in steaming against a head sea, and no later ship therefore mounted a bow gun below upper deck level. For saluting she was supplied with eight brass howitzers; and in 1878 two 14-inch torpedo carriages were placed on each side of the main deck. In 1886 she was fitted with a net defence. From first to last her heavy armament remained unchanged except for the replacement in 1892 of the 7-inch guns under the topgallant forecastle by a pair of 6-inch breech-loaders. Some light 6-pounder quick-firing guns were mounted on her bulwarks at the same time as an anti-torpedo defence. Her full complement was always 650 ranks and ratings.

Her armour was 9 inches thick on the sides and waterline, tapering to 6 inches at her extremities. She carried transverse battery bulkheads in addition to armoured screens covering the bow and stern as high as the upper deck, all of 6-inch thickness. But she had no conning tower, as these had temporarily gone out of fashion altogether. The truth was that the armoured 'pilot towers' tried in a few earlier ships were no use whatever, having neither steering wheels nor engine-room telegraphs inside. Everybody condemned them, and the argument that they would protect the captain from being shot by a sniper, like Nelson, was met by the sufficiently obvious reply that if he resorted to a position from which he couldn't possibly handle the ship, he might as well be shot before going into action.

It was always an open question as to whether the *Hercules* was the first ship of capital rank in the British service that could not hoist out her heaviest guns with her own appliances, ready for use at short notice. The 7-inch guns of her period could be put over the side by the ordinary boomboat tackles, as was done when the *Agincourt* was aground; and the 8-inch or even 9-inch by rigging the main yard as a well-secured derrick and using both sets of toptackle gear and both cat purchases, as was done when the *Lord Clyde* was aground. But a 10-inch gun weighing 18 tons was a problem which never happened to call for actual solution. Rope of sizes large enough for a purchase was available in the 6-inch and 8-inch hemp hawsers of the *Hercules,* and her main yard supported by a spare topmast would have been sufficient for a derrick, but her largest blocks were only tested to 12 tons working strain. Perhaps it may be said that the operation might have been attempted with a fair chance of success, but absolutely no margin of safety, in the event of an extremely urgent situation where risks had to be accepted.

Like so many of her consorts the *Hercules* was trunk engined; but Penn was now aiming at higher efficiencies and jacketed her cylinders, with successful results in preventing loss of pressure by radiation. The change increased their external diameter to 118 inches as compared to earlier designs, with added weight in proportion; but resulted in economy of fuel, and in conjunction with an increased steam pressure in the boilers, and better speed lines in the hull, enabled the *Hercules* to travel slightly faster and farther than the *Bellerophon* for an equal indicated horse-power and consumption of coal, though 1100 tons heavier. Her nine rectangular boilers were constructed to stand a pressure of 30 lb., and her two telescopic funnels had the novelty of an oval instead of a circular section.

Towards the end of the 'eighties a policy of attempting to modernise some of the older battleships found considerable favour at Whitehall. The expenditure involved was doubtfully justified if only because their armour was of the obsolete unhardened type with very low powers of resist-

H.M.S. *Hercules*, as Flagship at Malta about 1875. (NMM. HBB/1)

H.M.S. *Hercules*, as she appeared in 1887, with searchlights mounted on her poop and her fore bridge, and with light quick firing guns along her topsides. (NMM. HBB/2)

ance to high velocity projectiles; but in the case of turret ships such as the *Devastation,* which could be satisfactorily re-armed with breechloaders as well as compound-engined, modernising did increase their offensive efficiency appreciably. The Admiralty, however, extended the process in half measure to several of the central battery ships (which could not be re-armed satisfactorily without being almost re-built from the main deck upwards, as the experiment with the *Bellerophon* had demonstrated) by equipping them with most expensive new machinery while leaving them with their obsolete guns. Thus while the *Bellerophon* had been re-armed but not re-engined, the *Hercules* was in 1892 re-engined but not re-armed; although being twenty-four years old she had become quite outclassed as a battleship, and nothing could make her fast enough for a cruiser.

In spite of all this her old machinery was pulled out for replacement by inverted triple-expansion engines made by the Greenock Foundry Company, and her nine rectangular boilers were changed for eight steel cylindrical. By this the steam pressure was raised from 30 lb. to 140 lb., and the maximum number of revolutions attainable from 64 to 88. As the pitch of her propeller was reduced from 24 to 17½ feet the extra speed of the engines only added a bare knot to the speed of the ship, while the vibration was excessive as compared with that produced by the discarded machinery. The one definite gain was a great reduction in fuel expenditure, but as she only did one week's sea service during the remaining thirteen years in which she was still retained on the effective list, practically none of the money sunk in the new engines was ever recovered in saving coal; and the machinery was still almost unused when she was disarmed and relegated to harbour service in 1905.

As a new ship she had no machinery outside the main engine and boiler rooms except the combined steam pump and capstan, which was becoming a universal fitting in the new ships of the later 'sixties; but in 1874 steam-steering gear was added, of which the wheel was placed in front of the hand wheel under the break of the poop; and in 1892, when her yards were removed, she received a steam hoist for the substituted boat derrick.

Except in a few points of detail the spars of the *Hercules* were of the same dimensions as those of the *Achilles.* Her jibboom was shorter, but she carried a flying jibboom which the other vessel did not. The proportions of the lower masts also differed a little owing to her designer's partiality for shortened doublings. The reduction of the doublings, with its corresponding raising of the tops and lower yards, increased the drop of the courses (as already observed in the case of the *Bellerophon*), and as the *Hercules* was the only one of the six specially heavily yarded vessels in the Navy to carry the shortened mastheads, her mainsail was the largest sheet of canvas in the whole service in 1870; though, owing to the lesser height of the fore mast, her foresail was no bigger than those of the *Warrior, Black Prince* and *Achilles.* Her slightly higher maintop than that of those other three also made her the tauntest ship in the fleet from main truck to deck, with a total of 178 feet. Her sail area inclusive of stunsails was about 49,400 square feet, of which 24,400 were square sails, 9,700 fore-and-aft sails, and 15,300 stunsails. In spite of this great windtrap aloft, with its many miles of standing and running rigging, the *Hercules* responded indifferently to its purpose, her best record under sail being only 11 knots, which was only 1 knot better than the *Bellerophon*'s. But she rode admirably in lying-to under storm canvas; and, unlike the *Bellerophon,* she was reliable in tacking and wearing, doubtless because her finer run allowed a freer movement of water on the rudder. When supplied with triple expansion engines her whole sail plan was removed, and she was left with only two pole masts carrying fighting tops, a boat derrick and a truck semaphore.

The keel of this fine battleship — sometimes called Reed's masterpiece — was laid in the dock at Chatham in 1866, whence she was floated out in 1868. She was a complete success and source of well-justified satisfaction to her able designer and the officers who subsequently served in her. It would be going too far to say that she was faultless, but her faults lay in minor directions; and though she never had an exact sister vessel her general model was considered worth reproducing, with the drawbacks of a three-calibre armament and main deck chase guns eliminated. This provided the design for the *Sultan,* built immediately afterwards.

Being ready for sea the *Hercules* was commissioned at Chatham in 1868 for service in the Channel Squadron by Captain Lord Gilford, an officer with few equals in his day. For six years she belonged to that force, during which period it fell to her lot to be rather prominently associated with an event of sensational public interest, the grounding of the *Agincourt* on the Pearl Rock. When that vessel had been three days' hard and fast and another ship had failed to move her, Gilford was ordered to try by anchoring the *Hercules* close astern of the stranded ship, passing chain cables, and then going ahead with his engines and heaving in on his own anchor simultaneously. He was to keep two important considerations in mind, though their demands were rather contradictory. Firstly he was to be very careful not to get the *Hercules* aground herself as the tidal currents were tricky, and he was to handle the situation cautiously. Secondly time pressed and all delay was to be avoided. The calm weather, which had fortunately lasted for several days, showed signs of breaking, and if a swell set in (as it actually did a few hours later) a complete wreck of the *Agincourt* was inevitable.

What followed can only be fully appreciated by those who have personally handled a large single-screw ship in a restricted space, whether as captain or officer of the watch. Consummate judgment was required to drop the *Hercules'* anchor exactly right both as to bearing and distance from the *Agincourt*'s stern; and as the *Hercules* had no twin screws wherewith to manoeuvre into position, this had to be done entirely by very accurate use of helm. Watched by the whole squadron Gilford got under weigh and made his approach, dropped his anchor plumb on the right spot, slewed his ship partly by snubbing the cable and partly by helm and reversed his engines when he had got her stern pointed, till the two vessels were stern to stern only a few yards apart, and a stout hawser could be quickly passed to keep them so during the long job of hauling in and securing two chain towing cables; which had to be done by hand. When at length these were made fast on the main deck of the *Hercules* he hove in on his own cable, started his engines ahead — slowly at first but gradually working up to full speed — and eventually dragged the *Agincourt* afloat once more, cheered by all the other ships. The unhesitating directness with which the whole operation was carried through left a lasting impression on the officers who witnessed it. Steam handling was still in its early stages, and this was an object-lesson in its higher application.

At the end of the six years under various captains she was paid off at Portsmouth in 1874 to be re-fitted and prepared to relieve the *Lord Warden* as flagship in the Mediterranean. When ready next year she hoisted the flag of Admiral Drummond and proceeded to her new station, where she remained till relieved by the *Alexandra* in 1877. On returning to Portsmouth she was paid off once more, and after being specially commissioned for a few months in 1878, as flagship of a large force temporarily mobilised and assembled at Portland under Admiral Cooper Key, in view of the strained political situation existing during the Russian advance on Constantinople, she was transferred to the Clyde district as guardship of the Reserve for three years. In 1881 she was selected as flagship for the Duke of Edinburgh on his appointment as Admiral Commanding Coastguard and Reserves, and as such led the Reserve Fleet on a cruise to the Baltic. This sea had been unvisited by any British fleet since the close of the war with Russia twenty-two years before; but as the reigning Tsar Alexander was a brother-in-law of the Duke's, through the latter's marriage to the Grand Duchess Marie, the fleet was most hospitably received on anchoring off Cronstadt. As flagship of the Reserve Fleet, the *Hercules* afterwards carried Admirals Hoskins, Baird and Tryon, and continued to serve for nine years until relieved for a second time by the *Alexandra* in 1890.

Having by that time completed twenty-two years of continuous commissioned service with only two short breaks, her engines were worn out and the ship herself had become virtually obsolete for war requirements. Nevertheless she was taken in hand for the ineffectual modernising process already mentioned, and after much money had been spent on her lay for the next twelve years swinging round her moorings up Portsmouth harbour. In 1904 she was temporarily employed for the routine harbour duties of the Commander-in-Chief's flagship, but next year was removed from the effective list and sent out to Gibraltar, where floating accommodation had to be found for the largely increased number of workmen arising from the great extension of the dockyard, for whom there was no room on shore. As a barrack ship for the dockyard she lay alongside a jetty for nine years, being re-named *Calcutta* in 1909, as her name was appropriated for a new battleship.

* * *

By 1914 accommodation afloat was no longer required at the Rock, and the *Sutlej* was sent to tow her home, her own engines having been allowed to fall into ineffectiveness to save cost of upkeep. But her hull was in such good condition that the Admiralty selected her for indefinite further harbour duty as one of the four hulks forming the artificers' training school at Portsmouth where she became *Fisgard II* in 1915. Finally she was sold to Messrs Ward in July 1932, and having been stripped at Morecambe, arrived at Preston for breaking up on December 1st, 1932.

CHAPTER SIX

THE 'LORD CLYDE' AND 'LORD WARDEN'

H.M.S. *Lord Clyde* aground on the island of Pantellaria in the process of lightening ship prior to the attempts to refloat her. (NMM. 58/3699)

When, at the end of the 1850s, the conception of armouring ocean-going vessels had taken concrete shape, the British Admiralty had adopted iron hulls as their future choice on general principles, and were extending the machinery for their production; but two important considerations decided them nevertheless to continue timber construction for a time on a limited scale in conjunction with iron. In the first place the still rather limited iron-building resources of the United Kingdom were not capable of launching an iron-hulled armoured fleet keel for keel against the wooden fleet building for the tricolour. And, secondly, considerable quantities of oak lay in the Admiralty timber ponds as material already in government possession. From time immemorial it had been their shrewd practice to anticipate the requirements of national defence twenty years in advance; so that, after cutting the trees, a full period should elapse for seasoning the logs before taking them to the saw pits. Not having had any reason to foresee the change to iron this custom had continued until the change took place, with the result that a certain quantity of oak remained on their hands. In addition, therefore, to converting certain wooden ships already building into armoured types before launching, the Board ordered Chief-Constructor Reed to submit designs for new wooden armoured ships to use up the remnant of the timber; even though some of it had not been long enough felled to ensure thorough seasoning.

Reed had just completed his designs for the *Bellerophon* and would have liked to duplicate her, but no wooden hull of her length could carry that vessel's weight of guns, machinery and armour, especially in so concentrated a form. While adhering, therefore, to the same total weights he reduced the length but increased the beam of his new model, and spread the burden of armament over a greater number of weapons of lighter calibre. When completed he submitted the drawings to the Board of a larger, faster and more powerful wooden ship than had ever been laid down before, which were ordered to be used in duplicate for building the *Lord Clyde* and *Lord Warden;* while a certain residue of timber still left over after providing for these two was to go to a much smaller armoured ship, afterwards named the *Pallas.* Curiously enough, one of this pair of fine sister vessels turned out to be the very best investment Whitehall ever made in a wooden armoured hull, and the other the very worst.

To the *Lord Clyde* and *Lord Warden* belonged, in consequence, the distinction of being the largest, and under steam the fastest timber-hulled ships ever launched from British shores for purposes of either war or peace; a record which is safe now to stand in their names for ever. They represented the supreme achievement of the British shipwright who worked with the adze and the augur, the mallet and the saw, knee deep in chips and shavings, and surrounded by dockyard scents and sounds which vanished for all time when the spicy aroma of fresh planed oak was replaced by the fumes of the foundry, and the rumbling thunder of the mallets by the harsh metallic clang of the riveting hammers. Moreover, apart from their size, these two ships were unique in the history of naval architecture in certain features of their system of construction.

Their officially calculated displacement was 7840 tons, which was about 1000 tons more than the largest three-

H.M.S. *Lord Clyde*, at anchor at Plymouth about 1866, as originally completed. (NMM. HAO/1)

H.M.S. *Lord Warden* refitting in dock about 1875, showing the full lines favoured by Reed, her bronze-capped ram and the timber waterline belt outside of her armour. The head projecting over the stem, with its outflow on the waterline. All main rigging is down, and the mast stayed by runners and tackles. (NMM. PX. 21214)

deckers. Actually, however, the *Lord Warden* must have been the heavier of the pair by quite 300 tons, as when both were completed for sea she was found to draw 28 feet aft and 24 feet forward to the *Lord Clyde*'s 27 and 24. This was due in part to slightly heavier engines and a poop, but mainly to untraceable causes; because it was never possible to anticipate the displacement and draught of a wooden hull with the same accurate precision as an iron one. Theoretically the specific gravity of each kind of timber was a constant quantity, but in practice it varied a little with age and density of grain, and thus, though two wooden ships might be built from the same plans, their actual displacements when afloat were seldom precisely equal. Moreover, a wooden hull was sometimes insufficiently rigid to withstand the fluid pressure of being waterborne, without a slight loss of shape and increased immersion.

Unlike the converted wooden battleships, which were all straight-stemmed, the *Lord Clyde* and *Lord Warden* were armed with prodigious pointed rams of solid oak tipped with a massive bronze casting, somewhat in the style of their opposite numbers in France. Aft they carried the oval type of stern, with a counter below the waterline, and a wooden rudderpost joined at its foot to the keel by a stout bronze bracket. In entry and run their lines were even fuller than the *Bellerophon*'s, with a very square midship section; but in spite of that their maximum beam extended over a less proportion of their length, so that their waterline plan was less like a barge or pontoon. They had the *Bellerophon*'s defect of rolling like a pendulum in an even worse degree, from the same cause of an excessively low centre of gravity, but rode buoyantly to a head sea owing to their very bluff bows and light draught forward.

In actual structure their sides were unique, having no less than five alternate layers of iron and timber over the greater part of their length at the waterline. First came the main framework of 24-inch oak ribs; and not only were these more closely spaced than usual, but for the whole length of the main battery the intervals between the ribs were completely filled up with wood of the same stoutness from 6 feet below the waterline to the upper deck. Over that solid 2-feet timber shell was laid a skin of 1 ½-inch iron as a binding; and on that again a layer of 6-inch oak planking to receive the armour. Thus all over the central parts of these ships the armour rested on 31 ½ inches of unbroken structural solidity. Along the waterline the armour itself was 5 ½ inches thick, tapering to 4 ½ at the

Sail plan of the *Lord Clyde* to scale, exclusive of stunsails. Also of *Lord Warden* and *Bellerophon* except as regards bowsprit. They set 31,000 square feet exclusive of stunsails. (From Admiralty records.)

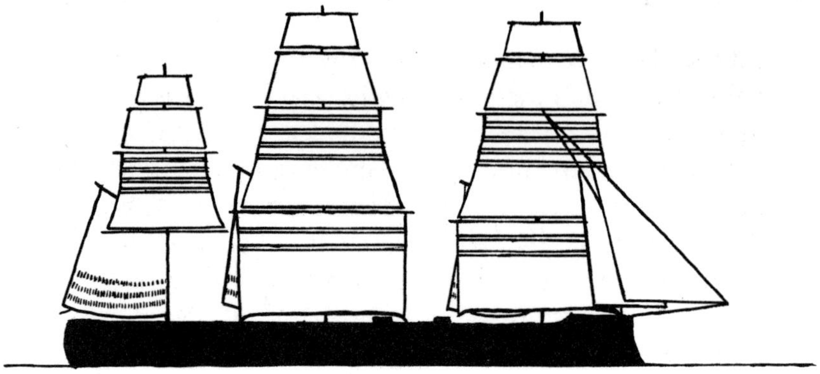

bow and stern, with 4 ½ plates on the battery and forecastle screen. But outside the armour at the waterline came yet more wooden planking (as in all the wooden-hulled battleships) to prevent the water from getting between the joints of the plates and under their lower edges and rotting the inner timber. A belt of 4-inch oak was carried with that object right round the hull as an outermost casing, from 2 feet above water to 6 feet below, and formed the fifth and outermost layer in the side structure of the *Lord Warden* and *Lord Clyde.* And finally this belt and the whole bottom were cased with a thin sheathing of the alloy known as Muntz[1] metal, to protect the wood from fouling and the ravages of the teredo.

The armour of these vessels not only covered every square inch of their exterior surface above water (except the ports) to the level of the upper deck — but was even carried up at the bows to the full height of the topgallant forecastle as a frontal shield. Through that extension they were the most completely covered ships in the British Navy in the matter of external protection; but they had no internal armoured bulkheads.

Up to this point in their description the pair were built exactly alike. But beyond it they differed in two points of external detail. The *Lord Warden* carried a poop and a light seamen's head outside the true bow like the *Bellerophon,* with small heads for the petty officers projecting outboard from the bulwarks in the fore part of the waist on each side. The *Lord Clyde* had no poop, and the heads for the whole ship's company were in the waists.

Internally they were constructed on the same general plan as all the contemporary British battleships, whether iron or wood, in so far as decks were concerned. But certain novelties in the structural application of material were introduced in their building, of a kind foreshadowing the composite type of naval architecture that found high favour at Whitehall a few years later for smaller classes of ships. For instance, although the cross beams and stanchions carrying the main and lower decks were oak, those carrying the upper deck were iron. And even more noteworthy was the division of the hold into five main compartments by four transverse iron bulkheads. These were not intended to be watertight, as it was scarcely practicable to partition a wooden-framed ship watertightly on account of the insufficient rigidity of her structure; but they strengthened the hull athwartships. Iron bulkheads also enclosed the coal, and the cable bitts and side bollards were cased in iron.

Of the five main compartments so created the central and largest held all the boilers, bunkers and machinery in one; without any division between the engine room and the stokehold. The compartment next forward of that carried as its chief contents the water tanks, chain lockers, sail room and cable tiers, and the one next forward of that again (which was the bow compartment) the forward magazines and shell rooms, and the gunner's, boatswain's and carpenter's main store rooms. The compartment next abaft the engines contained the provision rooms, bread room and slop room, and the stern compartment the after magazines and shell rooms, and the engineer's and officers' stores. In the sub-division of each main compartment iron-bar gratings were extensively used instead of bulkheads; so that they were partitioned into cages rather than sub-compartments, which economised structural weight.

Over the hold lay the lower deck right fore and aft. Being free from wing compartments and transverse bulkheads this deck was utilised (as in frigates) for berthing every soul on board except the captain, commander and navigating officer. At its after extremity stood the wardroom, in the run of the ship and therefore triangular in shape. It was spacious enough, but only the fore end received any illumination by the skylights lying one above the other in the two decks above, and the rest of it was too dark to read in without a lamp. Cabins extended the full length of the wardroom to starboard and port, and a wooden cross bulkhead enclosed it on the forward side and divided it from the adjoining steerage. On the starboard side of the steerage lay the engineer officers' and warrant officers' messes, and on the port the gunroom. In the centre stood the midshipmen's chests and here their hammocks were slung at night. Two steerage cabins cleared of all fittings were extemporised as steerage officers' bathrooms, with tubs and tin cans which they had to provide for themselves. The fore end of the steerage opened right on to the ship's company's mess deck without even a canvas screen. Here the long row of mess tables lying athwartships extended forward the whole remaining length of the deck on both sides, except for some bag racks in the bows. This deck was divided in the fore-and-aft line by the funnel casing, the engine room hatch and three ladder ways; but had no transverse partition of any kind, so that when the steerage hammocks were piped up in the morning the steerage officers shaved and tumbled into their clothes at their chests in full view of the watch below engaged in scrubbing the deck and mess tables.

Having neither guns, nor wing compartments, nor bulkheads, a lower deck of that kind was commodious as a berthing space, but badly lit and worse ventilated. A row of 6-inch pigeon-hole scuttles both sides — cut through the whole 3-feet thickness of the side including the armour — gave a little fresh air in harbour; but as they lay only a yard above water they were necessarily always plugged at sea. Such light as reached the messes came by the hatchways at secondhand from the gun deck, and was barely sufficient to enable the men to see their dinners. On this deck also were slung 550 hammocks at night. It was only natural, therefore, that the sick list was always heavier in the ships where the hands lived on the lower deck than in the iron-hulled vessels where they lived on the main. Another drawback to existence in a wooden hull was the difficulty of keeping down rats and cockroaches. Even in a modern ship these pests may be sufficiently troublesome in hot climates, though there is little woodwork for the rats to gnaw through or the insects to breed in. On board a wooden vessel they sometimes became a perfect plague.

One notable compensation, however, a timber hull did afford, which sometimes made a great difference to officers and men. It maintained an equable temperature in the living spaces of both during extremes of climate. The thin shell of an iron or steel vessel keeps out neither the heat of the equator nor the cold of high latitudes, but a timber side defied alike the rays of the fiercest tropical sun and the bitterest frosts of Newfoundland or Shantung. Furthermore, it was entirely free from condensation and sweating of inner surfaces which is so common in iron and steel ships. Add to this that the men always disliked berthing on a gun deck, and we find enough reason for the fact that, at first, they were by no means enthusiastic about the conditions of existence in the new iron-hulled vessels.

Except near the bow and stern, the main deck of the *Lord Clyde* and *Lord Warden* was practically clear of everything but the guns on each side and the funnel casings, hatchways and galleys amidships. The captain's three cabins were on the port side of the half deck with the ship's office, and the commander's and navigating officer's on the starboard. In the centre stood the half-deck wheel, and in the open space right at the stern a 9-inch chase gun with three ports. The sick bay was forward on this deck, and as these two ships did not carry messenger capstans the cables only came in a short distance from the bows.

For steering, their wooden rudders received an iron norman head, with one iron tiller on the main deck above the stern gun, and another in the after part of the wardroom. In the *Lord Clyde* the upper tiller was worked by a threefold wheel under the bridge, and the lower by a threefold wheel on the half deck. The *Lord Warden* had a fourfold wheel under the break of the poop, a double wheel on the half deck, and another fourfold wheel in the steerage. All three could be geared to work in conjunction on the same vertical spindle. The *Lord Clyde* was conned from the bridge just before the mizen mast, and the *Lord Warden* from the break of the poop. Neither had any charthouse, but the latter ship carried rifle-proof conning positions inserted in the upper-deck bulwarks each side abreast the wheel; each of which gave a view outboard on its own side through loopholes, though with only room for one officer.

When first completed for sea the pair were differently armed; but after her maiden commission the guns of the *Lord Clyde* were changed to correspond to those of the *Lord Warden.* The latter always carried the same armament throughout her long service, though it was very mixed both in composition and arrangement. On the main deck she mounted two 7-inch in the bows — useless in steaming head to sea — twelve 8-inch on the broadsides and one 9-inch at the stern. On the upper deck one 9-inch under the topgallant forecastle and two 8-inch on the quarter deck broadsides just abaft the main rigging. For saluting she used a pair of 20-pounder Armstrongs. At a late stage in her career she was equipped with Whitehead torpedoes and a net defence; being the only British wooden vessel ever so provided except the *Repulse.* Her Quarter Bill was made out for 608 officers and men.

The *Lord Clyde* began her sevice with twenty-four 7-inch guns; of which two were under the topgallant forecastle and two on the quarter deck broadsides. The remaining twenty were all on the main deck, and included a pair at the bows and a pair at the stern. Four years later she was re-armed on the same plan as her sister ship in the matter of ordnance, but did not last long enough in service to be given torpedoes.

Both ships carried nine rectangular boilers supplying steam at 28 lb. pressure, with thirty-two furnaces and two funnels; and both were peculiar among British wooden-hulled battleships of the period in having the engine room abaft the main mast as in their iron-hulled consorts, instead of before the main mast exactly half-way between bow and stern. But the engines themselves differed fundamentally in design with rather remarkable results. The contract for the *Lord Warden*'s was given to the Maudslay firm, who provided her with a three-cylinder set of the return connecting-rod type, that lasted out eighteen years' work. The contract for the *Lord Clyde* went to Ravenhill and Hodgson, who put in double-cylinder trunks, which became scrap iron in a single short commission. It was all a matter of foundation strains on the engine beds.

Owing to the comparatively inferior rigidity of even the best timber, one of the chief problems associated with the installation of powerful reciprocating machinery in a wooden ship was to steady the cylinder sufficiently against the strong reactions of the moving parts. These were somewhat complex, and varied with different types. Thus in trunk engines the up-and-down reaction of the connecting-rods bore on the cylinders themselves instead of on crosshead guides, and tended to produce a rocking motion of the bed plates. And in two-cylindered engines — whether trunks or otherwise — the horizontal reactions of the piston movements were especially severe; because the planes of the two cranks were set at right angles to each other on the shaft, in order that each might carry its partner over the dead centres at the ends of their respective strokes, and so maintain a combined uniformity of rotative effort all through a complete turn of the propeller. But under that arrangement both pistons are necessarily on the push during one part of the circle, and both on the pull

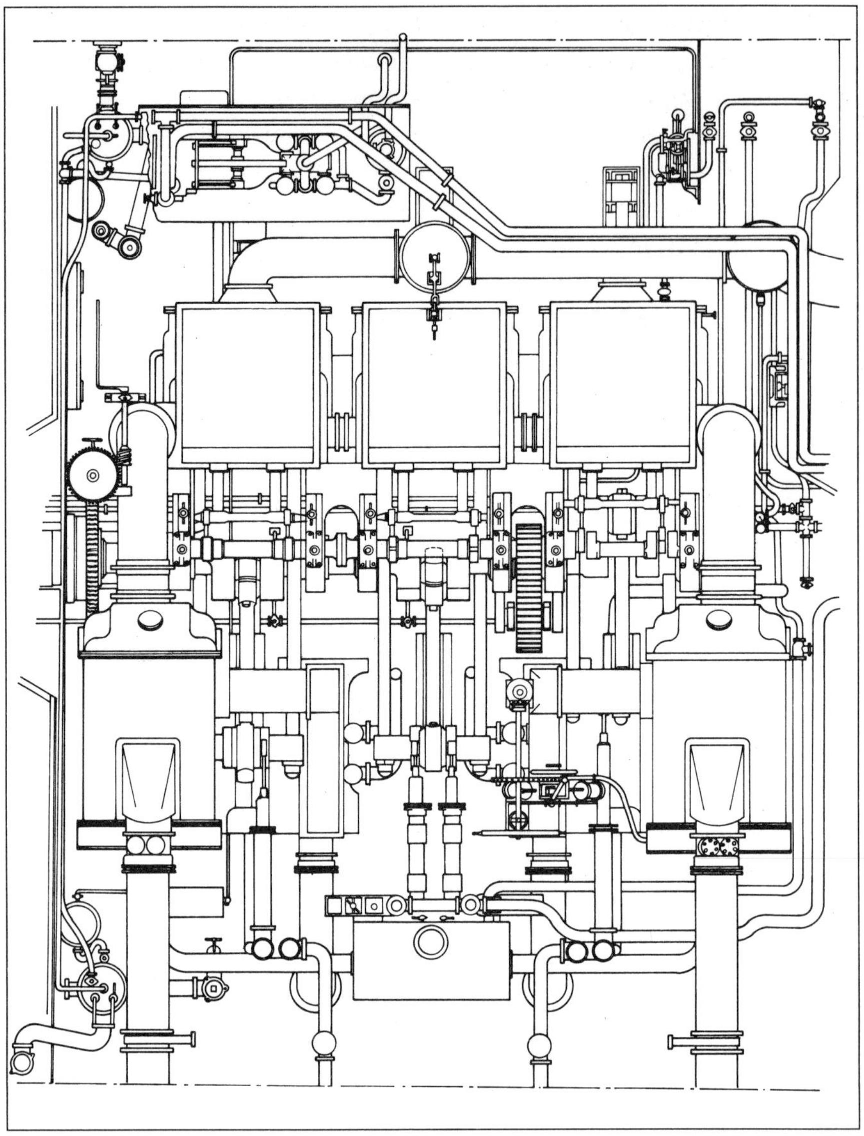

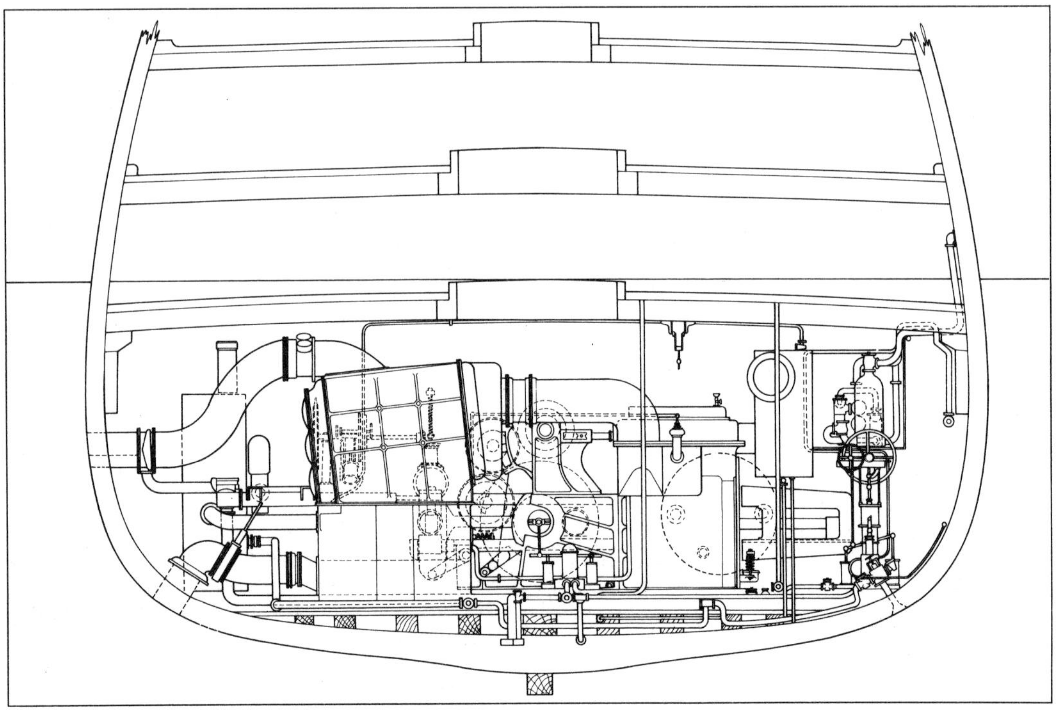

Above: plan of the triple-cylindered return connecting-rod engines of H.M.S. *Lord Warden*. Port is above in the plan, starboard below. Left: Athwartship section through engine room, looking forward.

during another; with the consequence that the reacting stresses on the cylinder foundations are alternately doubled in opposite directions in every full revolution. With three cylinders, on the other hand, the cranks are set at angles of 120° to each other; whereby the reactions of the piston movements are mutually balanced all through and the strain on the foundations from that cause reduced to a minimum, though such engines are rather heavier and more expensive in construction than others of the same power.

Maudslay attached importance to the above considerations, and equipped the *Lord Warden* with the triple-cylindered return connecting-rod machinery which of all kinds throws the least reactionary force on the bed plates. These engines proved to be not only the most powerful, but the most reliable ever put in a British wooden hull. She passed through her acceptance trials without a hitch, as well as the annual full-power tests that became a rule at a later period; and her engines would always take as much steam as the boilers could feed them. Incidentally they were the only triple-cylindered engines in the Black Battlefleet of 1870, though there is little question that if wooden shipbuilding had continued they would have established a general type for subsequent adoption.

On the other hand, the prominent Ravenhill firm and the Admiralty were both impressed by the undoubted success of the fine twin-cylindered trunk engines of the iron-hulled *Warrior, Resistance,* etc., and made insufficient allowances for the fact that only their iron hulls enabled them to withstand the reactions of that type of machinery, which are severer than in any other for the reasons already given, though it has advantages in other ways. Assuming that the wooden *Lord Clyde* should be able to stand these strains, the contractors furnished her with the largest pair of trunks so far made. These engines were a trifle lighter and cheaper than those of the sister ship and occupied rather less space, but proved quite insuitable. No actual breakdown occurred during the trials, and the contract speed of 13 knots was duly attained; but it was probably fortunate that, as those were the days before periodical full-power tests, the *Lord Clyde* was never driven hard again, for the rocking of her cylinders soon caused trouble. Within two years the oscillation caused by mere cruising speeds had worn them out so completely through the erratic travel of the trunks and pistons that the whole machinery had to be renewed except the shafting and condensers; and special precautions taken as to bolting down. Whether the fresh set would have lasted any better than the old can never be known, however, as the career of the ship herself came to an end within six months of hoisting the pennant for a second commission.

The *Lord Clyde* and *Lord Warden* were both masted on exactly the same plan as the *Bellerophon* with the same short doublings, iron lowermasts and steel yards and topmasts

H.M.S. *Lord Warden*, an early photograph taken about 1866 while her forecastle was painted black. (NMM. HAP/1)

which, in their case, were never changed for wood as in other ships. This meant that, except in their lowermasts, they were under-sparred for their size. Moreover, on account of the projecting head on her bows, it was not considered necessary to give the *Lord Warden* a jibboom, and for setting the jib she only carried a lengthened bowsprit. With an inferior spread of canvas and very full lines the pair were necessarily slow sailers, the best logged speed of the *Lord Clyde* being 10 ½ knots and of the *Lord Warden* only 10. But they were extremely handy in tacking and wearing, and through that, if in nothing else, the otherwise persistently unlucky *Lord Clyde* did leave a claim to distinction by the double record of being the largest ship of any class or kind ever to enter either Spithead or Plymouth Sound under sail alone; performances the more noteworthy from being accomplished under unfavourable conditions at both anchorages.

In all their general equipment of boats, anchors, cables and other gear, their store establishment was identical to that of the *Bellerophon,* with the same patent capstans and cat chains. Each stowed three months' salt provisions for 600 men.

Both ships were laid down in 1863, but the *Lord Clyde* took the water first. She was built in the old Admiralty yard at Pembroke, and seldom did the blocks receive the keel of a more chronically unfortunate vessel. Being ready for sea in 1866 she hoisted the pennant at Plymouth under Captain Dew for the Channel Squadron. Shortly afterwards disciplinary trouble occurred on board, followed by a court martial which disrated twelve petty officers and flogged several men. She then joined the fleet, which was cruising for a month in the north-east Atlantic to compare the seagoing qualities of some of its new ships. In his official reports on the *Lord Clyde,* Dew affirmed that though slow under canvas he had found her as handy as any frigate; which a little later he proceeded to prove. On the first occasion of being detached to go into Plymouth for stores he approached the Sound under double-reefed topsails in half a gale from the north-west, which compelled him to keep close-hauled to weather the breakwater end. On clearing it he rounded in on his weather braces and stood away for the anchorage, intending to make a running moor under sail; the most difficult but most beautiful of the old sailing movements, in which a ship kept her topsails set till the first anchor was let go and whipped off every stitch of canvas before the second. But the *Lord Clyde* carried more way than he had allowed for and marred the performance by parting the first cable before the second anchor was dropped, obliging him to remain content with a stern moor using bower and sheet instead of bower and bower. Never before, however, or since, did any ship of the *Lord Clyde*'s dimensions enter the Sound without steam.

Her next remarkable exploit was to beat into Spithead by the east entrance *against* the tide; an operation demanding even higher class seamanship than the first in a vessel of nearly 8000 tons and 27 feet draught. Roderick Dew was an officer with no lack of pretty severe critics at an earlier stage, but nobody ever accused him of professional incompetence or want of nerve. He was ordered to Spithead on some matter of minor service not calling for dispatch, and sailed up the Channel from Plymouth before a fresh south-wester, but found the weather very thick on nearing the Needles Channel before daylight, which (as more than one naval officer had had good reason to know) is a risky approach from the westward if the light is obscured. He stood on south of the island, therefore, still resolved to complete the passage under sail; though to reach Spithead from the east (with the wind as it then lay) he was now committed to an 8 miles beat as the last stage, through a narrow channel with dangerous shoals on either

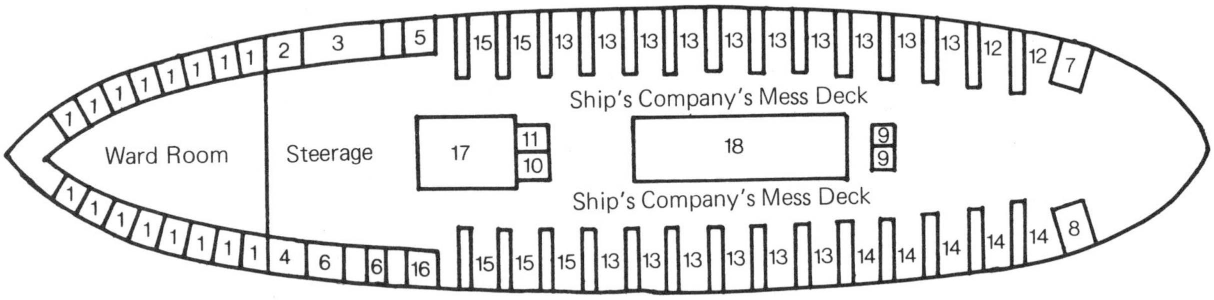

Scale diagram of general arrangements for berthing and messing officers and men on the lower deck of a British wooden-hulled battleship of 1870. Details differed with individual vessels. 1-1: Wardroom officers. 2: Wardroom pantry. 3: Gunroom. 4: Engineers' Mess. 5: Washing cabin. 6-6: Warrant officers' mess and cabins. 7: Chief Petty officers. 8: Engine-room artificers. 9-9: Master-at-arms and ship's police. 10: Sergeants of marines. 11: Officers' cooks and servants. 12-12: Petty officers' mess tables. 13-13: Seamen's mess tables. 14-14: Stokers' and day men's mess tables. 15-15: Marines' mess tables. 16: Ship's steward. 17: Engine room hatch. 18: Funnel casing. Engineers' and Gunroom officers' chests were in the steerage. Ship's Company's kitbags were stored in racks, between the funnel casing and the engine-room hatch, and in the forepart of the deck.

Cases of lower yards being carried away or sprung occurred in several mid-Victorian battleships besides the *Lord Clyde* and the *Lord Warden.* This sketch shows the usual method of hoisting in a new main yard with the ship's own gear. The spare jibboom was rigged as a derrick having the two mainstay tackles as topping lifts, of which the long tail pendants were hooked round the lower masthead to their own parts. Two sailtackles formed the guys, with the foremost one hooked far enough forward to allow the yard to pass underneath in swinging it athwartships. A luff was used as a martingale, with the lower block hooked to a strop round a large toggle across the inside of a main deck gunport. The heel of the jibboom rested on a wooden chock with a lashing to deck bolts. The derrick purchase consisted of two catblocks and one catfall with the hauling part taken through a leading block on the short leg of the masthead pendant and down to the deck, where it was manned by about 200 men. The yard was towed alongside lying fore-and-aft, having the chain necklace of the slings and the jeerblocks already in place. The purchase was then hooked to a stout lashing on the necklace, and the jeers rove and taken to the capstan. When ready, the yard was hoisted by the derrick high enough to clear the bulwarks, and swung athwartships by light steadying lines hitched to the yardarms. The weight was then transferred to the jeers by heaving round the capstan and easing the purchase till the yard hung amidships, when it was lowered on the crutches for rigging and fitting before swaying up into position. The sketch shows the point at which the jeers were just beginning to take the weight. (From a midshipman's log of 1877.)

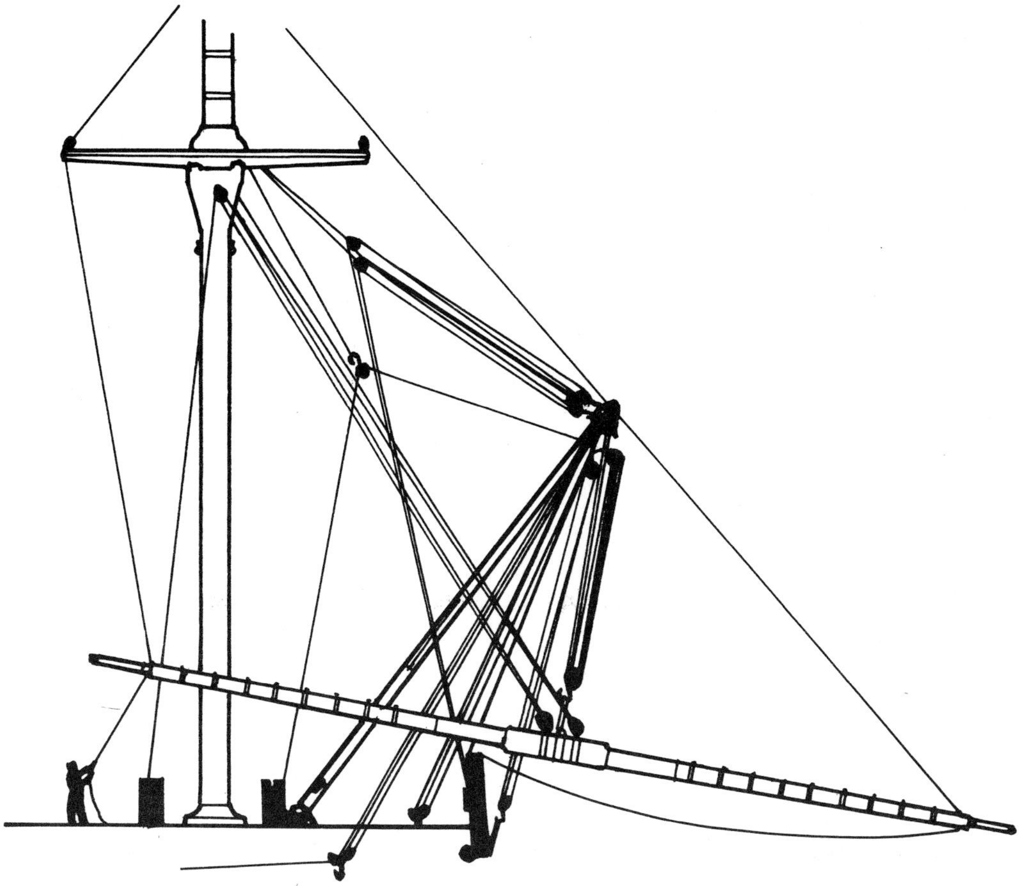

hand, and worst of all on a falling tide. But he kept to his purpose, and on passing St Catherine's shaped course for the Nab.

The ebb was running when he neared the lightship, and if the wind had been blowing straight down the fairway from Spithead no square-rigged vessel ever built could have tacked through the bottleneck between the Horse Sand and the Warner before slack water. Even with the wind about four points to the southward of dead foul — as it happened to be — five ships out of six would have anchored at the Nab to wait for the flood stream. But Dew calculated that the difference would enable him to stem the tide by long legs on the port tack; and on reaching the lightship, soon after seven o'clock, hauled to the wind to begin the long beat up. Never once did the *Lord Clyde* miss stays (she would have been on the mud if she had) and at ten o'clock she shortened sail and dropped her anchor at Spithead, the largest ship of all the thousands that ever arrived in that historically famous roadstead under the impulse of the wind alone. She carried all plain sail, less royals, till rounding the Nab, but the extra strain of bracing sharp up split the main topgallant sail, and from there she had to work through under topsails and courses only, with a wind of force 6 to 7.

By this time she had completed fifteen months' service in the Channel Squadron — including three as temporary flagship — and the inherent unsuitability of her engines had begun to produce constant trouble. In spite of this, however — or possibly because of it — she was now sent to join the Mediterranean, where at any rate she was not likely to be driven against quite such heavy seas as in the Bay or Atlantic. From that point onwards her history would have made good material for an article on the Soul of a Ship if treated by a writer with the imaginative gift of Conrad, or Kipling or 'C.F.S.'. It really seemed as if in spirit as well as substance the *Lord Clyde* belonged to the navy of the past and had no use for steam. She had shown what could be done with her under the motive agency of the old order, and apparently would rather perish than submit to the facile handling of the new. Perish at any rate she did in a very literal sense, at an early age.

Arrived on her new station she took part in one spring cruise with the squadron, in the course of which she carried away her steel mainyard in a violent squall. It was a torn and complicated fracture instead of a clean break, and the accident started the prejudice in the Navy against steel spars on account of the long time required for repairs. With a wooden spar her own staff of carpenters and shipwrights would have had the fishplates on in three or four days; but all the blacksmiths in the squadron assisted by engine-room artificers were unable to get the steel yard ready for sending up again under a fortnight.

Meanwhile the state of her engines became worse and worse, and on arrival at Naples a survey by the fleet engineers reported that they were no longer safe for use. On receiving this report the Admiral dispatched her under sail to Malta to see what the dockyard could do; and under canvas she entered the Grand Harbour without trouble or difficulty. There, after examination, the yard staff found that nothing short of new cylinders would be satisfactory, although they could patch up the machinery enough for a passage home. This was done accordingly by Admiralty orders, and she limped back to Plymouth to be paid off after only two years' service. Dew went to the *Northumberland,* a ship that declined to manoeuvre under sail as obstinately as his late command under steam; and a new set of engines was made for the *Lord Clyde* at the Keyham factory of Devonport dockyard, which was one of the earliest large-scale engineering jobs not given out by the Admiralty to private contract. Opportunity was taken at the same time to re-arm her like the *Lord Warden,* and to fit her with a two-bladed propeller, experience having shown that these had just as good a grip on the water as the four-bladers, with less cost, less weight and much less drag under sail.

When all this had been done she was kept in reserve for eighteen months awaiting orders, and then commissioned a second time for the Mediterranean in 1871 by Captain John Bythesea, one of the half-dozen officers then serving who had gained the Victoria Cross as juniors in the Crimean War. This commission was even shorter than her first. Six months after going out she was lying at Syracuse and received a wire from Malta to proceed at once to the help of a British steamer aground on the island of Pantellaria, on which service she got far worse aground herself. Anchors were laid out, coal jettisoned, guns, ammunition and stores hoisted over the side into small coasters in the island hired on the spot, and everything possible done to lighten her, but she remained a fixture except to sway in the swell from the open sea, strain her back, and wrench off her sternpost, rudderpost and rudder. It took some time to summon help, as that was a full generation before the days of wireless telegraphy and Pantellaria had no cables; but an officer was dispatched by a passing steamer to Malta, where the *Lord Warden* was lying as flagship and came at once to pull her crippled sister off the floor and get her to bed. This proved an extremely difficult job even when she was afloat again, as she yawed about so violently without a rudder when in tow, as to pull the sister in charge all over the place. It took three days to conduct her erring footsteps a distance that could ordinarily be covered in less than one, and all the while she was leaking at a steady rate of 2 feet an hour.

On arrival at Malta she had to be docked with great care on account of the badly damaged state of her bottom; and the yard reported that it would take six months to repair. The inevitable court martial meanwhile sentenced the gallant Bythesea and his navigating officer to be severely reprimanded and dismissed their ship, with the result that neither was ever employed again. On receipt of the report from Malta the Admiralty decided that only sufficient external repair for a passage home was to be carried out, but even that required half a year and cost a lot of money.

From what transpired soon afterwards this money might just as well have been thrown into the sea. Quite apart from accidental damage the *Lord Clyde* was by now a doomed ship stricken with a fatal and incurable malady. She was escorted home by the *Defence,* with a navigating party on board supplied from that vessel; and on arrival at Plymouth her engines, boilers and hold fittings were removed to ascertain what internal repairs were required.

This disclosed a disastrous state of affairs as a consequence of her insufficiently seasoned wood having become a prey to the germs of timber fungus from bow to stern. Thenceforward the account of the efforts made to deal with that malignant complaint reads exactly like a fight for the life of a human patient. Every remedy known to the expert science of the day, including chemical antidotes, was tried in turn, but all to no purpose. The disease had passed the stage of possible cure and its further and deeper spread could only be checked by measures rendering the ship useless for sea duty, such as maintaining an artificially high temperature by numerous charcoal stoves in the holds. It was hoped that perhaps she might serve for a time as a drill hulk attached to the Portsmouth gunnery school and her machinery was removed in 1876, but after nearly three years of ceaseless though unsuccessful curative treatment even that idea had to be abandoned, as she was sinking into decay so fast that she had to be sold before being too far gone to find a purchaser. She went to the shipbreakers accordingly for a mere song in 1884, and so ended a brief existence in which she failed to complete even one full commission at sea.

The career of the *Lord Warden,* on the other hand, was as long and prosperous as that of the *Lord Clyde* was short and full of trouble. From first to last her timbers were sound, her engines ran true and her navigation remained free from serious mishap; and when after twenty-four years the Admiralty eventually removed her with some hesitation from the effective list, it was only because she had become out of date in design and not because she was rotten or worn out. It has been said that happy is the nation without a history, and the same applies in some degree to a fighting vessel in times of peace. Year after year the *Lord Warden* led the routine life of a mid-Victorian squadron battleship, with its endless round of cruises, drills, steam tactics, firings and periodical trials; its occasions of full-dress ceremony and its intervals for refit and overhaul; and

H.M.S. *Lord Warden*, entering Portsmouth towards the end of her career, very little altered in appearance from her early days. (NMM. HAP.3)

year after year she was always ready and reliable for the requirements of national defence should the emergency arise.

She was built at Chatham at the same time and in close proximity to the *Bellerophon* and launched in 1865. When ready for sea in 1867 she joined the Channel Squadron for a few months under Captain Rolland, and then went to the Mediterranean, relieving the *Caledonia* there two years later as station flagship. As such she served till 1875, flying in succession the flags of Admirals Milne, Yelverton and Drummond. After eight years of continuous service in the Mediterranean, including six as flagship of the Commander in Chief, the *Lord Warden* was paid off in 1875, on relief by the *Hercules,* and after a thorough overhaul commissioned in the following year as guardship of the First Reserve in the Forth; serving as such for nine years more. In 1884 the ship was still in such good condition as to be considered worth fitting with Whitehead torpedo carriages and net defence; but in 1885 she was paid off finally and placed in the dockyard reserve, though not removed from the effective list even then.

For four years to follow the question of her sale came up annually with that of other old ships in preparing the Naval Estimates, but was always negatived on the ground that she continued stout and sound except for slight leakages in her seams. In 1889, however, the Board eventually decided that a battleship with no watertight compartments had no longer any place in a fighting navy, and sent her at last to the shipbreakers, who could scarcely find a rotten plank in her massive fabric of thousands of tons of seasoned English oak. All her wooden-hulled consorts of the battlefleet had gone before, except the *Repulse,* which was broken up only a year later.

No other ships of the Royal Navy ever bore the name *Lord Warden* or *Lord Clyde.*

CHAPTER SEVEN

THE 'MONARCH'

H.M.S. *Monarch* getting her torpedo nets out during manoeuvres at Queenstown in 1889. (NMM. 58/6653)

At the beginning of the year 1870 the Black Battlefleet under the White Ensign included two vessels in its composition, which though by no means sisters except in armament, were nevertheless widely different from, and superior to, the twenty-eight others in combatant design. These were the *Monarch* and *Captain;* the first sea-going capital ships, as distinct from mere coast defence craft, to be armed on the centre line principle in the whole of naval history. The *Monarch* in 1870 was the legitimate pride of the British service and the openly expressed admiration of those under foreign flags. The name of the *Captain* was associated before the end of that year with the worst disaster that befell the forces of the British Crown throughout the very long reign of Victoria, other than at the hands of a human enemy.

The story of the origin of these two forerunners of all existing heavy armament carriers is remarkably interesting, and centres chiefly on the inventive activities of an officer who met his untimely end before reaching flag rank, and yet left a name more familiar to the British public of his day than that of any contemporary flag officer through the converting inspiration of his masterful genius. For if that word may be taken as properly applicable to far-seeing vision, refusing to be trammelled by tradition, convention or compromise in its unremitting advocacy of measures necessitating the complete abandonment of an age-long and universally accepted practice in the teeth of great opposition, then genius can hardly in reason be denied to Cowper Coles. But it must in truth be added that with all the confidence and persistence of genius he combined in full degree its impatience of any opinion or outlook differing on the slightest point from his own. And therefore it was that although his main theories on armament received in the end full recognition not only in his own science but in all others, his insistence on attaching conditions to their application of a kind which many of his brother-officers instinctively distrusted, led to dreadful catastrophe involving his own death.

The first article of the faith which Cowper Coles preached with such emphasis laid down that as a fighting ship is solely built to fight, if so required, she should carry her arms in such a way that she can make the utmost use of them. Admitting that when a hundred guns more or less were in the case, no practicable alternative could be found for the immemorial and universal custom of disposing them along the two sides of the ship — even though under such an arrangement only half could be used at a time, unless engaged on both hands — he insisted that where only a dozen or less were to be mounted, as was then becoming the case, they should be so placed that all could be brought into action on either side.

In many cases the chase guns of existing ships answered to that already. So did the entire armament of small craft. But such guns were pivoted at the ship's side and fired through fixed ports; and the operation of traversing them right across the ship from starboard to port or port to starboard was a lengthy process at all times, requiring special care if the ship was rolling, in addition to always heeling her towards the engaged side and thereby reducing the maximum possible of elevation. Cowper Coles had mounted a gun during the Crimean War for bombarding purposes, with an all round fire from a turntable on a stout cask raft, and covered it with a shield like a dish cover, which revolved with the gun; and this was the original turret. Here was an arrangement which if placed on a ship's centre line on deck would enable a gun to be fired on either hand and trained from one to the other in a matter of seconds whether a ship was rolling or steady. The conviction that this was the best method of mounting the heavy armament of an armoured ship became in consequence the

H.M.S. *Monarch* at Malta about 1880, cleared for action, upper masts sent down etc. (NMM. HBR/1)

moving inspiration of the rest of his life. Less than two years after Coles had expressed his views before a public audience in England, his theory of a centre-line armament carried on top of a low hull was receiving a practical application by the Swedish engineer Ericsson, under the ordeal of actual war in the domestic feud of the United States; and up to a point proved very successful. The American vessel which provided the lesson — rather aptly named the *Monitor* — was little more than an elaborate iron raft in shape and proportions. Above her flat armoured deck she showed nothing except her funnel and a circular armoured turret mounting a pair of Rodman smoothbores with an almost all-round fire, which also served as a conning tower. Within a few hours of going to sea, she met and successfully engaged a far more heavily armed broadside opponent, and returned quite undamaged. Having thus demonstrated her powers in one respect, she gave an equally thorough display of her limitations in another by foundering with all hands on her first occasion of meeting a gale. But the success of the turret was undeniable, and as a first step therefore the British Admiralty accepted the system for coast-defence ships; which just at that period were being regarded as a valuable element in any navy through a precise inversion of the true teachings of the Crimean War. The result was seen in the 'flat irons' *Royal Sovereign* and *Prince Albert.*

To Cowper Coles, however, that was only a beginning. In spite of the fate of the *Monitor* he insisted that low freeboard was not incompatible with full ocean-keeping qualities, and offered the advantage of presenting a small target to an enemy; and to his way of thinking the turret system of carrying guns was properly inseparable from a deck lying comparatively close to the water, which seemed to remain an obsession in his mind through the memory of his early experiences of such a combination. He prepared a complete design accordingly for a low freeboard capital ship with an exceptionally powerful turret armament, and when the Admiralty suggested as a compromise that the armament he proposed should be adopted but mounted on the deck of a hull of normal height above water he declined to meet them.

Thereupon the Admiralty referred the idea to a picked committee of officers and members of the Chief Constructor's Department, who produced the plan of a vessel answering the Board requirements in addition to excellent sea-keeping qualities; from which was built the *Monarch,* the first blue water turret ship. She carried a fine entry, a pointed ram of medium projection, a stern of the same type as the *Bellerophon* and a balanced rudder. It would be hard to say that a more successful ship was ever launched in so far that she did everything expected of her and floated

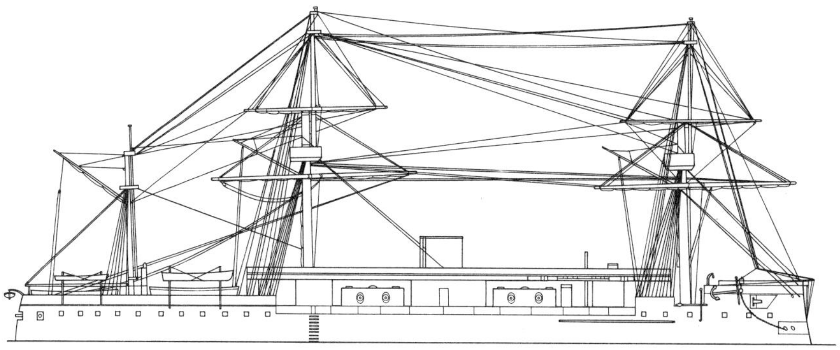

Outline sketch to scale showing the *Monarch* prepared for action in her barque rig. Topgallant masts and stunsail booms sent down on deck, sails unbent, mizen topmast housed, bowsprit run in, top screens rigged, waist bulwarks lowered outboard, yards braced up. Preparation for action aloft was usually a monthly drill evolution occupying from eight to twelve minutes.

exactly at her previously estimated draught. As tried after commissioning she proved a steady gun platform and buoyant seaboat, fast under both steam and sail, by the standards of her day, easy to manoeuvre, and comfortable in her living quarters for officers and men. Quite apart from the great constructional advance in power of attack which she represented, her general proportions gave a better combination of speed and handiness than those in any other British ship launched in the 'sixties; and exhibited a temporary reaction against the demand for an excessive shortness of model which was becoming increasingly prevalent at that period. Twenty-five years later, when the cult of the 'battle-tub' had passed, the *Monarch*'s ratio of length to beam became on a larger scale the accepted standard for British battleships with very little variation up to the Great War.

Internally she carried the same three decks as all capital ships of her period except the *Captain,* and was built under the same general system of watertight sub-division as the others with iron hulls. In its arrangements the partitioning of the main deck resembled that of a central battery ship with her guns omitted, as the armoured compartment enclosing the bases of both turrets occupied a position amidships precisely corresponding to that of the battery in such vessels as the *Bellerophon* or *Hercules*; with the whole of the rest of the deck outside it fore and aft used for berthing officers and men. All the ship's company except the marines lived forward of the central compartment, but the marines' 'barracks' came next abaft it and were separated from the officers' quarters by a light bulkhead. The captain's quarters were right at the stern; and most of the commissioned officers' cabins, as well as the engineers' mess and gunroom, were on the half-deck; but the old cus-

H.M.S. *Monarch* as she appeared in 1887 with quick-firing guns on her flying deck, with torpedo net booms stowed on the after end. (NMM.)

tom of placing the wardroom down on the lower deck, as in a frigate, was repeated.

The *Monarch* had a topgallant forecastle but no poop. Internally the construction of the former was peculiar; being divided athwartships by an armoured bulkhead into two compartments, of which the foremost was the ship's company's head and the aftermost a chase battery containing two bow guns firing through embrasures. This differed from the arrangement in the *Bellerophon,* where the head was an added external projection over the bows.

Though closely resembling a central battery ship in the interior plan of her hull, her upper deck was quite unlike that of any previously built sea-going vessel as a necessary consequence of taking in the armament platform. The two turrets stood close together in the central body, absorbing the space used in her predecessors for the boomboats and spare spars; and as the gun muzzles fired across the deck only three feet above it, the high bulwarks topped by hammock nettings universal in all classes of British sea-going ships in the Royal Navy at that time were impossible in the *Monarch,* where their place was taken by low iron plates, hinged to drop outboard for action or target practice. Some alternative stowage became necessary in consequence for the boomboats, the spare spars and the five hundred hammocks. The launch was placed therefore in the middle of the quarter deck, and hoisted in by a main yard purchase and two stay tackles between the main and mizen. A steam pinnace nested in the launch, and a sailing pinnace was stowed right aft between the mizen and the taffrail, which was hoisted in by using the spanker boom as a derrick. To carry the hammocks a light narrow superstructure like a fore-and aft bridge extended above the turrets from near the fore mast to the quarter deck, with the nettings forming parapets each side throughout its length, and ladderways leading right up from the mess deck through vertical trunks or casings near each end. Projecting angle-irons or brackets on either side of the fore part of this superstructure carried the spare spars. The large funnel and the main mast passed up through it.

For steering purposes the *Monarch* was equipped with a steam wheel in the conning tower — the first ever so placed — a threefold hand wheel on the quarter deck and another below, with a tiller on the main deck and a second on the lower. For ordinary peace purposes she was handled by the quarter deck wheel and conned from the small quarter deck bridge. Thanks to her balanced rudder the helm went over easily. Her cables were worked by a steam patent capstan on the fore main deck, and her anchors catted by the cat chain system.

In describing her formidable powers of offence the first point to be noted is that the *Monarch* was the first ocean-going vessel in which the three principal features of the great Victorian change in fighting-ship design were all combined. That is to say she fought under steam instead of sail; was built of metal instead of timber; and carried her main armament outside her hull instead of inside. All later developments in the battleships built since were merely advances along the lines thus established, except in the introduction of underwater appliances for attack and defence. Steam remained in use as the motive power, metal as the material of construction (though in steel instead of iron) and main armaments were still mounted above the hull until the end of the battleship era. But as being only the initial step in the process of evolution where the arrangement of guns was concerned, the influence of former practice showed itself in their grouping. Her main armament was closely bunched together amidships like that of a central battery ship, leaving the end-on fire to chase guns at her extremities; whereas in later turret ships the turrets stood well separated, and end-on fire was included in their arcs of training. As regards the fore-and-aft disposition of her ordnance she was in effect a central battery ship with the guns carried higher and moved back from the sides.

The two drum-shaped turrets had a diameter of 26 feet, a height of 7 feet above the deck and oval ports for a pair of 25-ton muzzle-loaders in each; these having a double historical interest not only as the first weapons ever to be fired from the centre line of a capital ship, but as the first of rifled make in the British service to throw a 12-inch projectile; which continued for half a century to be used in more than a dozen successively evolved marks and weights of guns from mid-Victorian days till after the Great War. The ancestors of all this family, with the same sized mouths but different voices, were the quartette that looked out from the turret ports of the *Monarch.*

They were mounted on high iron double plate carriages, travelling in and out on slides formed by fixed girders on the turret floor, and having three steps in vertical slots at their sides to take the gun trunnions at three different heights, so as to increase the angles of elevation or depression if required. Shifting steps was effected by hand-worked hydraulic jacks forming a part of the mounting. Like all muzzle-loaders they were run right back for loading till the muzzle was far enough inside the port to receive the charge and shell. If the gun had just been fired the recoil took it into that position, but at other times this had to be done by a rear winch with six handles outside the turret below deck. Loading, laying for elevation, and shifting steps were all

(1): Diagrammatic drawing of shot-trolley used on the main deck of the *Monarch.* The grab was slid on to the projectile as it lay on the deck or on a shot-mat. This arrangement was devised in the ship and was much quicker in operation than the alternative band and pram as supplied. By easing up with the handles the projectile was lowered on to the deck and thus anchored the whole apparatus. (2): Gear as supplied to lift projectiles from racks and to sling for hoisting into turret. For wheeling along a brass pram with four small wheels was supplied—a massive contraption, very slow in operation. The clip (3) was used in the flats and for lifting main-deck projectiles for cleaning, etc. The band was not used; its eye was the wrong way for the ship's carriage and the pram supplied was not satisfactory.

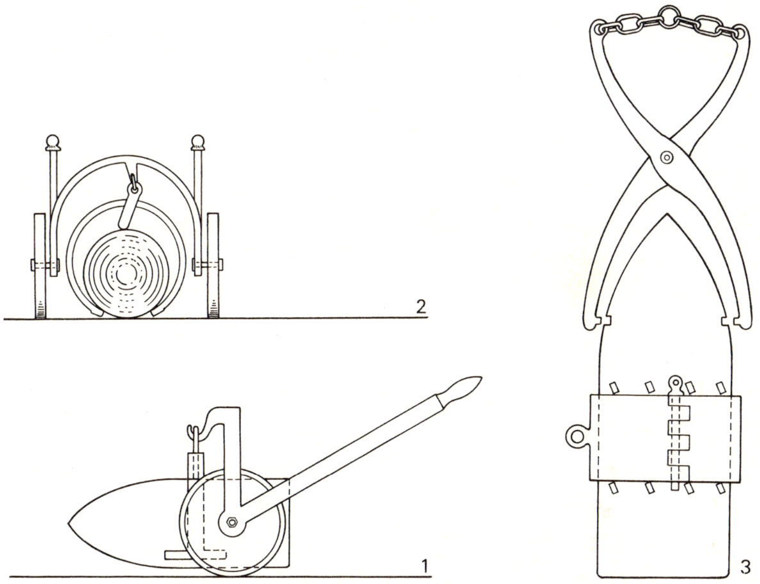

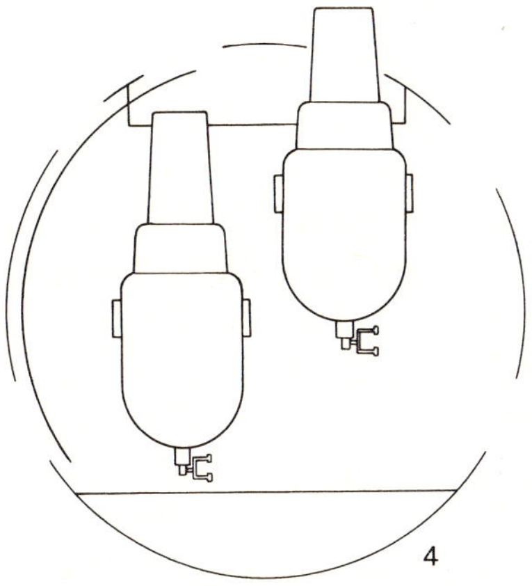

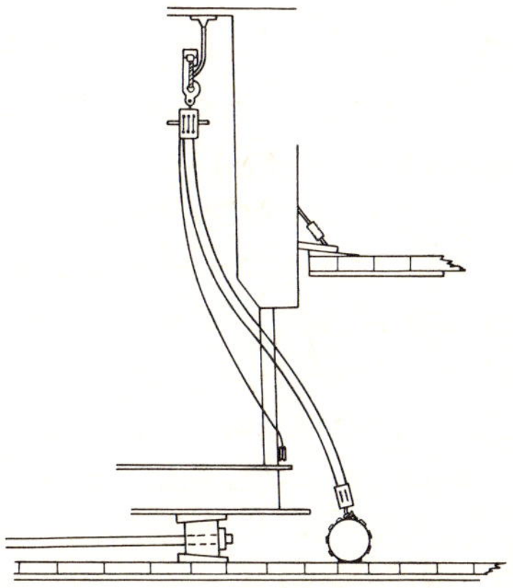

(4): H.M.S. *Monarch:* plan of turret (diagrammatic) showing entries and exits.
Right: Entry port to turret (section) through which the projectile was hoisted.

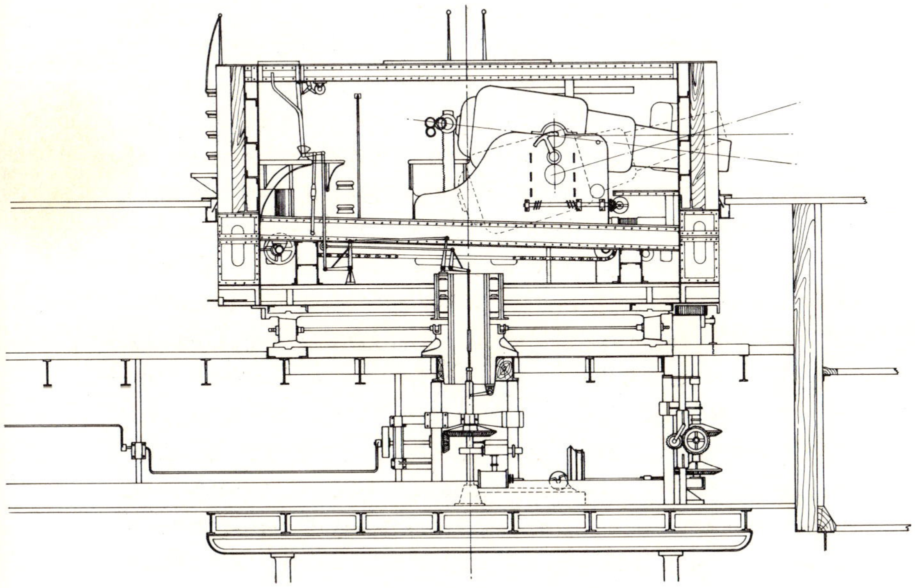

Left: Chatham Yard, May 5th, 1870. *Monarch.* Drawing of turrets.

carried out by hand, which in the confined space inside the turret entailed strenuous labour. Only a very smart turret's crew could maintain in 'simultaneous firing' a higher rate than one double round in two minutes. Steam was used solely for turning the turrets, with hand gear as an alternative. The guns were 17 feet above the water, which was 7 feet more than those of any other vessel in the battlefleet.

A pair of 7-inch guns were mounted under the topgallant forecastle, and a third at the stern (with three ports) which was the only gun on the *Monarch*'s main deck; but she was found to carry those forward so easily that they were replaced by 9-inch after her first commission. In 1897, when thirty years old, she received four 12-pounder and ten 3-pounder quick firers as an anti-torpedo armament, mostly mounted on the superstructure. Two torpedo carriages of the early moveable type were placed at a couple of main-deck ports in 1878; and net defence was added in 1885. Her guns being fewer than in ships of her size armed on both sides her complement was proportionately reduced, and though her broadside was heavier than the *Hercules*'s she only carried 530 ranks and ratings.

The armour of the main hull — as distinct from the turrets — was distributed on the same general principles as in the central battery ships; that is to say a complete waterline belt rose from about 6 feet under water to the level of the main deck for most of her length, but as high as the upper deck for a space amidships, with similar upward extensions at bow and stern as cover for the chase guns. The rise amidships formed the sides of the armoured box or citadel — with transverse armoured bulkheads for its ends — which in a central battery ship contained the main armament, but in the *Monarch* enclosed the turret bases as well as the major portion of the 12-inch shell supply. The side armour was 7 inches thick amidships, tapering to 4 ½ at extremities; the transverse bulkheads of the central enclosure being 4 ½ on the forward side and 4 on the after. A 5-inch screen round the bow guns on their forward side rose as high as the topgallant forecastle, and divided it internally athwartships through standing about 16 feet from the stem as already described. The wings of this screen extended farther aft on the upper deck than on the main; because on the former it was meant to afford side protection as well as bow to the guns, but below that level merely additional defence to the lower part of the hull against raking fire. The turret plating was 10 inches thick round the ports and 8 inches over the remaining circumference; and 8 inches also on the conning tower, which stood ahead of the forward turret.

The *Monarch*'s first set of machinery was made by Humphreys and Tennant, and consisted of a pair of 120-inch cylinders acting on the cranks by return connecting-rods with a 4 ½ foot stroke; each cylinder having four piston-rods instead of the usual two. Steam was generated by nine boilers working at 30 lb. pressure, with all the furnace uptakes heading to one large telescopic funnel of oval section. The ship attained a speed on trials of 14.9 knots at load draught, making her by about a quarter of a knot the fastest battleship afloat in 1870. Having a pair of turret engines in addition to a steam capstan and steam wheel, her equipment of auxiliary machinery was more extensive than that of any other vessel in the British service when she was launched. But her one serious point of inferiority to the *Captain* was the absence of twin screws.

Twenty-three years after her first completion for sea she was treated to the same half-measure of modernisation as her ancient squadron mate the *Hercules* — and in the same year 1892 — by being re-engined at considerable cost but left with her quite obsolete armament, though in her case re-arming offered no great difficulty. It would be a matter of historical interest to find on record the line of discussion that led to such a decision at the Board meeting (if any) where it took final shape; but official documents relating to the *Monarch* apparently go no further than to set forth the fact that the Maudslay firm was entrusted with the contract for new engines (specified in full detail) of the inverted triple expansion kind, and eight steel cylindrical boilers to work at 150 lb. pressure. Simultaneous instructions to the ordnance department merely approve the addition of light quick-firers for use against torpedo craft and leave the antiquated main armament of muzzle-loaders where it was. The new machinery at much expense raised her speed by half a knot and trebled her radius of action; but those advantages had no war value when tied to an armament that was unable to deal with even a light cruiser carrying up-to-date weapons.

Excepting in the lengths of her iron lowermasts the *Monarch*'s rig on first commissioning was one of the six of specially heavy scale in the fleet of 1870. All the remaining spars corresponded in full dimensions to those of the *Achilles,* and the lower and topsail yards and topmasts were steel. Her fore and main tops had a width of 26 ½ feet and her topmast crosstrees 18. In that rig she once logged 13 knots under plain sail and weather stunsails, which among British armoured ships was only exceeded by the *Royal Oak;* and her average under sail in crossing the Atlantic both ways equalled that of the wooden steam line-of-battle ships of the 'fifties and 'sixties. After her maiden commission however she was re-rigged as a barque, with topmasts reduced to 60 feet, topgallantmasts to 49, lower yards to 96, topsail yards to 68, and upper yards in proportion; all the steel spars being replaced by wooden. The original fixed iron bowsprit was also changed for a wooden

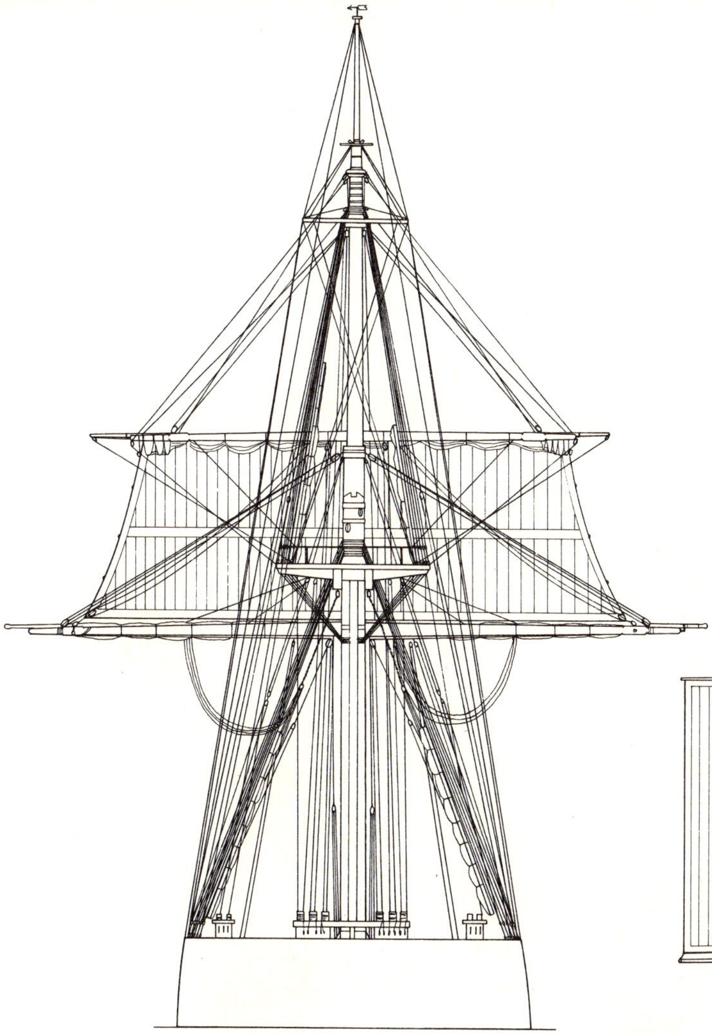

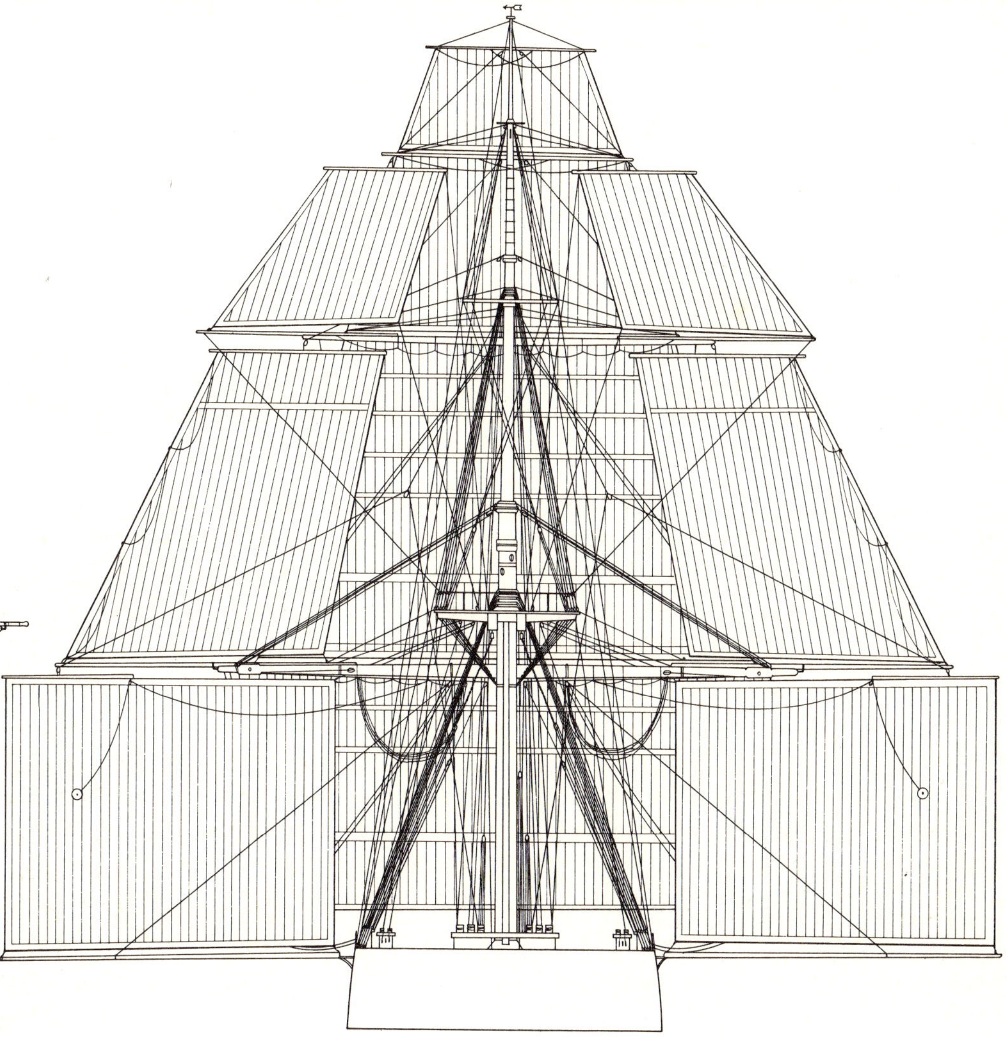

Left: Diagram on the same scale as that on the right, showing the minimum of square canvas on one mast in the British service rig of 1870. Fore mast of the *Monarch* with sail reduced to a close-reefed topsail and topgallantmast housed. A ship-rigged vessel in the Royal Navy of 1870 had eighteen square sails set altogether when carrying stunsails both sides, the mainsail being the only square sail clewed up. In shortening and furling when stunsails were set both sides, fifteen square sails were all taken in simultaneously, that is to say three royals, three topgallantsails, eight stunsails and the foresail. Twenty-six ropes were manned, and run away with all at the same moment on the single order 'shorten sail'; that is to say twelve upper clewlines, six stunsail downhauls, two topmast stunsail deck sheets, two lower stunsail tripping lines, two clewgarnets and two lower buntlines. Forty ropes were let go at the same order; that is to say six upper halliards, twelve upper sheets, eight stunsail halliards, eight stunsail tacks, four stunsail sheets and two fore sails. The stunsails booms were then rigged in by order, and the three topsails clewed up and lowered. The whole eighteen sails could be taken in under six minutes by a smart ship's company from the first order.

Right: Diagram to scale of the sail plan on the fore mast of the *Monarch* with stunsails set both sides in her ship rig; showing the maximum spread of square canvas on one mast in the British service rig of 1870. (Trysail and trysail gaff omitted, together with all rigging lying fore-and-aft.)

runner-in, lying on the topgallant forecastle. Though she sailed well with both rigs, the spacing of her masts was unsymmetrical in appearance by the normal standards. Owing to the position of the second turret the main mast had to be stepped considerably farther aft than was usual, and to maintain a balanced sail area the fore mast was moved proportionately farther forward; with the result that the span between the two was 155 feet which was 10 feet more than in the *Achilles,* though the latter was the longer ship by 50 feet. Through the height of the superstructure the mainsail and trysails had to be of special cut; and in order to minimise interference with the arcs of fire of the turrets the iron wire shrouds of the fore and main lower rigging were fewer than in other large ships, but individually stouter. Though not so handy in tacking and wearing as the converted ships, the *Monarch* was more reliable in stays than any vessel of her tonnage in the fleet of 1870. As the installation of triple expansion machinery in 1892 increased her steaming radius about three-fold, she was then stripped of all her canvas, and left with pole masts and a boat derrick with steam hoist.

Her equipment in anchors and cables was on a very heavy scale. Both bowers and both sheets were 6-ton iron stocked Rogers, in addition to which she carried a 50-cwt. stern anchor, a 35-cwt. stream, and a large and a small kedge, all iron stocked. With these were 575 fathoms of 2 ⅞-inch chain cable, 100 fathoms of 1 ⅝-inch, 100 fathoms of 1 ½-inch, one 18-inch hemp and two 12-inch. Cat and ground chains were used for stowing the bowers. The sheets were stowed immediately abaft the topgallant forecastle. As regards the boats, the placing of the launch and pinnaces has already been described. The regular complement of clinker-built boats was carried at iron davits in the usual positions.

The *Monarch* was laid down at Chatham in 1866, launched in 1868, and commissioned in May 1869 by Captain Commerell, V.C., for the Channel Squadron. Later in that year she was ordered on a special mission. A wealthy and philanthropic citizen of the United States, Mr George Peabody, had died in England after making a fortune in London business, and in official recognition of his colossal munificence towards London charities the British government offered the *Monarch,* as the finest ship in national possession, to take his remains to his native land. As an acknowledgment of this courtesy the American government sent the U.S.S. *Plymouth* to England to accompany the *Monarch* over; and the two ships left Spithead together on December 21st, under orders to touch at Madeira and Bermuda for coal on the passage.

The *Plymouth* was a wooden corvette of the old style that moved well under canvas, and it was generally assumed that the battleship would have to make considerable use of her engines or find the corvette shortening sail to wait for her. In light winds that might have happened; but the North Atlantic winter solstice is not noted for gentle breezes, and the reefed canvas weather they experienced suited the *Monarch* very well. When they reached Madeira it was the *Plymouth* that needed coal most. Still more was this the case on the seventeen days' run to Bermuda which followed. On that passage, before strong following winds, the two ships maintained the creditable average of 172 miles a day; but whereas the *Monarch* only used her engines for a day-and-a-half, and needed no coal at Bermuda (where she remained hove-to outside), the *Plymouth* had to go inside to fill up a second time. When she came out again the pair proceeded for Portland, U.S.A., but took another week to reach it owing to fogs. Here the funeral ceremonies of Mr Peabody were officially attended by the officers and men of both ships, after which the *Monarch* went on to Annapolis.

The interest awakened in America by this arrival of the first turret ship to cross the Atlantic fell little short of that caused by the arrival of the first British airship to reach New York nearly two generations afterwards. All ranks and ratings of the *Monarch* were received with most generous hospitality, and great crowds visited the vessel from Washington and the neighbourhood; including such prominent notabilities as the Secretary of the U.S. Navy, the Governor of Maryland, the Mayor and Corporation of Baltimore, and several of the principal naval and military victors in the then very recent domestic war such as Admiral Farragut and General Sherman. The British Minister at Washington brought down a very appreciative party of distinguished guests, who were taken out in the ship to witness some firing from the turret guns before she started for home. Leaving Annapolis on March 5th she reached Spithead on the 27th: the first eighteen days of the passage being made under sail and the best day's run being 242 miles; which was only one mile below the sailing record for a British armoured ship.

At the end of a further fifteen months' squadron service she was paid off for magazine alterations necessary for stowing the new pebble powder charges, and the opportunity was taken of her being in dockyard hands to change her rig as already described. In 1874 she was again commissioned for the Channel, but sent to reinforce the Mediterranean in 1876, at the time when the main fleet on that command were under orders to remain indefinitely near the Dardanelles during the progress of the Russo-Turkish war. In one of the frequent winter gales that sweep the open anchorage of Besika Bay (well known to many in the Great War) she nearly came to grief by dragging her anchors foul of the *Triumph* lying astern. In 1877 she returned home to pay off and refit, and in 1878 returned to the Mediterranean for seven more years of continuous service; a period in which she had her full share of the incidents attendant on naval duties. Her gunnery lieutenant was killed by the premature explosion of the primer of a spar torpedo which was being rigged for the naval peace time quarterly exercise in the steam pinnace — all the other officers and men in the boat being injured — and her whole main armament was employed for the first and last time on a war target soon after.

It was on July 11th, 1882 that for the first time in history a British armoured squadron went into action, and the *Monarch* was one of the eight armoured ships which on that occasion engaged the water-front batteries of Alexandria; a seaport ranking in those days as a first class fortress in virtue of its five miles of defence works mounting 250 guns. Four-fifths of these, or more, were smooth bores, unable to make any impression on their armoured hulls though hitting them often enough; but about forty-five were rifled ordnance of heavy calibres up to 10-inch, chiefly turned out at Elswick, on exactly the same principle of construction as those in the British service; and considering the

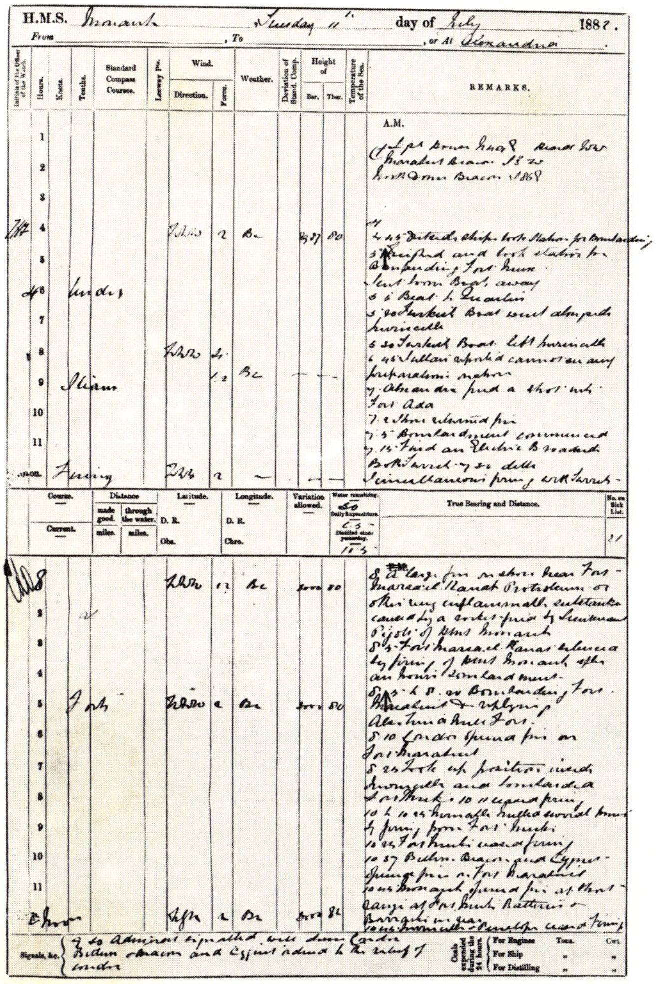

A page from the log of the *Monarch* for July 11th 1882, the day of the bombardment of Alexandria. In the 'remarks' column is the account of the action, beginning '7. *Alexandra* fired a shot into Fort Ada. 7.2 Shore returned fire. 7.5 Bombardment commenced'. Note also at the foot of the page '9.30 Admiral signalled Well done Condor'. (Public Record Office, ADM. 53/12157)

repeatedly demonstrated advantages which forts enjoy in conflict with ships, the attack looked like a sufficiently formidable undertaking on paper, even allowing for some larger guns on the water. But those on shore were manned by indifferently trained crews, and as the action progressed the marked superiority of the fleet gunnery completely overwhelmed the defence, though nearly the whole day passed before the return fire was silenced. This engagement was not only the first undertaken by British armoured ships, but the last in which British vessels carrying either muzzle-loading or broadside armaments ever bore a share.

In the nineteenth century Egypt was still nominally a vassal state under the suzerainty of the Sultan of Turkey; but in practice the Khedives at Cairo were independent rulers. The incompetent and extravagant Khedive Ismail having defaulted on the interest of his extensive foreign borrowings, the French and British governments established after some hesitation a joint control of Egyptian finances on behalf of the holders of Egyptian stock; and when Ismail kicked against such a restraint they deposed him and established his son Tewfik as Khedive. But the reforms and economies to which the Egyptian public officials and army were forced to submit under this arrangement aroused anti-European sentiment, and the discontent eventually took shape as an organised opposition by Egyptian regular troops under the leadership of Arabi Pasha, a brigade commander, who seized all power and compelled Tewfik to act at the army's dictation.

In this crisis the two Powers dispatched a joint note of protest to Arabi, and a combined Anglo-French fleet anchored off Alexandria. Shortly afterwards a change of government in Paris completely reversed French foreign policy, and the French battleships were recalled. To meet the altered conditions therefore the European Powers convened a special conference of their ambassadors at Constantinople to propose a plan of action to the Sultan and press for its execution; but the British Mediterranean squadron remained at Alexandria under the command of Sir Beauchamp Seymour.

As recognised heads of their profession the Ambassadors were powerful talkers, and for many weeks discussed the situation in punctilious French but without arriving anywhere. Meanwhile anti-foreign feeling in Egypt was breaking out in riots, a wholesale exodus of Europeans was taking place, and Arabi was strengthening the sea defences of Alexandria by mounting additional large calibre guns, repairing parapets and revetments, completing ammunition stocks and drilling his gunners; all being carried out in full view of the British vessels lying in the roads. Seymour could hardly remain a passive spectator of preparations for an attack on him, and at length notified the Pasha that these proceedings must stop, or by a certain date he would be compelled to stop them by bombardment.

Arabi ignored the warning and the ultimatum expired on July 11th. All foreign residents had already departed and the last foreign vessels left the harbour the evening before (including the men-of-war of five nationalities) with refugees. Seymour's dispositions for next day's engagement comprised an eastern group of three ships (officially described as the offshore squadron) to attack at close range the works on the arm of the land which forms the east side of the natural inner harbour; a western group of three others to deal with the defences lining the coast to the westward of the inner harbour at close range (officially known as the inshore squadron); and a third group of two as a central attack, to fire at long range over the breakwater on the batteries inside.

The *Monarch* belonged to the western group with the *Invincible* and *Penelope;* but as the two latter had a much lighter draught they were able to push in to far shorter range. Being the nearest targets they drew all the return fire from that section of the defences; with the result that the *Monarch* escaped from being hit altogether. In so far as her individual share of the action was concerned therefore it was virtually an unopposed bombardment; but she was kept very busy shelling the western forts for the greater part of the day, while steaming slowly up and down a line parallel to the shore at ranges of 3000 to 4000 yards. Towards noon it was observed that the work known as Fort Mex — which lay in the angle or salient where the land lines of the city joined the sea lines — was apparently abandoned though many of the guns seemed unhit. The ship anchored therefore at one o'clock to despatch a boat party on the adventure of spiking them; a sufficiently risky enterprise in the absence of definite knowledge that the work was evacuated, but completely successful, as it was revealed on closer approach that the holders had all retreated. This operation occupied about two hours, and on the return of the party the ship weighed again and opened fire on the forts inside the harbour, which was maintained till the close of the general action at about five o'clock. Altogether the *Monarch* fired 125 rounds of 12-inch, 54 of 9-inch and 21 of 7-inch. Her turret guns were the first 12-inch rifled ordnance ever used in war by land or sea. As they were entirely man-handled except in training for direction, the turret crews had an exhausting experience, especially as the weather was warm. Of the subsequent course of the war nothing further need be said here as it became a land campaign finished in three months.

For nearly three more the *Monarch* continued to serve in the same command, in the course of which she was the accidental means of discovering a pinnacle rock just outside what was the direct line of entrance to Malta harbour in those days before the breakwater was built. It might be supposed that no part of the sea bottom in the world would be better known than in the approach to the greatest sea fortress in the British Empire outside home waters, but this pinnacle had escaped all surveys. Before the construction of the breakwater the big swell set up by a 'gregale' used to run right into the harbour; and going in under these conditions the *Monarch* gave a heavy 'scend aft and dropped the heel of her sternpost full on this rock with a bang that shook the whole ship, but lifted instantly to the next sea without losing headway and passed on. The damage was only local and the obstacle was charted as 'Monarch' rock and now forms part of the breakwater foundations.

During the years immediately following the Egyptian war of 1882 it was the regular practice of the British government to keep a ship always at Alexandria and another at Port Said. In this the *Monarch* took her regular turn, and when the economically minded cabinet of the early 'eighties had reduced the Mediterranean battle squadron to six vessels, of which one at least was usually refitting at Malta, these spells of stationary duty came too often to be good either for the ships or those on board. With one of its number in dockyard hands, another at Port Said and a third at Alexandria, the disposable cruising force was reduced to three, or if any other ship was specially employed, to only a pair. Such was the condition of affairs when a very sudden political crisis arose in March, 1885, through an irregular but violent encounter between Russian and Afghan troops at Penjdeh on their vaguely defined mutual frontier; into which quarrel it seemed possible that Britain might be drawn as a pledged supporter of the independence of Afghanistan. Two Russian armoured cruisers were passing through the Mediterranean at the time on their way to Vladivostok; and the Russian Black Sea battle squadron was mobilised for its annual summer exercises. A wire was despatched therefore to the British Commander-in-Chief on the station, Lord John Hay, to concentrate his ships at Malta immediately, except the one at Port Said which was to remain as a Suez Canal guard. Hay himself happened to be at Alexandria in his despatch vessel at the moment — his proper flagship being partly dismantled at Malta for quadrennial survey — but left at once for Malta in the *Monarch,* which was on duty at Alexandria at the time.

What followed affords a good example of the perplexing uncertainty sometimes attendant on naval emergency movements in the dumb age before the wireless aerial. Hay had cabled to his Rear-Admiral at Malta before starting to announce his intended date of arrival there; and when he became two days overdue on a mere five day passage that officer took alarm. Two of the four battleships available for service were already at Malta and a third expected; and as he had no light craft at hand he sent off these two in search of the missing Commander-in-Chief, both of which of course were themselves lost to immediate touch in like manner. Thus the naval 'concentration' was reduced to one battleship ready for service. Meanwhile the *Monarch* was lying disabled in open sea nearly 300 miles from Malta; and as she had drifted off the straight track, the one searching ship ordered to proceed along it passed her at night a long way off and mutually unobserved, while the other was hunting in vain towards Crete on a false report given by a Levantine tramp. A torrent of telegrams was passing between Whitehall and Malta; although the situation was somewhat eased by intelligence that the two Russian cruisers had retired into neutral harbours in obvious anxiety for their own safety. Eventually news came in to Malta by a steamer that she had overtaken another steamer with the *Monarch* in tow the day before; and on receipt of this report the dockyard tug was sent out, which found her and took over the tow, but in such heavy weather that she was soon forced to slip the hawsers and leave the battleship anchored in sixty fathoms thirty miles from land. A large steamer at Malta was therefore chartered to bring her in and ultimately the Commander-in-Chief of the principal British battle squadron arrived at his station headquarters in this rather ignominious fashion. A week later the searching ships returned from their futile errand and the concentration thus completed at last was maintained till the crisis blew over.

That this breakdown of a battleship and all the trouble it caused at a time of acute political tension were due to preventable causes hardly admits of serious dispute. For some time previously Treasury pressure had cut down the coal allowance of the Navy to such a point that ships abroad had been degenerating into mere stationary drill sheds, of which the *Monarch* was an instance. So long was she kept swinging round her mooring swivel at Alexandria that to use a time-honoured phrase she was in danger of grounding on her beef bones, and on taking her departure at last proved unfit for sea. In a comparatively shallow and absolutely tideless harbour, such as that, a ship's underwater surfaces are apt to foul very badly, and in the case of the *Monarch* a minute but extremely hard shell growth spread into the interstices of the stern tube lining, which had become worn till the tail shaft was a loose fit. The single turn of the engines by hand gearing, which was a daily part of the engine room routine, was too slow to indicate that anything was wrong; but as soon as steaming began the tremendous friction in the tube caused such excessive propeller vibration and laboured running of the machinery that speed had to be reduced to dead slow.

Even so matters got gradually worse till the engines had ultimately to be stopped altogether and the ship was left helpless till picked up by a passing steamer. After being brought into Malta the damage was found to be so bad that she was paid off, and after three months in dock for temporary repairs sent home in charge of a navigating party under escort of the *Neptune*.

Being again made fit for service she was commissioned towards the end of 1885 for the Channel Squadron, in which she cruised for five years more without notable incident except Queen Victoria's first Jubilee Review at Spithead in 1887. In 1890 she was paid off once more for new machinery to be put in as already described, but not re-commissioned till she hoisted the pennant with a half-

H.M.S. *Monarch* at Simons Town about 1900 after she had been modernised. (NMM. AAP/2)

complement in 1897 to proceed as stationary guardship, to Simon's Bay naval base at the Cape which was then an unfortified harbour. A party of her men provided one of the contingents comprising the naval brigade all through the South African War.

* * *

In 1902 she was reduced still further to the status of a depot ship and removed from the effective list; and as such remained at Simon's Bay. She was renamed *Simoon* in March 1904 to free her name for a new battleship, and next year she was brought home and sold to Messrs Garnham on April 4th, 1905 for breaking up.

* * *

So ended the long and useful career of the first sea-going turret ship thirty-seven years after launching, of which twenty-six were spent under the pennant. And though Cowper Coles himself repudiated any responsibility for the design of the *Monarch,* as not being a true embodiment of his ideal of a battleship, the Admiralty insisted nevertheless that the main credit was his alone for the great advance in combatant efficiency which that design represented. In the words of the Board memorandum, issued as an official comment on the report of her comparative test cruise with the *Captain,* both vessels were due *'to the constant, persevering and talented efforts of Captain Coles to obtain the adoption of the principle of arming ironclad ships with revolving guns in shields or turrets in contradistinction to broadside armaments. This was the great idea of a mind highly practical, inventive and ingenious.'*

CHAPTER EIGHT

THE 'CAPTAIN'

H.M.S. *Captain* at sea, an artist's impression drawn in 1869. (NMM. 58/3142)

It has already been related how Captain Cowper Coles came to submit his design for a turret ship to the Admiralty, and how their special committee preferred Reed's design to Coles's. But it soon became apparent that Coles had a large following from the Parliamentary and Press agitation which ensued under the lead of *The Times,* whereby it was freely asserted that only official prejudice and jealousy prevented the recognition and acceptance of his far-seeing conceptions. In political circles criticism became vehement, and in some newspapers acrimonious.

The Admiralty were in a difficult position, and under intense political pressure they fell back on a compromise. Having ordered the *Monarch* to be built therefore, to teach the nation what a sea-going turret ship should really be like to the professional mind of the British Navy taken as a whole, they met the public demand at the same time by authorising the construction of another (with Treasury sanction) to represent the ideals of Coles and the small group of naval officers who agreed with him. The Board was divided on the issue; Gladstone's new First Lord Mr Hugh Childers, represented and led the political pressure for Coles, and the Controller the professional opposition. Admiral Sir Spencer Robinson was Controller — and his principal assistant was Chief-Constructor Reed, one of the ablest men who ever held that appointment. Robinson and Reed condemned Coles's design in the plainest language. Each side had the courage of its opinions, for Childers lost his own son by appointing him to the ship, and Reed carried antagonism to Childers to a degree which made his own position at Whitehall eventually impossible. But Childers was the head of his department, and backed by the Cabinet, so a Board decision to allow Coles to build his ship was taken accordingly, and the First Lord himself selected a name for this vessel which was to be laid down under such an unprecedented origin. His selection recalled to the fleet to which she was to belong one of its many great exploits in the past and the services of its greatest leader on that particular occasion. Many officers still serving were familiar with the last of the trophies of that famous day, for less than twenty years had elapsed since the breaking up of the old *San Josef* after service as guardship at Plymouth.

Captain Cowper Phipps Coles, the talented and persevering officer who advocated arming ironclad ships with guns on revolving mounts in turrets or shields, who so tragically lost his life with the sinking of the *Captain* in 1870. (NMM. 58/2056)

Sir Edward James Reed, a brilliant if outspoken man who rose to the position of Chief Constructor of the Navy, until recriminations with the First Sea Lord over the design of the *Captain* forced him to leave the service, turning his attention first to commercial projects, and later to politics. (NMM. 58/2643)

Though the *Captain* was to be built at the taxpayer's expense like any other regular fighting ship, it was considered desirable, in view of the Controller's attitude towards her design, that the work should be performed by outside contract instead of in a Royal dockyard; a view in which Coles concurred. Among other leading firms the Laird brothers were shown the plans; and the fact that after examining them they tendered for the contract, greatly encouraged Coles's adherents, and even to some extent modified the mistrust with which the design was regarded by the generality of naval officers; for no establishment outside the Royal dockyards had a sounder reputation in battleship construction than that of these two very distinguished Birkenhead shipbuilders.

The placing of an order to build under such unusual and debatable conditions in matters of main principle, was followed by an almost equally controversial correspondence between Coles and Whitehall as to responsibility for its proper execution, which caused several months of delay. The main point of dispute arose in connection with sponsorship for adequate constructional strength. Coles demanded the observance of the regulation arrangements whereby contract work had always been conducted under the immediate supervision of Admiralty overseers, and all responsibility for a soundly built hull had rested with the Controller's staff. But the Board replied that all vessels launched from private yards to national order had hitherto been built from the Controller's own drawings: and that in many points of scantling and structural specification Coles's design fell short of Admiralty standards. They considered that they had taken quite enough risk in allowing the *Captain* to be built at all in the face of the opposition of their own advisers, and insisted that the responsibility for producing a sound and strongly constructed hull must remain with Coles himself and the Lairds. The most they were prepared to do was to appoint a representative of the Controller to ensure the use of good material.

Preliminaries having at length been settled on these lines, the keel of the *Captain* was laid down in Laird's largest dry dock in 1867. Cole's design provided for a hull 320 feet long between perpendiculars, with the unusually restricted beam for a capital ship of those fore-and-aft dimensions of only 53 ¼ feet. At full load for sea Laird's estimate of her probable displacement was 6950 tons on an anticipated draught of 24 ½ feet forward and 25 ½ aft; which would have given a freeboard or height of side abreast the main mast of only 8 ½ feet at the most. Thus, even if that calculation had proved correct, her design would have made a very small allowance for the two prime factors of safety in carrying sail; that is to say, breadth of cross-section and height of gunwale. Had her mould answered to the standard of contemporary battleships of her tonnage in either of these particulars she would never have been lost; but it so happened that the worst of these defects in design proved to be worse still in the completed ship. Reed had foretold that she would float deeper in the water than her design intended; and Reed proved right, though his direst anticipations fell short of the actual result. With all weights on board to the last ton of coal, ammunition and provisions, her height of side was only 6 ½ feet instead of 8 ½; a discrepancy which Admiral Robinson described as a 'serious and unexampled error' in constructional calculations, which indeed it was.

Owing to Coles's obsession for a very low freeboard — to offer a reduced target and diminish the area and weight of armour — the *Captain*'s hull resembled that of a battleship in which upward construction has been arrested at the main deck; except for a section at the bow and stern where it has been carried to a normal height. She presented therefore the unique peculiarity among large types built for ocean duty of having only two whole-length decks instead of three, her upper deck, or hull top, being at the level above her keel corresponding to that of the main deck in all contemporary capital ships and frigates. Relatively to that deck the higher sections at her extremities were like a topgallant forecastle and poop, though actually standing no farther above water than the hull tops of her consorts. But instead of terminating in a square transverse break (like all other topgallant forecastles and poops) their enclosing bulkheads ran diagonally from the sides of the ships towards the centre; and were joined by a broad fore-and aft bridge or flying deck, passing over the turrets and resting on a long support between them amidships, which contained the lower part of the funnel casing, the well of the engine room skylight and the ventilators for the stokehold and central portion of the ship. This superstructure was a much more important feature than the corresponding erection in the *Monarch,* which had only been built to stow the hammocks as a substitute for the bulwark nettings of other ships. In the *Captain* it was not only lined on both

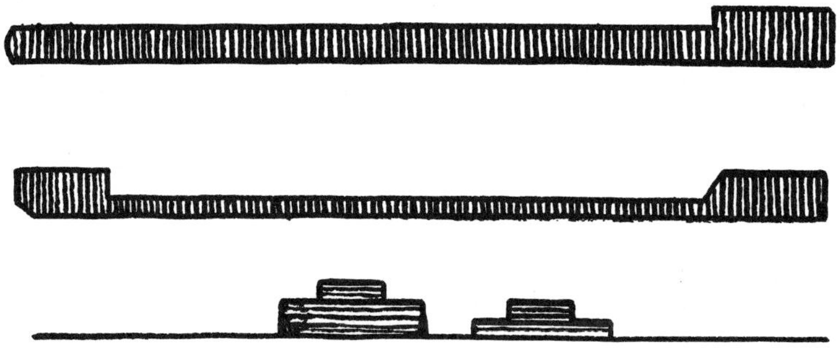

Side area above water (top, *Monarch*, centre *Captain*), and transverse section above water at fore turret (left *Monarch*, right *Captain*). (From Admiralty plans.)

sides with hammock boxes but carried most of the equipment and gear stowed or fitted on the upper deck of a broadside-armed ship; such as the boomboats, the conning tower, and the bitts and fife rails for the running rigging round the fore mast and main mast. It provided a working platform where all divisions, musters and small arm drills had to be carried out at sea when the deck underneath was awash (as was inevitable in any but very calm weather) and where all ropes were manned in every kind of weather for making, shortening or trimming sail, although the space was cramped for smart handling and gave no run for sheets and halliards. A light bridge crossed above it on the fore side of the funnel, with engine-room telegraphs and voice pipes to the threefold wheels on the poop and lower deck.

Except for a slightly projecting rounded ram at the forefoot the *Captain*'s stem was straight vertical; and, being the first capital ship to carry twin screws, she was the first steam battleship with no screw aperture under the counter. Being of the unbalanced type, her rudder hung therefore on the sternpost itself as in the old sailing days. Above the counter her stern rose in a semicircle like a half tower; which was a reversion to a style in some of the earlier armoured types that had gone out of favour. Her midship section was very square in the bilge, and her maximum beam, though small in itself, was carried over a considerably greater portion of her length than usual. This offered some small contribution towards improving her defective stability, though not conducive to good speed lines. Like many ships with a high centre of gravity, she proved a steady gun platform with a very easy and moderate roll, effectively minimised by long bilge keels. Under helm her manoeuvring qualities were indifferent, though, when helped by twin screws, necessarily better than those of other battleships lacking that advantage.

As the *Captain* had only one internal deck, that deck unavoidably fixed the accommodation level of the whole ship's company, and all the officers except the captain, commander and navigator, whose quarters were under the poop. All ranks and ratings therefore (except those three) were berthed at waterline height as in the timber-hulled battleships, though in every other iron-hulled ship the men enjoyed the advantage of messing and sleeping on a main deck as much as 8 or 10 feet above water. And as a great deal of space on a deck at waterline level in an iron hull, such as the *Captain*'s, was absorbed by watertight bulkheads and wing compartments, the space remaining available for slinging mess tables, stools and hammocks was considerably less than in a wooden ship where no such obstructions existed. But as she carried about a hundred fewer ratings than the wooden vessels of the same tonnage it sufficed without excessive overcrowding; and though worse ventilated it was better lit than in any other ship in the fleet. Having no ports anywhere on this deck it was lit fore-and-aft by large deadlights in the deck above; which of course admitted ample illumination in the daytime even during bad weather when the ports in other ships were closed and the messes in darkness. But the complete absence of all ports, and of all hatchways except those on the poop and forecastle, left very little movement of free air.

Except that both the upper and lower tiller carried the then newly introduced Rapson slide, the arrangements for conning and steering the ship had no novel features and marked no advance on those of the oldest vessels in the battlefleet. The *Captain* had neither a balanced rudder nor steam steering gear, nor even such a modest innovation as a charthouse. She was conned from the small bridge just forward of the funnel, whence helm orders were conveyed by voice pipe to the poop and lower-deck hand wheels, both threefold. Her cable deck, however, was equipped with the new inventions of a patent capstan and steam drive, having chain lockers abreast the fore mast instead of the main, and a hand capstan on the forecastle. As the ram was too low and too short to foul the anchors in weighing, cat chains were not required. Anchors, cables and hawsers were on the largest service scale. Her complement of boomboats differed from the service standard for battleships in the same detail as the *Monarch*'s; that is to say, instead of two launches and a pinnace she carried two pinnaces and a launch; one of the pinnaces being of the new type with steam propulsion alone. The stowage of the boomboats was unique, all three being on the flying deck with the launch between the funnel and foremast, the sailing pinnace resting inside her, and the steam pinnace abaft the main mast.

In armament the *Captain* had no equal in the world during her brief existence except the *Monarch*; and as regards that prime qualification for a fighting ship (next to seaworthiness) if in nothing else, the foresight of Cowper

The above-water armoured areas of *Monarch* (top-transverse citadel bulkheads not visible) and *Captain*.

Coles resulted in radical improvements which came to stay. His theory of revolving centre-line guns had been the starting-point from which the design of the *Monarch* was worked out in other men's minds, and formed the basis on which the design of the *Captain* took shape in his own. For each vessel it was made a condition on which everything else was to depend that the main weapons were to consist of four 12-inch M.L.R. guns with a weight of about 25 tons each. More than ever was the principle finding recognition that in the naval architecture of war the ordnance was designed first and then the ship that was to carry it, though in a less pronounced degree that rule had long been accepted.

The *Captain*'s 12-inch guns were carried, like those of the *Monarch,* in pairs in two circular turrets, which in her case were slightly higher sided but slightly less in circumference than the *Monarch*'s, and stood about three times as far apart. But the mountings were a special design. To give extreme elevation or depression the whole carriage and slide had to be lowered or raised, instead of the gun and trunnion box only, as in the *Monarch.* As the poop and topgallant forecastle restricted the turret fire to bearings between each bow and quarter, the sphere of engagement of this main armament did not differ from that of a broadside ship except in rather wider arcs of training. In other words, she was essentially a broadside and not an end-on fighter; and her equipment for conducting a chasing action was no better than that of the oldest and smallest battleships in the fleet, consisting as it did of only a single 7-inch gun on the forecastle and another on the poop, neither of which had any armour protection. The chief merit of disposition in her main armament lay of course in its application of Coles's first principle that its whole weight of metal could be thrown to either starboard or port. Its worst defect lay in its lack of command above water, from which the lower edge of the turret ports stood a bare 9 feet; that is to say, scarcely half the height of those in the *Monarch,* and even lower than the battery ports in the majority of the internally-armed ships. Even in a very moderate sea the turrets were liable to be flooded.

The armour completely encased the main hull from 6 feet under water to the upper-deck level, which obviated any necessity for internal athwartship plating like that of the *Monarch* and the central battery ships; but it did not rise high enough at the extremities to cover the chase guns as in many other vessels. It was 7 inches thick amidships, increasing to 8 abreast the turret bases and tapering to 4 at the bow and stern. On the turrets it was 10 inches round the ports and 9 for the remainder of the circumference, and on the conning tower — standing on the forward part of the superstructure — it was 7.

In provision for tactical movement the design of the *Captain* marked the first notable advance since the original application of steam propulsion to the British line-of-battle, through her possession of twin screws. In that respect therefore she was the best-equiped battleship in the armoured fleet when completed for service. By the adoption of this novelty her risk of a complete breakdown at any time whatever was of course reduced by half; and thereafter only three more capital ships were ever built for the British service with single propellers.

As incidental to her twin-screw machinery it happened that she was the first British armoured ship to be engined by her builders. During the early years of the use of marine steam engines their manufacture was quite distinct as an industry from hull construction, requiring as it did such an extensive special plant that for a long period only some half-dozen at most of the largest regular engineering firms were prepared to invest the capital required for its erection on a scale commensurate with the output of battleship machinery. But in a few shipbuilder's yards the production of engines of medium horsepower was undertaken as an additional branch to their hull construction, and of these establishments Laird's was one. With the introduction of twin screws the individual size of the two separate engines was naturally less than that of a single engine developing the same power; which brought the castings and forgings for the *Captain*'s machinery down to weights that Laird's machinery workshops could turn out. Two sets of trunk engines were made for her there accordingly, rotating a pair of outward turning two-bladed propellers. Steam was generated in eight rectangular boilers at a pressure of 30 lb. She carried 600 tons of coal nominally, but, owing to her excessive draught, only filled bunkers right up once. As a

The forward gun turret of the *Captain* during her fitting out, showing both the design of the ports, and the shield to protect the flying deck from blast. (NMM. 7110)

Looking forward from the poop of the *Captain* during her fitting out period, with work still proceding on her rigging. (NMM. 7111)

steamer she was not found to be so economical as the *Monarch,* though a considerably smaller vessel.

And it was in this matter of the extravagant fuel consumption, and consequently short radius of action under steam of the early armoured capital ships, that we find the first link in the chain of causes leading up to an awful disaster. As not one of them could carry enough coal in their ordinary bunkerage to count on covering even 1500 miles if spells of head winds were encountered, all were masted to provide an auxiliary equipment for oceanic voyaging; and in addition usually stowed extra coal in bags on the main deck if despatched to a destination exceeding that distance. The *Captain* had neither the space nor the buoyancy to carry her nominal bunkerage, let alone extra, and it was essential she be rigged for sail. Coles's genius sometimes found expression in an impulsive over-confidence which carried his collaborators along with him. Nothing on a second-class scale was good enough for the ship of his dreams, and so he endowed her with the maximum standard spar plan of the British Navy, with a sail area of 50,000 square feet including the stunsails.

The Admiralty were surprised when they were shown the plan, and said so; but the Lairds replied that they were satisfied that the ship could carry it; and at that it remained. One of the indirect results was rather paradoxical. The *Hercules* and *Monarch,* then building, were to be masted on a second-class scale; but the idea that the *Captain* might outsail these Admiralty planned ships was too much for Reed, who straightway obtained Board approval to increase their canvas to the *Captain*'s spread. But Reed could afford the extra gear in those two ships because their maximum stability angle was over 70°, whereas that of the *Captain* was calculated at only 21°; being less than a third of that in any other capital ship in the British service.

Nor was the mere magnitude of Coles's sail plan the only indication of a confidence apparently without limits. Its position as a whole relatively to the hull threw a heavier tax on her stability than was imposed in any other ship, because it was erected with the line of the superstructure as a base instead of the top edge of the freeboard. This admittedly obviated the trivial loss of a few feet in the drop of the courses, but only at the expense of raising the centre of wind pressure by quite 10 feet, whereby its capsizing effect was increased by many tons in a gale. As a direct consequence of this arrangement the *Captain*'s main lowermast had the tremendous height of 96 feet from cap to gunwale level, that being 16 feet more than the corresponding measurement of the tauntest of her consorts. Her fore and main yards and fore and main topmasts were 2 feet shorter than the maximum as carried by three other vessels in the fleet; but against that her topsail and upper yards were the squarest in the service.

A more symmetrical spacing of the masts than in the *Monarch* gave a better longitudinal sail balance. This was obtained by stepping the main mast on the forward side of the after turret, which in the *Monarch* would have been impossible. But the positions of all three masts relatively to the turrets was such that if supported by the usual array of shrouds a considerable arc of fire in both turrets would have been masked. In the *Monarch* this interference was much less, and provided against by halving the number while increasing the size of the shrouds, and fitting them with slips. But in the *Captain* the shrouds were abolished altogether and their places taken by iron tubular tripod legs, of which the pair on the fore mast sloped forward to the forecastle and the pairs on the main and mizen aft. Tripods were a novelty then in the British Navy, and in certain respects offered advantages to a stiff sail carrier quite apart from narrowing the blank sectors of fire. A shroud or stay only reacts to a pulling strain, and when a ship is sailing with the wind at any point before the quarter the lee shrouds are idle. But a tripod leg, being rigid, reacts to a push as well as a pull, and a mast so reinforced has at all times a double lateral support wherever the wind may come from. That adds to the safety of the mast itself in a proportionate degree, but thereby reduces the safety of the ship; which is possibly the reason why tripods were never adopted in any merchant service. Though it may be doubtful whether so crank a sail carrier as the *Captain* would ever have been reasonably safe, even with shroud-held masts, it is at least certain that tripods removed her last insurance.

Her gear was peculiarly designed in several other details. For example, the topgallant masts were fidded on the after side of the topmast crosstrees, and struck abaft all; which possessed the advantage that they could be sent down without lowering the topsails (if under sail), but, as it spoilt the set of the topgallantsails and caused excessive belly chafe, the experiment was not a success. Another unique feature was a boom for the main trysail as for a spanker, which enabled the foot of the sail to remain full spread on any point of sailing, but could only be fitted because there was no mizen stay to interfere with its swing or gybe. And that itself was a novelty. For the main and mizen masts the tripods were considered sufficiently a fore-and-aft as well as an athwartship support to obviate the use of stays as well as shrouds. Stays were used on the fore mast, but with that exception the only standing gear below the tops consisted of five dummy shrouds a side to each mast, leading down to the superstructure as ladders for the men to run aloft by.

As no rigging was allowed to mask the fire of the turrets, it was impossible to set up the topmast and upper backstays in the regular places at the ship's side. But, on the other hand, they could not be set up on the superstructure with a sufficient lead aft unless by interfering with the use of the trysails; nor was any method of tripod support feasible as regards topmasts and topgallant masts. To overcome this difficulty stout iron outriggers were hinged to the masts to give the necessary angle of spread; those for the topmast backstays being 15 feet long and projecting aft in pairs from under the tops, while for the topgallant masts they were single, 6 feet long, and projecting aft from the topmast caps. None were required on the mizen, where the splay of the poop allowed of the usual leads. All hauling parts of running gear, whether rove through sheaves in the bitts and fife rails or through lower leading blocks, and all lower purchase blocks for halliards, lifts, trusses, etc., were on the superstructure, except those for the main tacks and fore and main sheets, which were necessarily at the ship's side to permit a proper set of the courses when braced sharp up. The fore and main toptackle falls and jeers could be taken forward and aft respectively to the hand capstans on the forecastle and poop. Alone among ships under the white ensign in her day the *Captain* carried main topmast stunsails; a superfluous addition to weight aloft at best. She never had the opportunity in her brief career to show how she could move under canvas in a strong quaterly breeze; but all twin-screw vessels were dull sailers even under the most favourable conditions, on account of their double drag of propeller and broken line of run.

She was delivered by her builders as ready for sea in January 1870. While yet in an early stage of her construction the Admiralty overseer at Birkenhead had reported his conviction to the Controller that she would prove to be far deeper in the water than intended; and when floated out of dock this anticipation proved correct, for she was already down to her designed full load draught though about 2000 tons more weight still remained to be put on board in machinery, armament and stores. On arrival at Portsmouth from the Mersey her immersion was 22 inches in excess of original estimate, which was increased to a full 2 feet when ammunition had been embarked. This was too much on the wrong side altogether for the Admiralty, who protested in

Captain Burgoyne and officers of the *Captain.* (NMM. A.2544)

very plain terms and declined to pay the final instalment of contract price due on delivery. Laird's, however, insisted on payment, and on the advice of the law officers of the Crown a compromise was reached whereby Whitehall agreed to settle the account if the sea-going tests of the *Captain* were found to be in all respects satisfactory.

Though her design did not commend itself to the generality of seamen, Coles had his followers among them nevertheless; and, as usual in the case of such leaders, the more he was criticised by the profession as a whole the greater was their ardour in his support. Prominent among them was Captain Hugh Burgoyne, V.C., an officer with a very fine record and the only son of the field-marshal whose statue stands in Waterloo Place. On him fell Mr Childers's choice to command the ship which had been the cause of such endless controversy, in the natural expectation that, as an enthusiastic follower and ally of Coles, he might be relied on to make the most of her. Fate decreed that he should have no successor. She hoisted the pennant at Portsmouth on the last day of April 1870, and it was a persistent lowerdeck belief for long afterwards (quite possibly well founded) that the signalman at the ensign staff that morning accidentally bent the colours union down.

Owing to the wide attention and curiosity aroused by this extremely novel battleship, large numbers of officers visited her while lying at the north jetty, and great discussions took place between Burgoyne and other senior members of his profession. Among these was Captain Sherard Osborn, than whom certainly no man stood higher in the estimation of all ranks and ratings in the mid-Victorian Navy. Osborn had commanded the *Royal Sovereign,* which at Coles's instigation had been converted when half built[1] as a three-decker of the old style into a mastless low freeboard coast defence turret ship. His opinions therefore were founded on practical experience, and though he was an intimate friend of Coles and Burgoyne alike he was not at all happy in his mind about the *Captain.* Burgoyne, however, persistently expressed his entire confidence in her seaworthiness; and it transpired afterwards that among the standing instructions for officers of watches in his night order book was a reminder of the importance that she should compare favourably with other ships under canvas as well as under steam; which was not improbably regarded by them as in the nature of a warning never to shorten sail on their own responsibility unless in very pressing emergency. Meanwhile the officers (who were mostly of his own choosing) remained satisfied with his assurance, and the men took their cue from the officers. In any case both officers and men had to serve where they were sent, but in their evidence at the Court Martial after the disaster the handful of survivors declared with remarkable unanimity that they had never entertained any doubts about the safety of the ship.

When the usual commissioning trials were over the *Captain* was sent on a fortnight's cruise in company with the *Monarch,* in order to see how she compared with a turret ship whose sea-going qualities had already been thoroughly tested in Atlantic crossings. In a sense the comparison was hardly fair, as the *Monarch* was appreciably the larger and more expensive ship. Her superiority to the *Captain,* however, extended to every qualification for a blue-water vessel, including those in which the smaller competitor ought to have held an advantage, such as economy of fuel at all steaming speeds and a lesser turning circle at all angles of helm. In the relatively unimportant matter of movement under sail the *Monarch* had the best of the trials on every point of the wind, though the sail areas were equal; but the *Captain* was necessarily handicapped by her twin screws. It was in the supreme question of combatant efficiency during heavy weather conditions, however, that the greatest difference in the war value of the two ships existed; for while the *Captain*'s decks were swept by the sea, and her turret ports frequently flooded in a moderately fresh breeze of force 6 to 7, the decks of the *Monarch* remained bone dry.

After this cruise the *Captain* joined the Channel Squadron for another fortnight at sea and thereby came in for a full gale off Ushant, which she rode out lying-to under easy steam like the other ships with a close reefed main topsail and storm staysail set; awash fore-and-aft below the superstructure like a half-tide rock, but apparently safe; though in the light of subsequent events it was afterwards believed that her margin of security must have been much narrower than anyone on board realised. However, it was then regarded as a final proof of her sea-keeping sufficiency.

On returning to Portsmouth she was acclaimed therefore as a great success by the elements of the public and Press which had been committed from the first to a vociferous approval of her design; and it was freely contended that she had shown herself superior in all respects to her Admiralty rival. But even when all this exaggeration and over-statement had been discounted Coles's critics were bound to admit that the fact remained that the *Captain* had spent a month in open ocean and faced full Atlantic weather with every seeming success, which they had believed to be impossible. Alone among her detractors the Controller's department still refused to regard the vindication as being complete.

Before going to sea again she was detained three weeks at Portsmouth for some minor adjustments and alterations to her machinery, and during this interval an incident occurred in connection with the question of her stability, which gave rise to much discussion at a later date and some rather unmerited censure. Prior to the mid-Victorian period the matter of stability in ships had been dealt with under rule of thumb methods; but with the introduction of armour naval architects began to treat it as a subject for scientific attention in order to ascertain maximum angles of safety. These may be determined more or less accurately by prolonged calculations based either upon the purely theoretical data of hull design, or upon the practical results obtained through heeling the hull over to certain degrees of inclination by moving weights transversely from the centre

H.M.S. *Captain* at Plymouth on the eve of her departure of what was to be her last voyage. (NMM. HBQ/1)

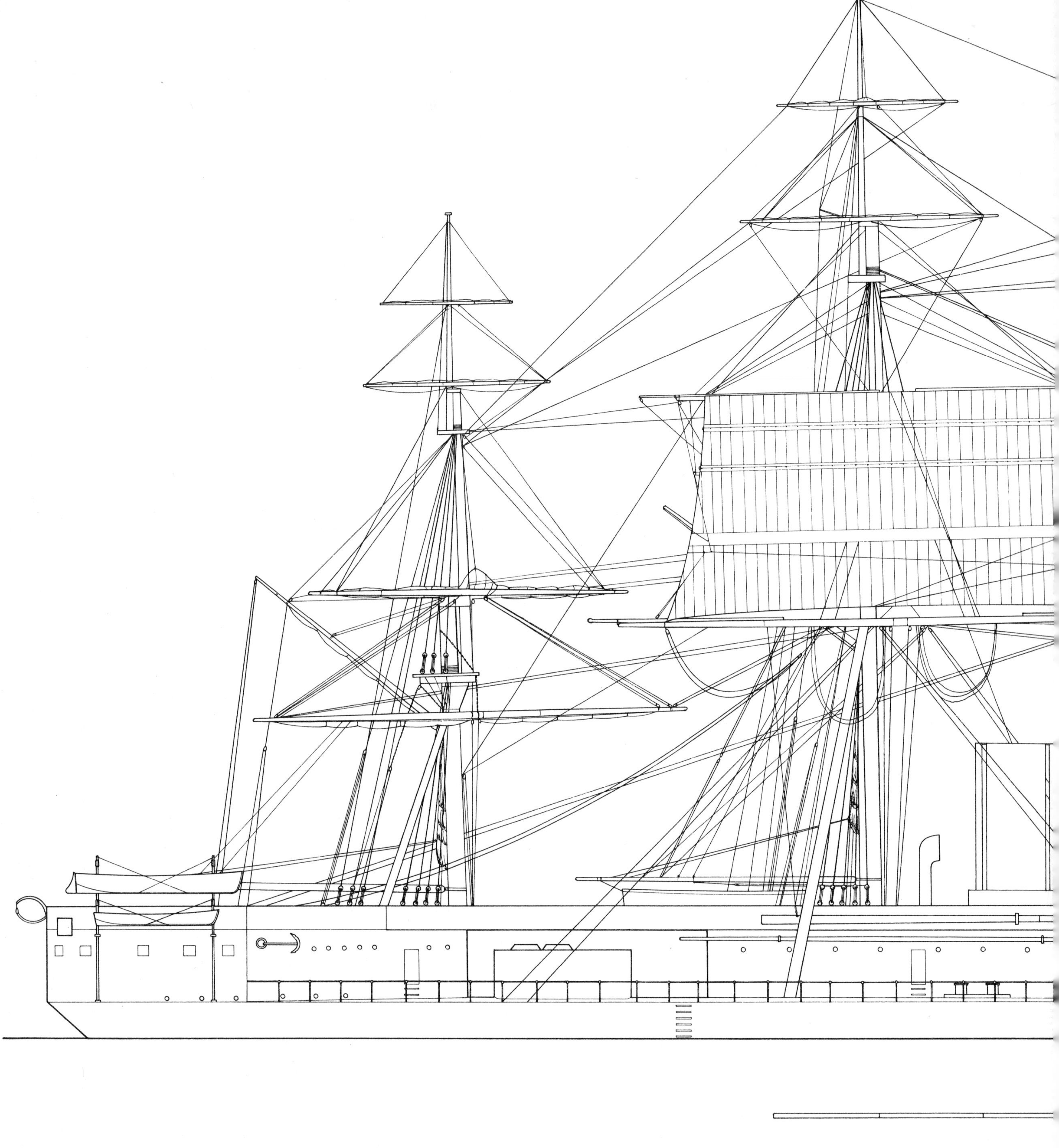

H.M.S. *Captain*, showing hull and rigging, and also the sail she was carrying at the time of her loss.

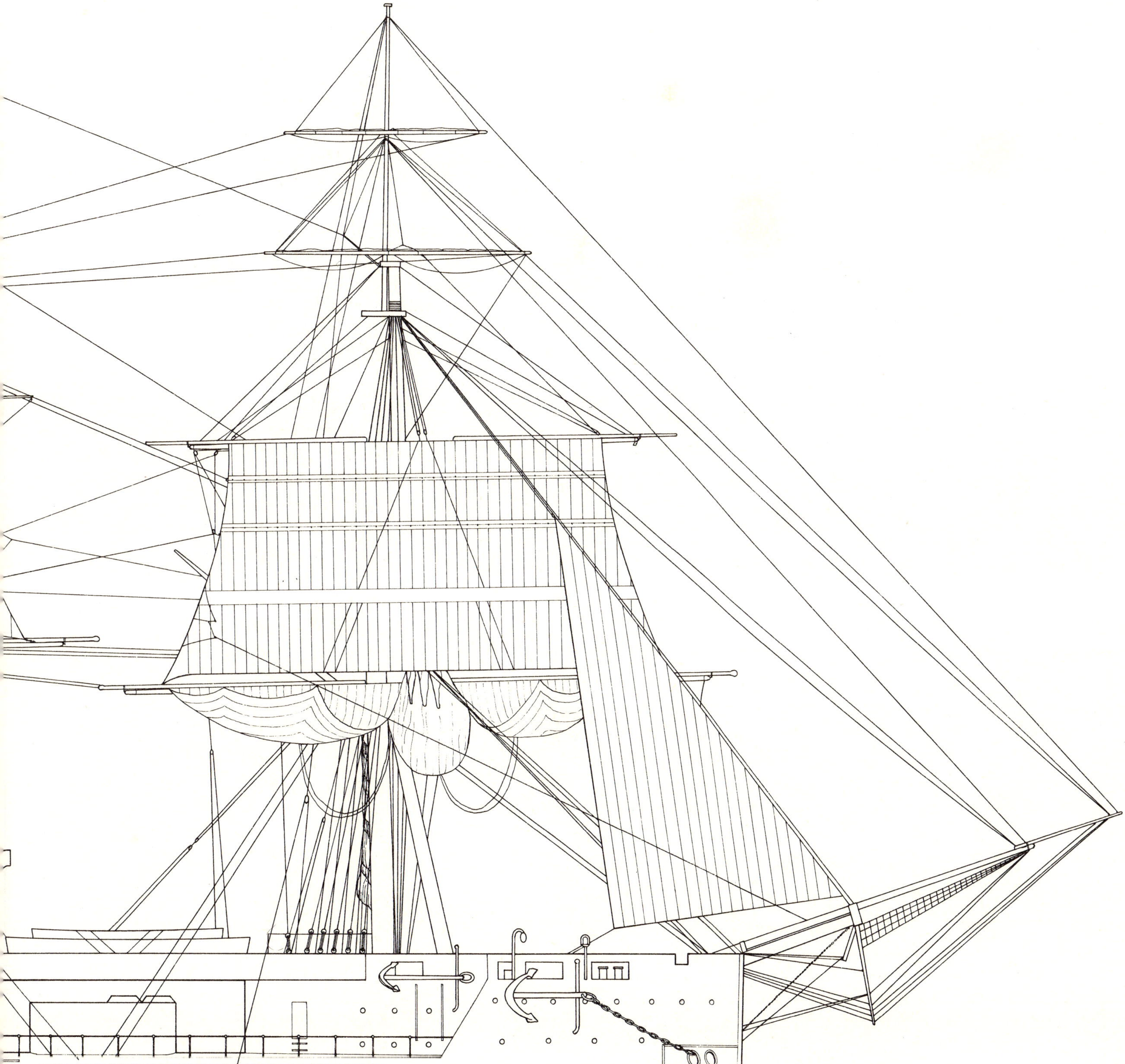

Admiral Sir Alexander Milne, who as Commander-in-Chief, the Mediterranean Fleet, had the ill-fated *Captain* under his command, and when he stepped aboard his cutter after inspecting her, was the last commissioned officer to stand on her deck and survive. (NMM. A.1218)

line. Both methods are dependent on a knowledge of exact immersion, and when it was found that the *Captain*'s draught far exceeded their anticipations, Laird's worked out her maximum angle of safety by the purely theoretical method for their own information and found the answer to be no more than 21°, which happened to be a very correct estimate. Apparently, however, they mistrusted their own calculations — or else were uneasy at the lowness of the figure and wanted the Admiralty to discover it for themselves — for after the ship had been delivered they endeavoured to get this verified by asking Whitehall to ascertain her stability by the practical test system, though without suggesting that there was cause for anxiety. The Board assented and directed the Controller to use the favourable opportunity of the ship being in a dockyard basin to carry out heeling experiments. That was undertaken accordingly and the results sent to the Chief Constructor's staff as the basis for the necessary mathematical investigation; but as the order from the Board was not marked 'urgent' or 'immediate' and as Laird's previously expressed confidence in the *Captain*'s ability to carry her sail plan had never been contradicted or modified, the report of the heeling tests was placed on office file to await attention in its turn, because the experts had a heavy pressure of other labours in hand already.

Meanwhile the minor adjustments to the engines were completed and she was despatched on a three weeks' gunnery cruise by herself, at the conclusion of which the Admiralty decided that her sea-keeping qualities might be considered as sufficiently established. Having ordered her therefore to take a regular place in the columns of the Channel Squadron they approved payment to Laird's of the final instalment of the contract price. For a brief period the triumph of Cowper Coles and his disciples, including Mr Childers, was complete.

In point of fact the *Captain* had barely six weeks longer to float. She left England with the squadron for Gibraltar to participate in an ocean cruise with the Mediterranean battleships. The outward voyage was made at easy speed in fine weather without noteworthy incident. When the assembly was complete, and all had coaled, the combined force sailed northward on August 19th under Sir Alexander Milne, the Mediterranean Commander-in-Chief, whose flag was flying in the *Lord Warden;* the other vessels being the *Minotaur, Agincourt, Northumberland, Hercules, Warrior, Bellerophon, Monarch, Captain, Inconstant* and *Bristol.* Many of the 7000 officers and men present had not seen the *Captain* at sea, and her behaviour was incessantly under critical observation from all the fleet, her station being in the weather column next astern of the Commander-in-Chief. Milne himself had never seen her till she joined his flag; and as she was now for a time under his orders he was in some respects the most directly interested spectator in the gallery, and notified his intention of inspecting her at sea before the force dispersed. During the earlier part of the cruise he was too much occupied in other ways, but on the morning of September 6th, when the combined squadrons were in the Bay of Biscay, he boarded the *Captain* to witness target practice and remained till late in the day.

In the course of the afternoon the weather began to look unsettled and the glass to fall, and when he signalled to the *Lord Warden* about an hour before sunset to close and send the cutter for him, a swell of ugly appearance was setting in from the westward. The *Captain*'s deck was already well awash and the *Lord Warden* was wallowing after her wont. Under such conditions boatwork is attended by an element of risk and Burgoyne pressed Milne to remain on board for the night. But Milne held the view that with bad weather prospects overtaking a fleet at sea the proper place for the admiral is on board his flagship and adhered to his decision. The fleet hove-to by signal accordingly, while the cutter was despatched and came alongside under the *Captain*'s lee, where she rose and fell to the level of the turret ship's deck. In accordance with the custom of the service the flag-lieutenant then got into the boat, and the admiral followed after taking the gangway salute. He was the last officer of commissioned rank ever to stand on the *Captain*'s deck and survive. The boat returned and was hoisted up, the two ships resumed their stations in the line, and the fleet filled and proceeded on its course W.N.W. with yards braced sharp up on the port tack, and easy steam in use for keeping station.

Half an hour later it was nearly dark, and as the wind was already strong and steadily freshening, upper sails were furled throughout the lines and topsails double reefed. All through the first watch the weather grew rapidly worse and before eleven o'clock accurate station keeping had become impossible. By midnight a whole gale was roaring and hooting through the darkness, and soon afterwards the last reef had been taken in the flagship's topsails, while her log recorded a 'very heavy and confused sea'. At about a quarter past twelve the wind suddenly increased to a blast of extreme violence, which swept over the combined squadrons and either split or blew clean away no fewer than twenty-three of their sails in all, including fourteen reefed topsails. Not a single ship escaped loss of gear except the *Captain,* where unhappily the hemp and canvas was too stout for the cranky hull beneath. Her number was up, together with that of 472 ranks and ratings.

When her relieving watch were called at midnight the other watch were all crowded on the cramped space of the superstructure, trying amid great difficulties to clew down the topsail yards, having previously taken in the foresail and furled the other sails. The halliards had been let go, but the ship was already lying over too much for the sharply-braced yards to come down by their own weight, and not an inch of the weather braces could be gained. Burgoyne himself was personally taking charge from the bridge just above; and judging by the evidence of some of the survivors had come up in haste, being scantily dressed in spite of the rain. He was giving a rapid succession of orders, and among others a call to the officer of the watch to take the angle of heel by the bridge clinometer, to which the latter answered eighteen degrees. The incident suggests that Burgoyne's professed confidence in the ship had for some reason given way to doubts, for in no other ship in the service would so small a list have attracted the least notice. The senior midshipman of the relieving watch had just finished mustering them on the forecastle when the above-mentioned squall struck the whole fleet. It threw the *Captain* down almost immediately to her maximum angle of safety and then past it. She was gone now beyond hope of any recovery.

The 200 hands on deck were all hove into a raging sea, while the 300 below were trapped — mostly in their hammocks except the stokers — and one of the most terrible sounds in those awful moments were the agonised shouts heard from below through hatchways and ventilators. For a few tragic minutes she lay on her beam ends, till the inrush of water through the funnel and superstructure openings destroyed her buoyancy. Then she rolled completely keel up and disappeared under the heaving Atlantic, where its depth is nearly 4000 feet, about twenty miles to the westward of Cape Finisterre; the only British armoured vessel ever lost through inability to outlive a gale in the open. In complete ignorance of the catastrophe the *Inconstant* swept past within fifty yards of the spot a bare ten minutes afterwards, but the extreme darkness of the night and the noise of wind and sea prevented anybody on her decks from either seeing or hearing the drowning crowd in the water, though to these unfortunate men her lights were plainly visible.

When the ship was thrown on her beam ends the three boomboats on the superstructure fell overboard out of the crutches, and if all three had floated upright many of the stronger swimmers might have found temporary refuge at least; but that was not to be. The launch — which was of course much the largest and could have taken sixty men or more even in that weather — had the pinnace inside her as usual; and the captain's galley happened to be inside the pinnace instead of at the taffrail davits, in order to permit the firing practice of the day before with the stern gun. Under this double burden the launch was quickly swamped and sank, but left the pinnace floating, a 36 foot boat of the 14-oared service pattern, into which last hope for hundreds clambered seventeen hands, who very wisely threw out the galley at once. The third boomboat (a 37 foot steam pinnace) capsized in falling into the water, and in such a seaway her wet and slippery undersides afforded little hold and no shelter to the few who succeeded in straddling her keel for a brief period, among whom were Burgoyne himself (partially disabled) and Mr James May, the gunner. The other pinnace was driven rapidly close past this upturned boat by the wind, and May actually succeeded in springing and saving himself by catching the gunwale, while the men tried to save Burgoyne by thrusting him a boat-hook. For a few seconds he gripped it, but then let go and called to them to keep it for their own safety, after which the boats drifted fast apart in the darkness and Burgoyne was seen no more.

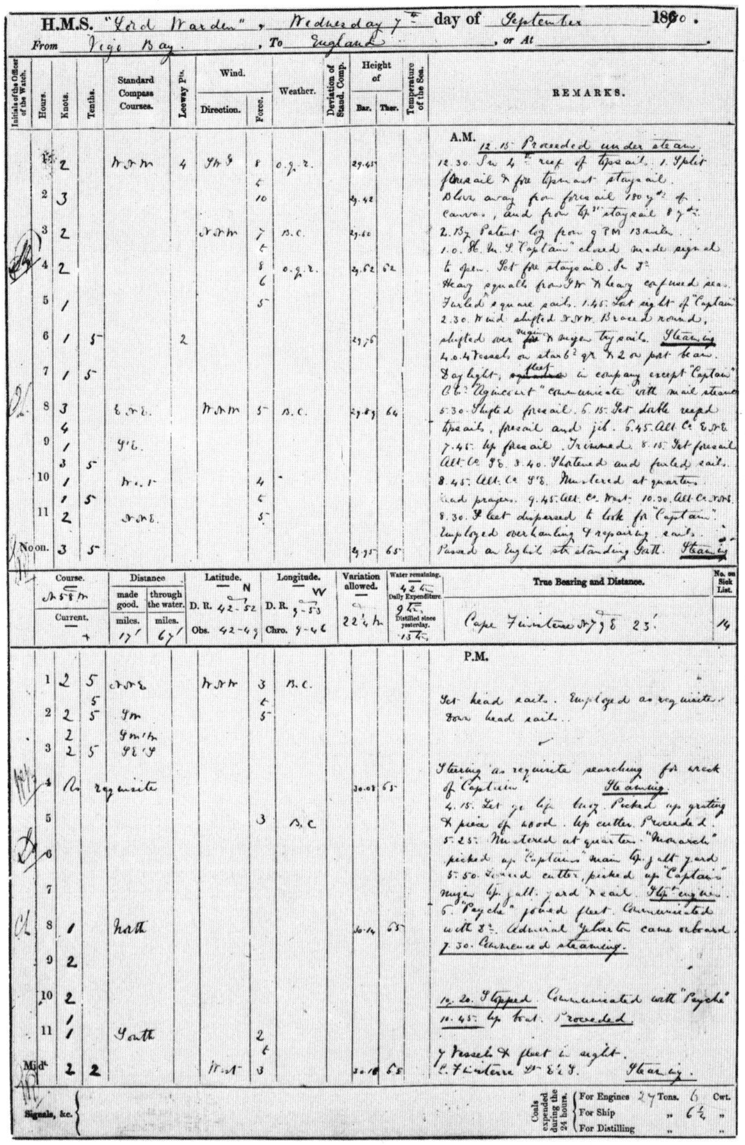

The log of the flagship *Lord Warden* for the night of the *Captain*'s loss and the following morning. It begins at midnight and in the 'remarks' column at one o'clock is: 'split foresail and fore topmast staysail'. This was the gust which suprised the *Captain*. Later it reads 'H.M.S. *Captain* closed made signal to open' (that is, to increase her distance). This must in fact have been the *Inconstant* just after passing the survivors of the *Captain*. Next morning the 'remarks' column has 'Steering as requisite searching for wreck of *Captain*'. (Public Record Office, ADM. 55/9641)

May was the eighteenth in the pinnace and the last. He owed his life to attention to duty. Having turned in for the night about an hour earlier, he was awakened by the violent rolling of the ship, and left his bunk to make certain that the turret guns were properly secured. Engaged thus in the after-turret when the ship capsized, he just managed to get through an opening in the turret roof in time to escape being carried down. Except for these eighteen souls, everyone in the *Captain* that night perished, including all the marines, all the non-combatant ratings, and more than nine-tenths of the seamen complement, the total loss amounting to 472 officers and men. Cowper Coles was among them, having been on board the ship as an observer by special Admiralty arrangement ever since her delivery from the builders. Of his last moments nothing was ever known. None of the survivors had seen him on deck, but they were all forward, whereas he as a passenger without active duties would almost certainly have been aft on the poop if on deck at all, where nobody else was stationed but the hands at the wheel and the sentry on the lifebuoy. For him the tragedy touched its lowest depths, denying the opportunity of seeing the complete recognition of his theory of centre line armaments yet allowing existence to last just long enough to know the failure of the vessel which he had designed to put it into practice.

Fortunately the pinnace with the small band of survivors was a stout and buoyant boat extremely well handled; as she needed to be all that in the circumstances; and, most providentially for them, land was dead to leeward at no great distance. Being blown straight down wind, they sighted the light on Cape Finisterre before the morning and early next day ran into the shelter of Corcubion Bay on the coast of Spain, where they were afterwards embarked by the *Monarch,* which had been sent to look for any sign of boats possibly escaping from the disaster.

The admiral had spent the whole night on the poop of the *Lord Warden* and in his report stated that the *Captain*'s lights had been visible till after one o'clock. But from the evidence afterwards forthcoming these must have been the lights of the *Inconstant,* which had been the *Captain*'s next astern before the lines were broken up by the weather. Through the thickness of the night half the fleet were temporarily lost to view from the flagship, though all were within signalling distance at daybreak except the missing vessel. Her absence caused immediate and general dismay, course being at once altered to follow back the track steered for the previous twelve hours and all ships extended in search. Ample signs of the worst description were soon discovered. It is remarkable over how wide an area the wash of a gale will spread the debris of a foundering in a very short time; and soon after the search commenced ships far separated were signalling to the admiral finds of spars, oars, breakers, deck lockers, gratings and all sorts of other floating gear. But no traces of survivors were to be seen, and for a time it was feared in the fleet that not a soul had escaped.

When the news reached England the nation was aghast at the loss of so many of its servants, and in a lesser degree only to that at the loss of a ship which had been so exceptionally under public notice and cost about three times more to build than any other ever lost in British history under any circumstances of war or peace. But even then some of Coles's supporters refused to admit that the calamity was due to defective design. Lack of careful seamanship was suggested as at least possible, and *The Times* in a leading article demanded that the keel of another similar vessel should be laid down at once as a memorial to Coles. In responsible quarters, however, the vindication of the Controller's department was admittedly complete, and all Press interference in the matter fell thenceforward on absolutely deaf ears. Nor has a name recalling so tragic an occurrence from every standpoint ever again been revived in the British Navy despite the lustre of its earlier history, save a hired trawler during the First World War.

The regulation formality of a Court Martial on the survivors followed in due course under the compulsion of the Naval Discipline Act. Its proceedings lasted for many days, during which every man that found safety in the pinnace was exhaustively questioned, and the only knowledge that could ever be forthcoming either then or afterwards as to the circumstances of the *Captain*'s loss was derived from their evidence. Their acquittal from blame was practically a foregone conclusion; but the statements of certain other witnesses, called before the Court and examined, produced information upon which the Court added some pretty strong comments to its finding, and these in turn started a long process of subsequent recriminations and denials of responsibility. Representatives of the Controller's department, as well as both the Laird brothers in person, were summoned for examination. It transpired that the stability calculations founded on the basin tests had been completed and submitted to the Board just after the ship had sailed for the last time, and that they confirmed the theoretically obtained estimate of the Lairds in regard to the angle of safety. Coles and Burgoyne almost certainly knew the results of Laird's deductions, but neither those nor the results of the Admiralty tests were ever officially communicated to any of the flag or senior officers under whose orders the *Captain* was placed during her short career. This omission supplied the basis for a paragraph in the finding of the Court in which, by implication at least, the Board of Admiralty, the Controller and the Laird brothers were all involved in a general expression of censure. To this the Lairds could legally plead that it was no part of their contract to notify results of work undertaken for their own information, and that they had at any rate suggested that the Admiralty should ascertain the stability of the ship for themselves. The Board for their part could say with perfect truth that Lairds' original assurance as builders of the ship that they did not regard the masting plan of the *Captain* as a source of danger had never been contradicted or modified, and was in fact still adhered to, since the Lairds persisted, even at the Court Martial, in attributing the disaster to some undiscovered cause. Other possible causes did come before the notice of the Court, such as suggestions that the superstructure acted as a dangerous wind trap, or that the starboard bilges were flooded, but all were rejected after close investigation. And lastly the Controller could point out that the order to determine the stability of the ship was not accompanied by any instructions as to urgency or priority of treatment, though the Board knew that his department was working overtime on the designs of new ships in preparation for the next estimates.

Mr Childers, the First Lord, was a man of autocratic temperament, notoriously given to differences with his naval colleagues and subordinates, and suffering from a heavy personal bereavement through the loss of his son as a midshipman in the ship. He resented and repudiated the strictures of the Court and tied to put the blame on Robinson and Reed for not calculating and reporting the results of the stability tests more expeditiously. It was true that both of them had from the first regarded the *Captain* as being dangerously over-masted; but all their representations to that effected had been turned down, and Childers's action was looked upon in the Navy as neither just to the Controller's department nor creditable to himself.

Elsewhere the finding of the Court was considered on the whole as a fair, impartial and not unduly severe summary of the causes which led up to this tragic page in the history of the British senior service. The exact wording of that historic pronouncement, which bore the signature of some of the most distinguished officers of their day, ran as follows:

H.M. Ship Captain *was capsized by pressure of sail assisted by the heave of the sea; and the sail carried at the time of her loss (regard being had to the force of the wind and state of the sea) was insufficient to have endangered a ship endued with a proper amount of stability.*

The Court find it their duty to record the conviction they entertain that the Captain *was built in deference to public opinion expressed in Parliament and through other channels, and in opposition to the views and opinions of the Controller and his Department, and the evidence all tends to show that they generally disapproved of her construction.*

It further appearing in evidence that before the Captain *was received from the Contractors a grave departure from her original design had been committed, whereby her draught of water was increased by about two feet and her freeboard diminished to a corresponding extent, and that her stability proved to be dangerously small, combined with a sail area under those circumstances excessive; the Court deeply regret that if these facts were duly known and appreciated they were not communicated to the officer in command of the ship, or that if otherwise the ship was allowed to be employed in the ordinary service of the Fleet before they had been sufficiently ascertained by calculation and experiment.*

CHAPTER NINE

THE 'PRINCE CONSORT', 'CALEDONIA' AND 'OCEAN'

H.M.S. *Prince Consort* depicted during her first voyage to the Mersey when she was in danger of being swamped. (NMM. 58/2216)

In the description of the *Lord Clyde* and *Lord Warden,* mention was made of certain conversional work in the building programmes of the early 'sixties, whereby some of those timber-hulled vessels under construction at the time, which had been originally laid down as unarmoured capital ships of the old type, were transformed before completion into armoured classes of the new.

Eight ships in all were converted. Of these only one had been begun as a three-decker, and she was finished as a mastless coast defence turret ship; which never went further from England than across to France, having no canvas and only enough coal for three days at sea. But the seven originally laid down as two-deckers were all eventually launched as armoured ocean-going vessels carrying broadside batteries under a sail area sufficient to take them round the world. Five of the seven were lengthened while still on the stocks and took the water all alike in external mould and dimensions; and three of these five were completed internally as well as externally on an identical plan at first, though modified differently in detail afterwards. These were the *Prince Consort, Caledonia* and *Ocean.*

The object of lengthening was to provide more space for an armament mounted entirely on one deck except for the chase guns. Such extensions are carried out by a process well recognised in naval architecture; which is quite possible at any stage in the building of a ship, not by adding to her extremities, but by cutting her in two at the point of greatest beam and separating the halves sufficiently to insert a new section between them. This is usually done by erecting a launching cradle round the stern half, moving it down the slipways on the cradle as far as required and building up the gap. In such fashion the *Prince Consort, Caledonia* and *Ocean* (as well as the *Royal Alfred* and *Royal Oak*) were each drawn out 23 feet when already partially in frame, all having been laid down in 1860 and ordered for conversion the year after. But as the midship portion of a hull is always the part subjected to the greatest hogging and sagging strains in a head or following sea, special care is necessary to ensure that an inserted section is sufficiently strongly built in; particularly if it has to bear the burden of the engines as it did in the *Prince Consort* class. Exceptionally stout timbering was therefore used for the purpose in those ships; though even with that their central structure sometimes exhibited signs of stress giving rise to trouble in a minor degree.

In general construction they were less complex than the specially built *Lord Clyde* and *Lord Warden;* having no iron bulkheads, no iron skin between ribs and planking, and no filling between ribs; though with the same system of iron beams and stanchions to carry the upper deck and the same iron cable bitts and bollards. In completing the bow the wooden knee of ancient tradition was discarded, but no ram was added, and the forward profile ended in a vertical stempost with a very slight outward curve. Aft the overhang of the counter was narrowed to an oval termination foreshadowing what became the normal battleship type of stern thirty years later, except that the counter was above water instead of below. As these were the first vessels with that novel design of stern they were often called 'double-enders' until the design became so common as to suggest no peculiarity.

Although the added central section had a full bilge throughout, their floors rose in the old fashion forward and aft, because their under-water lines had taken shape before conversion was ordered and could not be altered without an almost entire rebuilding of the ship. This reduced the carrying capacity of the hull as compared to that of a full bodied vessel such as the *Bellerophon;* but retained a good model for easy movement and contributed to their excellent steering qualities alike under steam and sail. The upper parts of the wooden rudder and screw aperture showed above water, and the rudder was of the old narrow unbalanced type with a bronze neck socket to receive an iron double-tillered norman head. As the first to be completed, the *Prince Consort* was copper bottomed in the prevalent style for all timber hulls; nobody in that age of inexperience in metals having foreseen that this would set up electrical action with the lower strake of armour, whereby the iron would become eroded. But as that was discovered when she was docked for general examination a few months after commissioning, Muntz metal was substituted for the copper and used in all later wooden battle-

H.M.S. *Prince Consort*, after 1866 showing her second change of rig, still as a barque, but with topgallants and royals to her fore and main masts. (NMM. A.2884)

H.M.S. *Caledonia*, anchored at Malta about 1868, after both her poop and topgallant forecastle had been added. (NMM. HAI/1)

ships. A series of large external scupper pipes leading right down to the waterline formed a conspicuous peculiarity in these vessels. Originally the scupper pipes had been carried down internally as in other ships; but the experience of the *Prince Consort* on her maiden voyage led to additional outside shoots being fitted to the whole class, which were found to answer requirements though extremely unsightly to the eye.

In their process of transformation the seven which became armoured broadside battleships were all cut down one deck from their original building plan; so that in their finished form the original main, lower and orlop decks became the upper, main and lower respectively. The three *Prince Consort* class were launched with flush upper decks; but at a later stage poops were added to the *Caledonia* and *Ocean* and topgallant forecastles to the *Caledonia* and *Prince Consort*. This had the effect of increasing the maximum draught of the *Caledonia* by six inches and of the *Ocean* by nine; although by bringing the *Prince Consort* on a more even keel hers was reduced by three.

All officers and men were berthed and messed on the same general plan as that for the *Lord Clyde;* except that in the absence of main deck chase guns the captain's accommodation lay right at the stern instead of the port side of the half deck. In the *Ocean* the sick bay was situated on the fore main deck, as she had no topgallant forecastle. The ship's companies' heads were external structures in all three of these vessels.

Beneath the lower deck these three sisters shared a marked difference of internal arrangement from that of all other British armoured ships ever built, in the disposition of the boilers and engines, which affected their whole internal plan by allowing a larger proportion of machinery space in their design than was allotted in the other converted types. What remained available for ammunition, stores, provisions and water was necessarily congested as compared to the space in the others; but in general arrangement the storage was on the same plan, on a reduced scale, as that of the *Lord Clyde;* though divided only by wooden bulkheads and iron bar gratings.

In most details of their hull appurtenances the three ships were alike. Each carried a threefold hand wheel on the quarter deck, a double wheel on the half deck and another double wheel down in the after part of the wardroom. The first and second connected with the upper tiller and the third with the lower; though as a rule only the tiller ropes of the quarter deck wheel were kept rove and the others reserved for emergency. Both tillers were in the captain's after cabin; the upper one close to the beams overhead and in general use; the lower close over the floor, but unshipped and clamped down when not required. Four hand pumps with bilge and sea suctions were fitted to rig on the main deck and work with about a hundred and fifty hands at a spell when fully manned. One steam pump was situated in the engine room. For working cables the *Prince Consort* and *Ocean* carried quarter deck capstans with chain messengers, but the *Caledonia* was fitted with a patent capstan in the usual position abaft the fore mast. Though the ships were timber-hulled the boat davits were iron; in which respect they differed from many of their iron-hulled contemporaries with the wooden quarter davits that most officers of the period preferred as carrying the boats higher. These three carried their boats too low and frequently lost them in heavy seas.

In their sea-going qualities the *Prince Consorts* were admirably handy under steam or sail in all conditions of weather, riding well in a head sea and yawing very little in a following; but shared the common defect among ships of the early 'sixties of an extremely low centre of gravity, with the usual result that they were deep rollers and unsteady gun platforms. Various expedients were tried to remedy this fault but without much success, as the main weights were necessarily fixtures. From the same cause however they were stiff as a church under canvas and could easily have stood up to a much larger sail area than they carried.

As an instrument designed solely for decisive contact with the main force of an enemy, the premier qualification for a capital ship — apart from absolute seaworthiness — has always been a highly developed power of attack. And to that ideal the five lengthened and converted vessels named above came nearer for their size than any others in the battlefleet of 1870; because with them the ratio of the weight of armament to total tonnage was the greatest of all. Next to offensive power in importance for a battleship comes defensive power; and here also they were remarkably well equipped for their date. Though considerably smaller than the *Warrior* or *Achilles* their broadside fire was not appreciably inferior and their armouring much more complete. Their speed was a little less and they had no watertight division; but in those respects they were no worse off than their foreign contemporaries; by which standard they had in the end to be judged. Even, therefore, if a few extremists of the old school on the one hand, and the new school on the other, chose to decry them as hybrid compromises of naval architecture, their title to a place in the front rank remained unchallengeable on any reasoned point of criticism for quite ten years after they were launched. In tonnage they were only middle-weights of the ring, but in ability to inflict and withstand punishment the equals of many units in the line of heavier body.

Like the majority of the capital ships launched during the first three years of the building of a British armoured fleet, the *Prince Consort* and *Caledonia* began their service with a combination of rifled breechloaders and smoothbores. The former carried seven 7-inch Armstrongs (three being upper-deck chase guns), eight 100-pounder and sixteen 68-pounder smoothbores; while the latter carried ten, eight and twelve weapons of corresponding sizes, of which four were upper-deck chase guns. All were on wooden mountings of the pattern representing the first step in evolution from the gun truck to the iron carriage and slide. But before the end of the first commission in both this armament was entirely removed to make way for rifled muzzle-loaders; of which four 7-inch were placed as chase guns on the upper deck, while four 8-inch and sixteen 7-inch formed the main deck broadside battery. Twelve of the main-deck guns however remained wooden mounted.

As the last completed of the three, on the other hand, the *Ocean* was armed with rifled muzzle-loaders from the time of first hoisting the pennant, when she carried twenty-four 7-inch; and except for the substitution of four 8-inch for an equal number of 7-inch on her main deck, only a few months later, she carried that armament all through her service. By the year 1867 therefore all three of those ships had exactly the same weight of metal as fighting units, with a Quarter Bill made out for a complement of 605 ranks and ratings. Having no lower-deck shot lockers, they stowed a larger proportion of projectiles in the main deck racks than the iron-hulled battleships; and being provided with no light saluting battery they saluted with their heavy armament, which sounded impressive enough at Spithead or the outer Nore, but was apt to cause a mixed sensation in a confined harbour among people on shore.

A complete armour casing covered every square inch of their sides fore-and-aft from the upper-deck level to six feet under water, with a broadside thickness of 4 ½ inches; but, like all the other vessels with complete external protection, they had no interior armoured bulkheads, nor were the upper-deck bow guns behind a forecastle armour shield as in the *Lord Warden*. In order to provide what was intended as a conning position for the captain in action a rifle-proof half tower large enough for one officer was inserted each side in the waist bulwarks, with voice pipes to the quarter deck and half-deck wheels and the engine room; but the ship's own guns rendered these cubby holes untenable when firing herself, which it may be presumed was not improbable if encountering an enemy. They were generally used as wash deck lockers in consequence. As flagships these vessels had no provision for protecting the admiral in action, who had to take his chance on the poop, after the ancient tradition, if he wished to keep an eye on his fleet.

The *Prince Consorts* were all three engined by Maudslay and Sons on an identical plan, and, as already remarked, the arrangement of their machinery was very unusual, though not entirely without precedent in wooden-hulled steamers. When horizontal reciprocating marine engines were being installed it was necessary as a first step to select a position for the cylinders with enough athwartship level to give a proper length of stroke. Some latitude of choice in the matter was to be found in the iron-hulled battleships and in the specially built *Lord Clyde* and *Lord Warden*, for in their cases the transverse sections were full for a considerable distance abaft the midship point. In those vessels therefore the engines were usually placed at roughly one-third of the ship's length from the stern to obviate excessive length of shafting; while to balance their weight, and keep the ship on a moderately even keel, the boilers and bunkers were grouped so that their centre of gravity as a whole lay a little forward of the ship's centre of flotation. But as the *Prince Consorts* had been shaped with the under water run of the old style before being ordered for conversion, and as that could not be altered without a wholesale reconstruction entailing a great and expensive waste of already shaped material, their floors rose considerably abaft the midship point; which meant that the only practicable position for the much larger engines they were to carry after conversion was in the added central body halfway between bow and stern.

Having consequently no choice as to the position of the cylinder beds, it only remained to deal with the problem of placing the boilers, and here too much was attempted by aiming at a more powerful installation than the ships could properly carry. Boilers and machinery sufficient to develop 1000 nominal h.p. were ordered; whereas in the other two lengthened ships — the *Royal Oak* and *Royal Alfred* — the Admiralty were content with 800 nominal h.p. only; which proved in the end to give just as good results in speed at less contract price and consumption of coal, because imposing 300 tons less burden on the hull. As too little space was available on the forward side of the cylinders in the *Prince Consorts* to fit in all the steam generating plant required for their heavier engines, Maudslay had to look elsewhere to place the excess; and, having grouped four large boilers of five furnaces each on the forward side installed four smaller ones of three furnaces each right aft in the position occupied by the cylinders of an iron-hulled vessel. Each group had its own funnel, and these three battleships were double funnelled therefore, like the ten others already described, with this difference that, instead of the funnels standing close together, one was in the forward half of the ship and the other on the quarter deck. As each group of boilers had its own coal bunkers abreast it, a considerable burden was imposed on the after part of these ships, partly accounting for their heavy trim by the stern. The boilers worked at 20 lb. pressure.

Return-connecting-rod engines were installed; a type in which Maudslay's specialized, very suitable for wooden hulls. Though well designed in itself, however, the machinery was really too heavy for the ships, and needed frequent adjustment to meet hull distortion — particularly as the shafting had to be exceptionally long — the worst case being a permanent keel sag of 3 inches developed by the *Caledonia* before being two years waterborne. As the cylinders lay forward of the main mast and the after stoke-

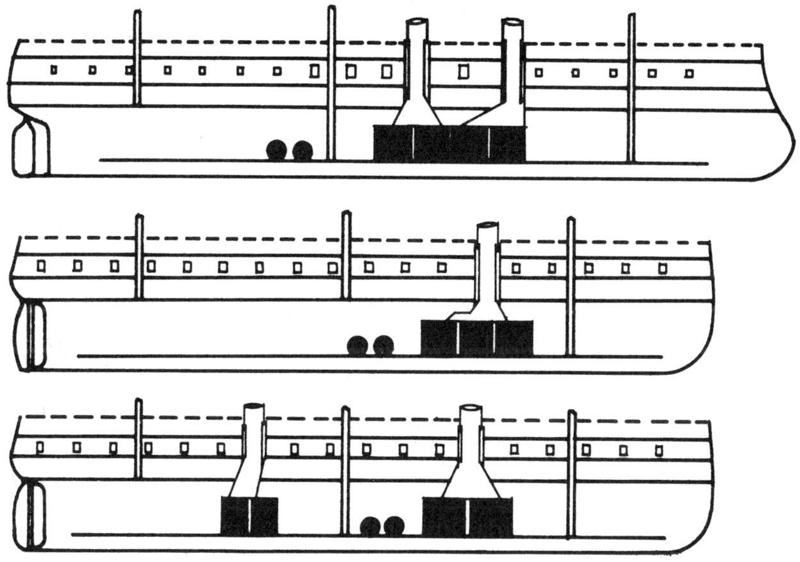

Diagrams to scale showing the three different systems of placing the boilers and engines among the British battleships of 1870. Top: General arrangement of boilers and cylinders in the iron-hulled battleships, also in *Lord Clyde* and *Lord Warden* (with three cylinders). Engines abaft the main mast, boilers in one group, from four to ten in number. Centre: Boilers and cylinders in *Royal Oak*, also, with minor variations in detail, in *Royal Alfred, Repulse* and *Zealous*. Engines before main mast, boilers in one group of six. Main mast stepped on lower deck. Bottom: Boilers and cylinders in *Prince Consort, Caledonia*, and *Ocean*. Engines before the main mast, boilers in two groups of four each. Main mast stepped on keel with shaft passing through it.

hold, the shaft had to pass through both. To allow it to lie through the mast the latter was slotted to five feet above the step, making it double heeled to straddle the shaft; and, running as it did right down the middle of the after stokehold on bearings between the two rows of furnace doors, the shaft inevitably cramped the work of slicing and raking the fires.

Taken as a whole, the steam propulsion of the *Prince Consort* as compared with that of the *Royal Oak* provided a good illustration of the principle that given certain lines and dimensions of hull — and especially a certain ratio between length and beam — no appreciable advantage results from raising the power of the machinery beyond a certain point; unless it can be effected without a proportionate increase in weight, and even then the gain in speed is never great. When additional power is only obtained by using heavier boilers and engines — as in the case of machinery of the same general type working at the same steam pressure but on a larger scale — the resulting increased immersion of the hull absorbs all the extra force of the drive, and the greater horse power developed does no more than pull its own weight, though more expensive in initial cost and upkeep and burning more fuel. Moreover, by imposing more burden on the ship, it is detrimental to her qualities as a heavy weather seaboat and reduces her speed considerably under sail. In all those points the *Royal Oak* proved superior to the *Prince Consort;* both having the same external dimensions, lines and speed, though

H.M.S. *Ocean*, after yards and a topgallant mast had been added to her mizen in 1866. (NMM. HAJ/1)

the latter carried machinery of 25 per cent. greater power, contract price and consumption of coal.

But if the *Prince Consort* class were rather over-engined on the one hand, they were certainly under-masted on the other, for distant service in the days when coaling stations were very few and far between outside the North Atlantic and Mediterranean. Had they been completed as two-deckers of the old type their sail area would have measured more than 33,000 square feet exclusive of stunsails; but when the order for their conversion was issued a double topsail barque rig was designed at the Admiralty as a substitute; in which the original area of canvas was cut down by nearly a quarter and the lowermasts alone remained on a line-of-battleship scale. Of the three vessels however only the *Prince Consort* — as the first to be ready — ever actually carried that sail plan for a brief period. Before the others were ready for masting, the generally unsatisfactory reports from the earlier battleships in which such a rig was already under trial had led Whitehall to abolish it in the Navy; the chief ground of condemnation being that extra topsail yards meant more top hamper aloft when steaming head to wind if left up there and more lumber on deck if sent down. In consequence the *Caledonia* and *Ocean* were completed for sea as single topsail barques with fidded topgallant masts; and the *Prince Consort* was altered to that rig soon after. By 1866, moreover, yards had been added on the mizens of all three, and the resulting full ship rig was retained to the end of their career. Though a moderate spread at best for oceanic passages, it proved quite sufficient for quick tacking and wearing,

Having been made in the pattern required for double topsails and stepped without alteration, the iron lowermasts carried their tops about five feet lower down than if constructed for a single topsail rig; which gave them very long mastheads and reduced the height of all gear above by the same amount, as well as the drop and area of the courses and gaff sails. The topmasts, topgallant masts, yards, gaffs and booms were all wood; the dimensions being on the same second class scale as the corresponding spars in the *Minotaur* except that mizens were necessarily different in a five-master and a three-master. At first these ships had fixed iron bowsprits stepped on the main deck with a considerable steeve; but in the *Prince Consort* and *Caledonia* wooden running-in bowsprits on the topgallant forecastle were substituted when their topgallant forecastles were built. They lost many spars through their being too light; and owing to the heavy drag of their great four-bladed propellers (even when disconnected) and their reduced sail area, these vessels were indifferent movers under sail at all times; the best ever reached by the *Caledonia* and *Ocean* being 11 ½ knots and the *Prince Consort* no more than 10. In spite of that, however, to the *Ocean* belongs the distinction of creating and holding for ever by just one mile the best record for a day's run without steam by a British armoured ship; though it can scarcely be doubted that at least five of the others would have done better in similar conditions.

Like all mid-Victorian battleships, they were provided with seven anchors and nine boats; the anchors corresponding in respective weight and description to the heaviest made for the British service previous to the building of an armoured fleet; that is to say, two bowers and two sheets of 95 cwt. Admiralty pattern with wooden stocks, one iron-stocked stream of 24 cwt. and two kedges. For use with these the standard cables were 525 fathoms of 2 ¼ inch chain, 100 fathoms of 1 ⅜ inch stream chain and a 14 ½ inch hemp. Chain messengers were also supplied to the *Prince Consort* and *Ocean;* and as none of the class had a ram, all used the old type of cat purchase, but the fish davits were iron fixtures. For hoisting out the boomboats the reduction of the lower yards to only 90 feet in length gave barely sufficient clearance when braced inwards for the purpose in these vessels of 58 ½ feet beam; which remained always a source of official complaint by their captains. In such heavy rollers moreover the substitution of iron davits for the old and high straight wooden pattern caused the loss of many quarter boats at various times in stormy weather.

The *Prince Consort* was laid down at Pembroke Dock in 1860 as the 90-gun line-of-battle ship *Triumph,* but ordered to be converted in 1861 and launched in 1862. Before completion her name was changed to commemorate the then recently deceased husband of Queen Victoria. It was an age when the atmosphere of foreign politics was charged with electricity from the domestic thunderstorm in North America. The failure of the British Governement to prevent the Southerners building their famous raiding sloop *Alabama* at Laird's yard in Birkenhead, and sailing her to ravage northern commerce, caused Downing Street no end of trouble, and as Laird's was also completing a pair of small armoured turret ships for the same prospective owners, the Cabinet decided to notify the Mersey Customs to detain them.

As the detained vessels were armoured and heavily gunned for their size, the Cabinet deemed it expedient to back their instructions with an adequate show of force, which meant an ironclad. Of such ships Britain only then possessed five, all at sea or refitting at the moment except the *Prince Consort;* which remained the only one immediately available, having just finished her trials. A temporary crew was placed on board at Plymouth accordingly, drawn from ships in dockyard hands, and she was ordered to Liverpool; on which service she fell in with weather of a kind threatening worse consequences than any likely to arise from an encounter with the potential enemy. A tremendous gale, which swept the Irish Sea and compelled all shipping to run for shelter, caught her on her way, while passing the Welsh coast just where no deep draught shelter lay anywhere near.

The *Prince Consort* was not only a brand new battleship, but the first of a kind entirely untested at sea, about which some seamen of mature judgment had predicted disaster. Moreover, with the exception of the Chief Engineer and the Warrant Officers, not a soul on board had been more than three days in the ship, or ever had any experience of handling this novel type under stress of violent weather. When the storm broke the captain tried to keep her to her northerly course, intending to make for the lee of Angle-

H.M.S. *Caledonia*, after 1866, ship rigged, with both poop and topgallant forecastle, and a running in bowsprit. (NMM. 58/6186)

sey; but her unknown rolling qualities in a beam swell were now disclosed at their worst, and as the wind increased every sea poured green over her seven foot bulwarks. That would not necessarily have been a source of genuine anxiety if the freeing arrangements had proved equal to the occasion; but the scuppers were too small to carry off the water as fast as it was coming in, and after a time her upper deck was like a swimming tank with cataracts escaping down the hatchways and funnel casings, until at length it became imperative to bring her round and try lying-to.

Once head to wind she behaved well enough, and though pitching heavily from her half drowned condition and still shipping seas over the bows, any immediate danger would have ended but for an unforseen accident. As her new auxiliary machinery had not had time to settle itself, the one steam pump broke down completely through over driving just when most required, and the situation at once became critical. Assisted by the four hand downtons, the steam pump had been reducing the well soundings since the turn to windward; but by themselves the hand pumps were barely able to compete with the water still coming in forward, much less get rid of what was already below. The weather showed no signs of moderating, nightfall was at hand and the long wall of the Welsh cliffs lay to leeward. Hour after hour therefore the hand pumps were kept swinging round without a moment's rest, though at one period it looked as if the men were fighting a losing battle. The water rose till the stokehold plates were submerged and threatened to extinguish the fires, in which case nothing could have saved the ship. But that fate was averted by an uncommonly good effort, and happily the main engines continued to work well. Gradually therefore, but surely, the *Prince Consort* drove through the gale all night towards the shelter of the Irish coast; where she anchored in Kingstown outer roads at eight o'clock next morning, badly down by the stern with a couple of hundred tons of water in the bilges, but safe.

It had been a tough night for all hands. Incessant care had been necessary on the bridge to hold the ship head to sea, and incessant care in the engine room to prevent a breakdown through propeller racing. The competence of the officers, however, had been well seconded by the unremitting labour of their subordinates; seamen and marines sticking to the pumps without slackening an instant from sunset till morning, while the stokers fed the furnaces knee deep in hissing water. For a raw ship's company, in which all ranks and ratings were strange to each other, the work of all was afterwards officially commended as highly creditable, and if any branch had failed the efforts of all the others would have been in vain.

Meanwhile great uneasiness had prevailed at the Admiralty, owing to the extreme violence of the weather and the absence of any news of the *Prince Consort*. She had been due off the Mersey the day before, but had been reported neither from there nor anywhere else, and, as that was long before wireless telegraphy, a ship out of sight was a ship of whose proceedings or welfare nothing could be learned. As usual on such occasions, all kinds of rumours spread rapidly, and even when she was in sight from Kingstown the first news received about her in London was disquieting. She anchored very far out so as to be well clear of shoal water in case of a shift of wind, and, the weather continuing rough, did not communicate with the shore for some time. Since she appeared to be out of trim, the sensation mongers started a story that she was in a sinking condition, which found its way over the wires before any official report. Eventually, of course, the truth was known and promulgated from the Admiralty, but the public anxiety was sufficiently indicated by the fact that Queen Victoria sent a personal telegram congratulating officers and men in bringing the ship safely through. She was ordered back to Plymouth when the weather moderated, to repair damages; but the publicity of the whole episode served to advertise the fact that the British Governement were carrying out naval movements in support of their announced intentions, which had a good effect; and the problem of disposing of the two vessels at Laird's was soon afterwards solved by their purchase for the British service under the names *Scorpion* and *Wivern*.

Moreover the behaviour of the *Prince Consort* afforded very useful indication of the good and bad points of this new class of vessel as heavy weather seaboats. Where details of construction or fittings had proved defective they were remedied, and the freeing arrangements in particular were largely augmented by external scupper shoots capable collectively of discharging six hundred tons of water a minute. The battening down and pumping equipment was also improved. Finally, as regards the design of the ship herself, the experiences of the occasion had shown that even if she was apt to roll excessively when off the wind she could face any weather lying-to, and if properly handled might navigate any sea in the world with confidence.

Her later career was uneventful. A few months after the incidents described above she was commissioned by Captain Willes, and flew the pennant under that officer and his successors for seven continuous years, of which the first three were in the Channel Squadron and the remainder in the Mediterranean, where for a time all three of her class served together, as well as their half sister the *Royal Oak*. This quartette in fact formed the backbone of the station fleet during a somewhat unsettled political period. In 1871 she went back to Plymouth to pay off; by which year she had answered the original purpose of her conversion to 'armoured rank', as a sufficient number of iron-hulled new battleships were now afloat. She was laid up accordingly but retained for emergencies till March 1882, when being no longer required and getting leaky she was sold for breaking up to Castle's. No other ship of the British Navy ever bore the name *Prince Consort.*

The next to leave the slipways was the *Caledonia,* built at the historical government yard at Woolwich, and launched in 1862, only a few months after the *Prince Consort* had taken the water at Pembroke. But she was much longer in completing for sea, having to wait for her guns; and it was not till 1865 that her orders were issued for hoisting the flag of Rear-Admiral Yelverton as second in command in the Mediterranean; an incident of some interest in the annals of the British service as she thereby became the first British armoured ship to carry an admiral. In 1866, after her poop was added, she took the place of the two-decker *Edgar* as flagship in the Channel Squadron, of which Yelverton had been placed in temporary charge to conduct a series of experimental cruises with an assembly of new armoured types. Thence the *Caledonia* went back to the Mediterranean to relieve the *Victoria* (the last of the three-deckers) as flagship to Lord Clarence Paget, the Commander-in-Chief. Paget formed a high opinion of her fighting powers and especially her ease in manoeuvring. In 1869 she herself was relieved as flagship by the *Lord Warden,* but remained on station for another full commission under the pennant. After coming home in 1872 and paying off, her next duty was as Forth guardship, in which after three years' further service she was relieved a second time by the *Lord Warden* and paid off finally at Plymouth; where, after lying out of commission for ten years more and showing signs of decay, she went to Castle's breaker's yard on September 30th, 1886. It was the triple distinction of the *Caledonia* not only to be the first British armoured flagship but the first in the two most important British sea commands.

The last of the three to be launched was the *Ocean,* to which ship distinction attaches in quite another line, as the only armoured ship ever to double the Cape of Good Hope under sail. She was built at Devonport, launched in 1863, and commissioned by Captain Stanhope for the Channel Squadron in July 1866, but transferred to the Mediterranean before the end of that year, where her sisters were already on duty. In 1867, however, a second and very different change of station awaited her. It was an age of highly important political evolution in the East affecting the interests of several flags, and the arrival of the French ironclad *Belliqueuse* rendered it imperative that an armoured battleship represent British interest in those waters. Their Lordships' selection fell on the very appropriately named *Ocean;* because the complete absence of any dry docks in the Far East gave a metal sheathed bottom a great advantage over an iron hull in keeping free from fouling, and of the sheathed ships she was the newest and last out of hand.

As the Suez Canal was still far from complete, vessels outward bound to China in 1867 were obliged to double the Cape of Good Hope. At that period anything in the nature of through passages under steam had scarcely come into existence on account of the long intervals between coaling stations, the extravagant fuel consumption of the early types of engines, and the high cost of coal throughout eastern ports. The worst stage in the direct steam route outward lay in those days in the 4000 miles stretch between Cape de Verde Islands coaling station and Table Bay, about half of which had to be covered driving against the strong and perennial south-east trade winds, of which the zone lies as a natural obstruction to south bound traffic right athwart the South Atlantic. Such very few merchant steamers therefore as went to the Cape were obliged to carry a large amount of extra coal in their usual cargo spaces for their own use in crossing this stage, but in ships-of-war it was only possible to stow coal elsewhere than in the bunkers in comparatively limited quantities. The sailing traffic crossed the south-east trades zone by a long stretch on the port tack, which unavoidably took them to the opposite side of the South Atlantic, near the Brazil coast, before they were far enough south to pick up the westerly winds that carried them on a Great Circle round the Cape into the Indian Ocean; passing three or four hundred miles to the south of the Cape itself.

A decision was required therefore as to which route the *Ocean* was to take. She was primarily a steamer and a very full powered one for her day, though with the short steaming radius of her day as well. But she was perfectly navigable under sail also, if slow as compared to the old class of unarmoured vessels. If sent out under steam she would have to creep all round Africa at five knots with colliers specially dispatched to meet her at almost every sheltered anchorage on the coast beyond Sierra Leone; which would mean not only a very long but a very expensive passage. Following the sailing route, on the other hand, she would probably arrive in the China seas just as soon at far less expense, and that route was selected accordingly. It was familiar enough to the mid-Victorian Navy, and incidentally no man carried the confidence of the whole service more than Captain Stanhope, who would take the *Ocean* out. But to get as much help as possible from her engines she was ordered to call in for coal not only at Madeira and the Cape de Verdes but at Rio de Janeiro.

With these sailing orders Stanhope parted company from the Mediterranean fleet at Gibraltar on June

18th, 1867, and as no other armoured ship ever followed the track of the *Ocean* — or ever will — it would seem to be deserving of some description in detail. She first steamed to Madeira, arriving on the 23rd for coal. Leaving on the 26th, she picked up the north-east trades at once as a direct fair wind and made a sailing passage of nine days to St Vincent in the Cape de Verdes. Thence she started for Rio on July 6th, crossing the line on the 15th and arriving at Rio on the 28th. Of this passage of twenty-two days, ten were under sail. At Rio she made a stay of twelve days, partly because coaling was a lengthy process there in old times and partly to make some preparation aloft for the long heavy weather now in front of her. Being the month of August it was the very depth of the southern winter, and, as the *Ocean* had been already a year in commission without refitting, Stanhope considered it advisable to reeve some new running gear, besides sending up lashing backstays and fitting a second reef to the courses. When ready she took her departure on August 10th, and nothing further was seen or heard of her till she turned up off Java Head more than two months afterwards.

On leaving Rio she followed the arc of a great circle, but experienced a fortnight of shifty winds before picking up the true westerlies, during which stage she put in a couple of days steaming at five knots. On August 24th she passed Tristan da Cunha, a lone speck in the vast void of the South Atlantic, and on the 28th — running before a gale — she reached the fortieth parallel and so entered the 'roaring forties'; the girdle of stormy waters on which the Southern Cross never sets, once an ocean highway of chequered memories for those who had to sail it, but now almost abandoned to the whale and the albatross in these days of steam and trans-isthmian canals.

A ship in the southern forties was 'eating longitude' nearly all the time, breaking off from a great circle on reaching her selected parallel to steer due east as much as the wind would allow for 1500 miles or more. And, unless it blew so hard that she was forced to lie-to, the wind generally helped her with both hands, coming strong from the westward five or six days in seven, often indeed in gale force for weeks together. The *Ocean* experienced her full share of roaring forty weather, although as it was the winter she did not go south of the fortieth parallel. For fifteen days she rolled along it pretty closely, under reefed topsails most of the time, with all main deck ports closed and preventer backstays on every mast. Under such conditions it was pitch dark on the mess deck except for candle lanterns, and a good deal of the pea soup found its way into the scuppers between the galley and the mess tables. But she was putting the mileage astern often logging double figures even if not travelling as fast as a frigate or clipper, and on

H.M.S. *Ocean*, as converted to ship rig, and with fixed iron bowsprit. (NMM. 58/3009)

August 26th established the record twenty-four hours run for an armoured ship under sail of 243 knots, the ship being at the time under double reefed topsails, whole foresail and foretopmast staysail. With a wind of force 6 to 9 two points abaft the port beam, and a disconnected propeller spinning about 28 to the minute, she logged 65 knots between 6 p.m. and midnight on 25th. On September 3rd she crossed the meridian of the Cape of Good Hope — another sailing record for armoured hulls — being then twenty-four days out from Brazil and 320 miles south of the African continent; and on September 13th, having traversed fifty-two degrees of longitude on the fortieth parallel, she resumed a great circle track in 49° east and entered the outer limits of the Indian Ocean. Only twice in the forties did the wind fail her, and then for a few hours she was under steam at economical speed.

Two days after entering the Indian Ocean she ran into a cyclone, which is rather unusual so far south, but proved to be the genuine article. The *Ocean* was steering E.N.E. and the first squalls struck her from a little on the port bow, veering after a time to starboard and becoming a continuous blast of swelling force as it shifted round the compass. These indications showed that the storm centre lay to the northward of her course and was travelling eastward; which meant that the centre and the ship were drawing together on convergent tracks. Since that morning the failure of the westerly wind, combined with the behaviour of the barometer and the threatening appearance of the sky to the northward, had combined to put Stanhope on his guard; and the *Ocean* was ready for the fray when it came, with steam up, propeller connected, topgallant masts on deck, storm canvas bent, hatches battened down and main deck tiller ropes rove. Until it was practically certain that a cyclone lay somewhere on his port hand he kept the ship to her proper course, but in time grasped the fact that this would mean crossing the path of the centre. He broke off, therefore, first to the south-eastward and then to the south-west, as the wind changed in direction and increased in violence all through the scale to its maximum figure of hurricane force. At first he met it under close reefed top-sails, but soon had to come down to storm trysails and then to run under steam and bare poles. Throughout the after-noon the cyclone raged in explosive fury, driving a rushing succession of mountainous seas before it; but the constant veering of the wind confirmed Stanhope's anticipations, and by corresponding alterations he knew that he could eventually regain his proper course by passing behind the centre of the cyclone.

With a steering party of twelve hands divided between the quarter-deck and half-deck wheels, the *Ocean* was doing her best under a merciless ordeal, though she rolled 35° each way and shipped hundreds of tons of water over her six-feet hammock nettings both sides. Fortunately she could spew it out again as fast as it arrived; but her exces-sive straining and labouring caused uneasiness by working loose the ends of the main and lower-deck beams round the engine room hatches, which were the weakest point in her structure, and the four boats at the side davits were all washed away. By the evening, however, the wind had veered to south and begun to abate, and during the night she crossed her track of the previous day, passing behind the now retiring storm centre and ultimately heading on her great circle again for Java Head after making the com-plete round of the compass. A second very rough day fol-lowed, in which she had the misfortune to lose a man over-board without the remotest chance of rescue, and then the weather settled down to the normal for the remaining month of the voyage. For a fortnight she held steady wes-terly winds and made good progress under sail; but on September 30th ran into light south-easterly trades and for the remainder of the passage used steam at half boiler power with fore-and-aft canvas until entering the Straits of Sunda, and so arriving within the official boundaries of the China command. On October 15th, being 133 days from Gibraltar (inclusive of stoppages) and 66 under weigh from South America, she anchored at Batavia and saluted the Dutch flag. Her further passage to the station headquarters at Hong Kong was made under steam.

Space does not permit here of a full account of her ser-vice in the China seas, which lasted for five years. In 1869 she relieved the old 90-gun ship *Rodney* as station flagship, and hoisted the flag of Vice-Admiral Kellett; a new ship's company being sent out to her round the Cape in the *Done-gal,* her new captain being Hewett of the V.C. with bar, and her new commander the most publicly known naval officer in the world forty years later, being none other than John Arbuthnot Fisher. At the end of her commission as flag-ship she was relieved in her turn by the central battery ship *Iron Duke,* one of the latest additions to the service, which had been sent out by the now open Suez Canal.

As at that period, however, the canal could only take ships of 24 ½ feet draught, the *Ocean* had the prospect of another Cape passage before her in order to get home. Owing to the direction of the trade winds, it is always an easier matter to cross the southern tropical zones under sail homeward bound from China than outward; but, as the ship had been five years without docking, her sheathing had at last became extremely foul, and her rate of move-ment under canvas very sluggish in light breezes. That was a strong consideration in favour of making the voyage mainly under steam; which received added force from the fact that since the opening of the canal the price of fuel in the east had been halved, whereby one of the most import-ant points in choosing the sailing route out no longer counted. With orders therefore to proceed under steam at an average speed of about 4 ½ knots, she anchored in Ply-mouth Sound on June 5th, 1872. On being docked after paying off at Plymouth her under water planking was found to be in a bad state through the loss of a great deal of sheathing and leakage along the lower edge of the armour. She was placed in the 4th Division of the Dockyard Reserve in consequence and never hoisted the pennant again, though not sold for breaking up by Castle at Char-leton till 1882, nineteen years after her launch.

The *Ocean* was the most widely travelled vessel among the thirty forming the Black Battlefleet of 1870. In mid-Victorian days the main forces of the great maritime Pow-ers were permanently stationed in European waters, and no battlefleet proper was to be found anywhere else. All but a few of the world's capital ships were kept within easy reach of their own territorial waters, while service in more distant seas was left to the frigates and small classes with only two or three battleships from each navy as foreign sta-tion flagships. Two-thirds of the British armoured vessels of 1870 never went farther than Madeira; of the others a pair did alternate duty as flagships in the Pacific, and five crossed the North Atlantic. Only two besides the *Ocean* ever sailed east of the Mediterranean, and both of them travelled by the Suez Canal. But the *Ocean* navigated the three main waterspaces of the globe, entered the ports of four continents, doubled the Cape each way, and never once dropped an anchor in English mud between the day on which as a new vessel she first hoisted the signal asking permission 'to proceed in execution of previous orders' till the day she entered the Sound at the end of her sea career.

Sail plan of H.M.S. *Ocean*.

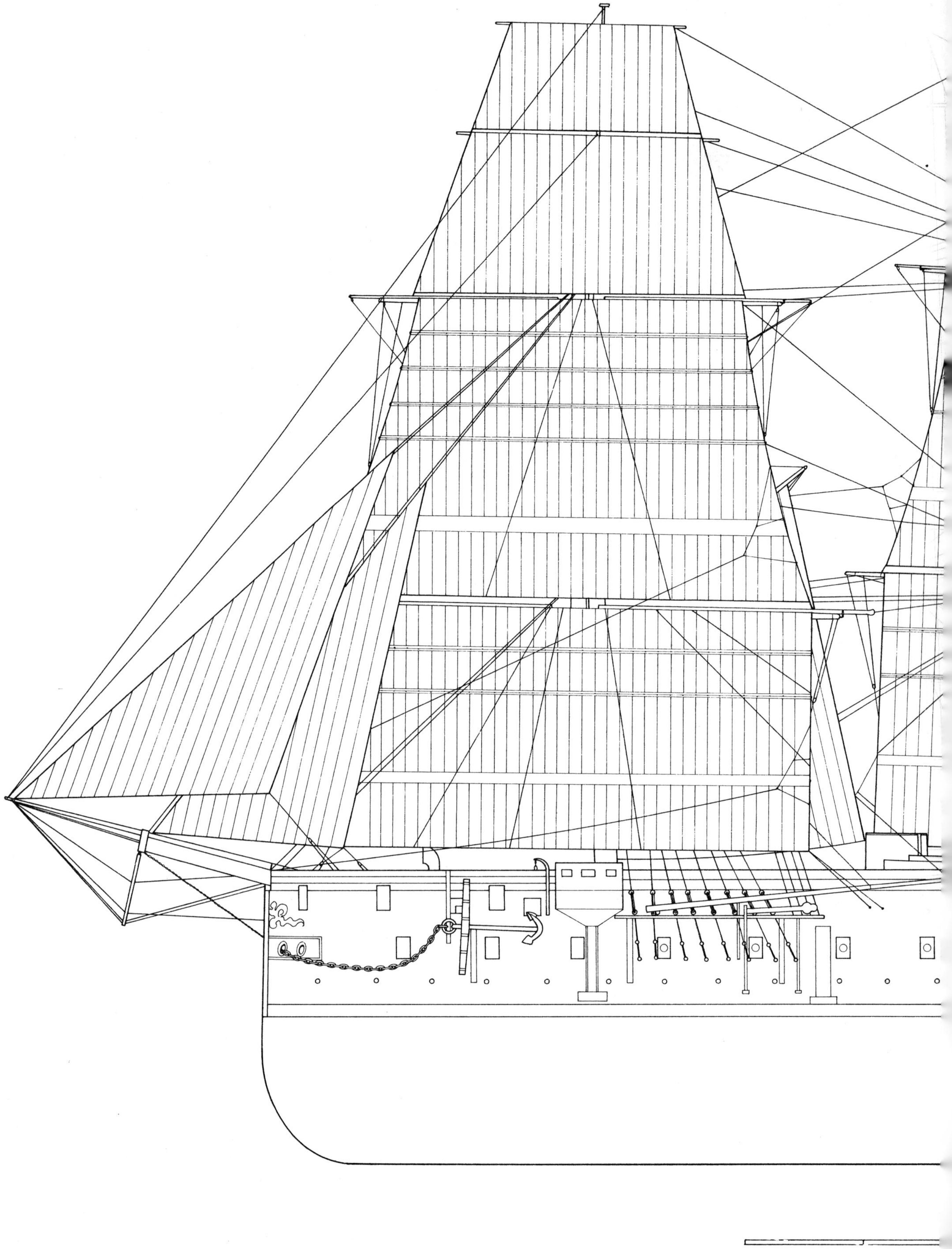

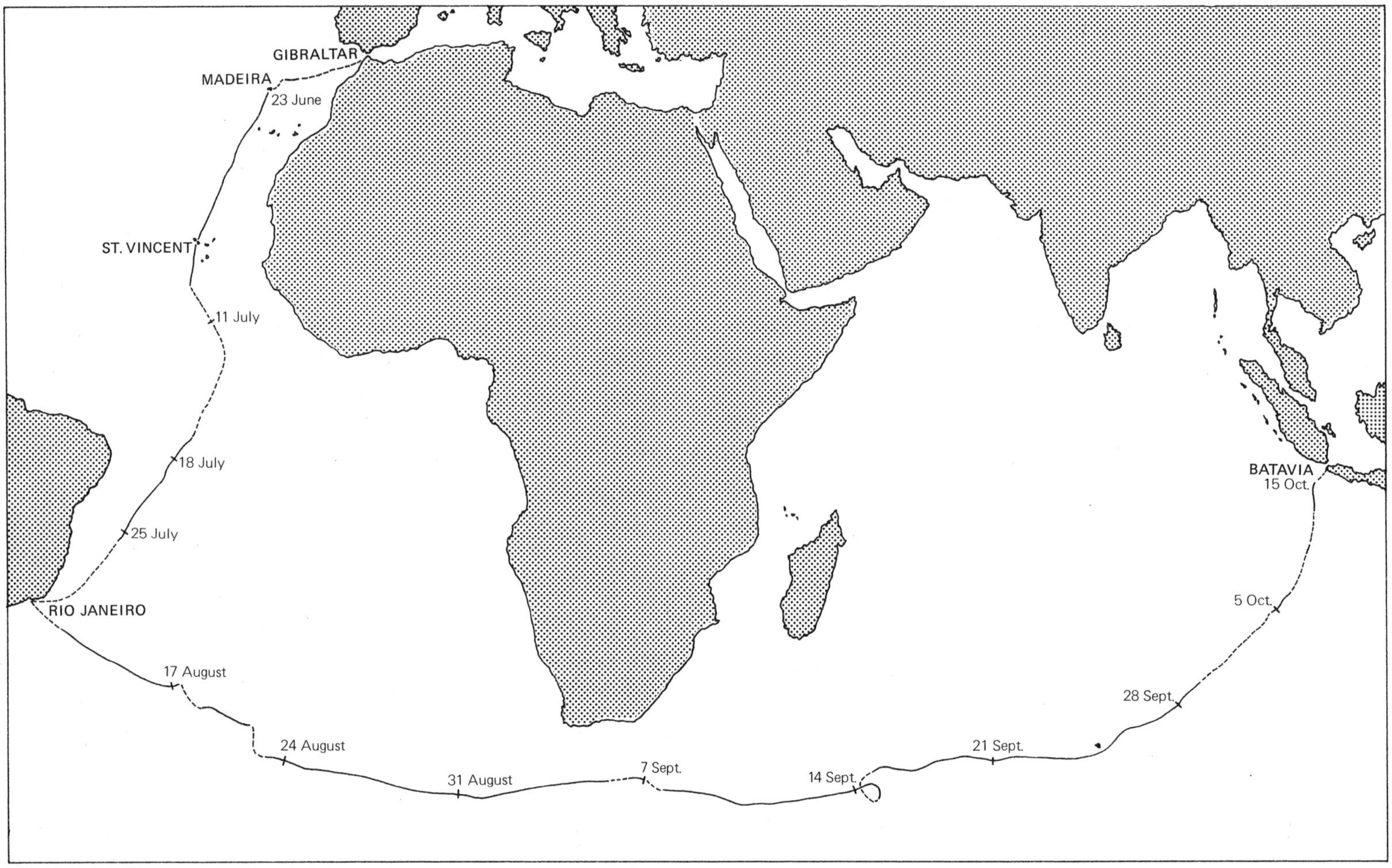

——— under sail

– – – – under steam

Track chart of the *Ocean* on her voyage from Gibraltar to Java in 1867, indicating weekly positions when at sea. Distances covered under steam shown in dotted lines.

CHAPTER TEN

THE 'ROYAL OAK' AND 'ROYAL ALFRED'

H.M.S. *Royal Alfred*, as Flagship, North American Station about 1870. (NMM. HAM/1)

H.M.S. *Royal Alfred* anchored in the Narrows, Halifax Harbour about 1869. (Photograph by courtesy of the Nova Scotia Museum, Halifax)

Of the five wooden battleships building in 1861 which were lengthened in the process of conversion to armoured designs, the three of the *Prince Consort* group have been described. The other two so treated were the *Royal Oak* and *Royal Alfred,* which were completed to the same external mould as the *Prince Consort* but on a different internal design. In the case of the *Royal Oak* the dissimilarity was confined to the scale and placing of the machinery, though in the *Royal Alfred* it extended to armament and protection so that neither of the 'Royals' had a full sister.

At the time of the issuing of the converting mandate to the Crown dockyards three of the seven 90-gun ships under construction had been a year longer on the stocks than the rest, and in their cases a stage had been reached in which a good deal of completed structure needed demolition as a preliminary. As that entailed additional expense the Admiralty decided to delay all work on these three until enough experience had been gained from the conversion of the others to determine whether the experiment — for such it was — justified the cost. Of the other four the *Prince Consort* at Pembroke and the *Royal Oak* at Chatham were at the point of building in which completion to the revised design could be most expeditiously accomplished; and though in their altered form they were to be finished alike as regards ordnance and armour, it was thought advisable to provide them with different standards of engine power to obtain practical guidance in connection with the motive machinery of the three postponed vessels. In the case of the *Royal Oak* therefore the engines already designed for use in her originally intended form as an unarmoured ship were installed unaltered except for a necessarily longer screw shaft; which gave her a nominal 800 horse-power, with six boilers and a hoisting propeller. For the *Prince Consort,* on the other hand, the machinery contractors — who were the Maudslays in both ships — were instructed to furnish a nominal 1000 horse-power, with eight boilers and a four-bladed non-hoister.

The results were instructive, and proved that the more ambitious conception had been a mistake, since by overloading the ship it added nothing appreciable to speed, while actually reducing radius of action by consuming more coal. The difference of scale of course affected the internal plans of the two ships in their allocations of space beneath the lower deck; but the principal immediate effect of adding 250 tons to the weight of the machinery was an increased draught, and a tendency to hull distortion. Thus although both vessels had the same length, beam and lines; the same weight of broadside; the same thickness and area of armour; the same speed under steam and the same bunker stowage, the maximum draught of water of the

Royal Oak was only 25 feet to the *Prince Consort's* 26 ½ feet. Her actual displacement proved to be only 6360 tons, though it was officially given as 6830 as in the *Prince Consort* class.

After the same fashion as in the other wooden armoured battleships a limited measure of iron work was introduced for completing the *Royal Oak's* frame by using it for the beams and stanchions supporting the upper deck. Elsewhere her hull fabric was well-seasoned timber throughout. Having no boilers abaft the main mast the space in the after-hold allowed of much more capacious shell, store and provision rooms than in the *Prince Consort,* and her decks aft were not cramped by funnel casings. A threefold wheel stood on the quarter deck, a double wheel on the deck immediately below it and another in the after-part of the wardroom. The wooden rudder had the bronze neck socket with iron norman head of all the armoured timber battleships of the period, with two tillers. The ship was conned from a small bridge resting on the quarter deck bulwarks, without conning tower, rifle tower or chart house. As first built she was flush-decked, but after her first commission a topgallant forecastle was added. Having the old-fashioned straight wooden quarter davits, she carried her cutters much higher than the *Prince Consort* class and never lost them even in the worst weather; but she had two pairs of iron davits also. Her original messenger capstan was altered to a patent at a later date, but without steam. As a gun platform she rolled less deeply than the *Prince Consort* — having less weight of machinery — but being on a lighter trim by the stern was more apt to dip in a head sea; a fault which was so much increased by the building of a topgallant forecastle that it was found desirable to transfer the contents of the forward shell room aft. The berthing and messing arrangements for officers and men were all on the lower deck except for the captain, commander and navigator.

As in all the British armoured ships under construction during the early 'sixties, the *Royal Oak's* armament warrant was constantly changed before she was launched, owing to the unprecedentedly rapid and extensive developments in the contemporary manufacture of naval ordnance. Eventually, after half a dozen alterations on paper, she received the customary mixture of smooth-bores and rifled Armstrong breech-loaders, with which all the British capital ships completed for sea between 1860 and 1864 started their career. In her particular case the precise combination consisted of eleven 7-inch Armstrongs and twenty-four 68-pounder smooth-bores. Excepting three of the Armstrongs which were upper deck chase guns — two being in the bows — all that armament was in main deck ports on the broadside. In 1867, at the end of her maiden commission, these were all removed to make way for a rifled muzzle-loader combination of exactly the same description as in the rearmed *Prince Consorts;* that is to say four 8-inch and twenty 7-inch, all on the main deck broadsides except four of the latter calibre as chasers on the deck above. The *Royal Oak's* armour consisted (as in the *Prince Consort*) of a complete hull casing to upper deck level with a maximum thickness of 4 ½ inches and a total weight of 935 tons. Her full complement was 585 officers and men with both armaments, having 15 fewer in the engine room than the higher-powered converted ships.

The engines were laid precisely half-way between the two extremities of the ship, in the added central section of her hull. In accordance with their usual practice the Maudslay firm installed return connecting-rod drive, with the cylinders on the port side and the condensers on the starboard. The six rectangular boilers were ranged in two fore-and-aft lines with the furnaces facing inwards in conformity with the invariable custom of the period, and supplied steam at 20 lb. pressure; engines and boilers having no separating bulkhead or partition between them. It was a peculiarity of the *Royal Oak* and *Royal Alfred* that they had no encircling bulkheading round their propeller wells between decks. The hoisting holes in the lower and main decks through which the screw and banjo frame came up were kept closed therefore by circular covers or lids; and as the absence of any surrounding trunk between them would have exposed the lower deck to the risk of a wholesale flooding if the lower cover had been removed in even a very moderate seaway, the propeller was hardly ever raised in practice except for occasional harbour inspection or repair. When the ship was under sail it could be disconnected by a clutch in the tail shaft as in nearly all other vessels of that date. Her coal stowage of 550 tons gave her about 200 miles more steaming endurance than the *Prince Consort.* She had only one funnel.

The *Royal Oak* was one of the eight British armoured ships originally rigged with double topsails and pole topmasts. When converted afterwards to single topsails all these vessels were sparred to the same dimensions as the fore mast and main mast of the *Minotaur.* As regards the double topsails, their chief advantage (apart from ease in handling) lay in the reduction of wind strain on the topmast, and on the topsail and lower yards, by the introduction of an additional yard between the latter. The area of canvas pulling on the topmast head was thereby halved, as the whole pull on the lower sail was transferred to the lower cap; while at the same time the aggregate expanse of canvas to be kept spread between the top and the crosstrees was held extended by three yards instead of only two. How great a relief that proved to be may be judged by

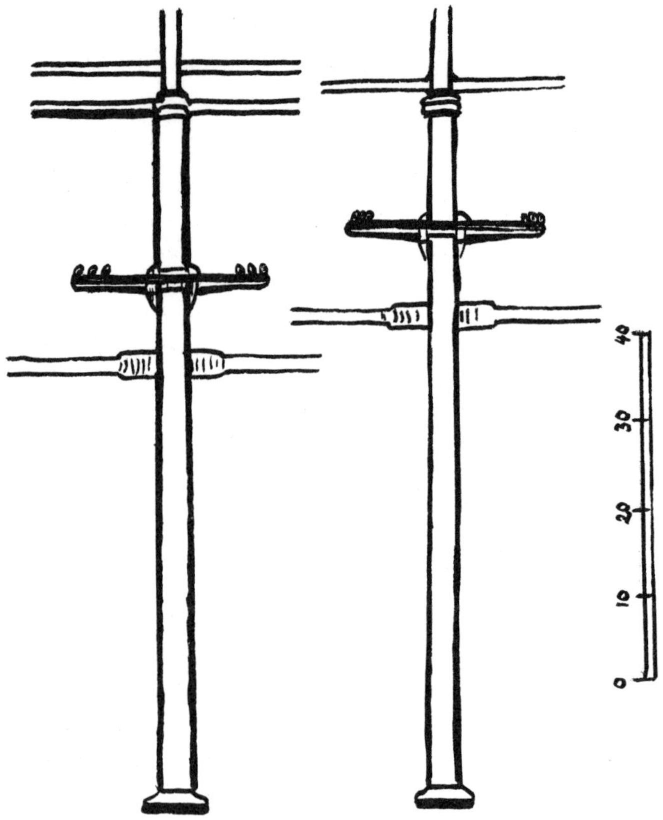

Sketch showing the design of lowermasts for double topsail and single topsail rig in two British battleships with lowermasts of the same total length from cap to deck. Left: *Royal Oak*, right: *Bellerophon*.

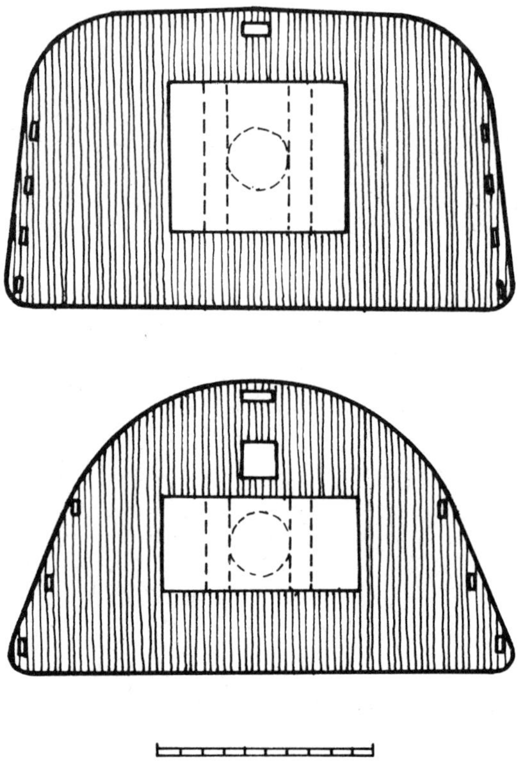

Sketch showing the shape and size of the tops for a single topsail (top, *Bellerophon*) and double topsail rig (bottom, *Royal Oak*) in two British battleships with lowermasts of the same length from cap to deck.

some of the entries in the logs of the armoured ships composing the Black Battle fleet of 1870; which amongst them in the course of their sea service carried away a matter of a dozen lower yards at one time or another when rigged with single topsails, but none at all when rigged with double. Moreover as the reduced pull on the topmasts diminished their necessity for strong lateral support, the tops could be cut away diagonally at the forward corners, and the spread of the topmast rigging thereby narrowed enough to brace the cap topsail yards up much more sharply than the yard of a close reefed single topsail without shrouding; which was an advantage when steering full-and-by in heavy weather or lying-to. Even with a single topsail hoisted right up and the bowline taut with the yard braced at its sharpest, the extra bellying of the larger sheet of canvas was apt to force weather helm more than with double topsails.

In the British Navy the lower sail of the pair was the larger and deeper, though in the merchant service the opposite was usually the case. The area of the two combined was slightly more than that of a single topsail for the same scale of masting; but the total wind pull was no greater in practice on account of the air escape between them. In cut the upper sail followed the shape of a single topsail from head to fourth reef, and the lower sail the shape from fourth reef to foot, but with more spread above the belly band. The head of the upper sail therefore was about 12 feet shorter than the head of the lower sail in a ship with the spar plan of the *Royal Oak;* and the sail was fitted with two rather deep reefs, whereas the lower had none. Both had drop enough in proportion to width to allow the clewline blocks to lie well in on the quarters of the yards, and the sails to be furled on the leeches from the yardarms inwards, with the high bunts which were universal in all navies of the period. No bowlines were necessary as they were slats of canvas rather than squares; and the weight on the reeftackles in reefing was so much reduced that they were only rove single. With the men double topsails were naturally popular. Having each only half the depth of single sails their reefing and furling came very easily to hands accustomed to deal with windbags having an area of anything up to 4000 feet, and a hoist of perhaps as much as 60 feet.

The topgallant sails in the *Royal Oak,* and other similarly double-topsail rigged ships, were mere strips of canvas with an area of less than 700 feet each and a hoist on the short pole-headed topmast of scarcely 15 feet. For all the use they were in fair weather or foul they might just as well have been left out of the sail plan as they were in the *Defence.* But the courses and gaff sails were all on the same size as for a single topsail rig. Only the lower stunsails were carried in these ships, and none had square sail on the mizen; all of which was changed, however, when the subsequent alteration to single topsails took place.

The main drawbacks of double topsails in a battleship were extra weight aloft and increased wind resistance in steaming against it. In the *Royal Oak*, the two 75-feet cap topsail yards with their fittings and running gear added about 6 tons to the tophamper but nothing to the sail area. And though wind resistance did not come into the question where a sailing ship pure and simple was concerned, it mattered a great deal to a mid-Victorian steamer. Making laboured progress at best when steaming head to sea, the battleships of that period always struck topgallant masts, which was easy enough; but to strike cap topsail yards in a heavily pitching vessel was a strenuous evolution, and moreover lumbered up the deck, which bore upon the all-important consideration of efficiency in action. To counterbalance in some degree the added weight of the yards the Admiralty abolished fidded topgallant masts in these ships; but that necessarily decreased the light upper canvas, and in the *Royal Oak* the area of working sail in her double topsail plan was quite 8000 feet less than the quantity she would have carried if completed to her original design.

For service in European waters that discrepancy was not of really serious consequence; but by the time that the *Royal Oak* was half through her maiden commission the Admiralty had been forced to look ahead as to the possible necessity of despatching armoured vessels to distant stations such as China, the Pacific and the West Indies, which would have long oceanic trade wind passages to make, requiring all the sail they could carry within reasonable limits unless their movements were to be unadvisably slow. In 1864, therefore, double topsails with their accompanying pole topmasts were abolished in the Black Battlefleet, and replaced by single topsails and fidded topgallant masts carrying upper sails of full hoist. Ships nearing the end of a commission waited till they were paid off before making the change, so that it could be carried out by the dockyards; but those in the early stages of a commission were re-rigged by their own hands after the necessary spars had been drawn from mast-house stock. New tops were fitted, but in order to save expense no constructional alteration in the length of the lowermast heads took place.

H.M.S. *Royal Oak*, after conversion to single topsail rig in 1864. (NMM. HAN/1)

In the case of the *Royal Oak* the ship was transformed by her own men into a single topsail barque as ordered. But in the following commission they added yards on the mizen and she thus received full ship rig on exactly the same scale as the *Prince Consort.* It was a little curious, however, that while still in her original dress of a double topsail barque she not only made her own best record under sail, but attained the highest speed ever reached under sail by any British armoured ship. The scanty area of this sail plan was not of course affected by a freshening wind in the same degree as that of a vessel with a full proportion of upper canvas; and under weather conditions in which the latter would be shortening sail after sail the *Royal Oak* could still afford to hang on to everything, till by the time it came to reefing topsails both would be under an equal show. Thus it came to pass that when running before a half gale on passage from Gibraltar to Malta, being at the time reduced merely to single reefed upper topsails, with a clean bottom and good sailing trim, the *Royal Oak,* on February 9th, 1864, logged 13.5 knots; a speed equalling the best of any full rigged old type ship-of-the-line then in existence. It was the sole occasion on which a British armoured vessel ever moved faster under sail than her own best trial speed under steam. The force of the wind was 7 to 9, giving a total pressure of about 12 tons on the area of sail in use. Her gear being comparatively new everything held. She was rolling 18° to 20° each way, but well under yawing control without connecting up the main deck wheel.

The *Royal Oak* was laid down at Chatham as a 90-gun two-decker in 1860, ordered for conversion with all similar ships then building in 1861, launched in 1862, and ready for sea in March, 1863. In April of the last-mentioned year, she was commissioned by Captain F. Campbell, being the first of the seven converted 90-gun ships to hoist the pennant. After a few months' service in the Channel Squadron she followed the *Resistance* to the Mediterranean; and as these were the first armoured British vessels to appear on a foreign station they were objects of considerable interest to the Mediterranean maritime Powers, whose own armoured fleets were newly coming into existence. Though not called upon for any service outside the normal routine of fleet duties in times of peace, her long commission of more than four years and a half happened to coincide with a period of marked political activity on the shores of the middle sea; including the unification of Italy and that country's war with Austria which occasioned the battle of Lissa, wherein the success of the Austrian ram affected the theory of naval tactics, and in one respect thereby, the practice of naval architecture for a generation to follow.

After returning home to pay off at the end of 1867 she was recommissioned almost at once for a second spell in the Channel Squadron. A few months later on a thick stormy night the squadron was cruising off the Scillies in open order under double reefs with a port beam wind, when the *Royal Oak* was partly taken aback in a short squall, and losing some headway was over-run by the *Warrior,* her next astern. Always unhandy at any time, the *Warrior* was above all slow in paying off under sail, and though her helm was hard up she took so long to answer it that she struck the *Royal Oak's* lee quarter and crashed along her side, shearing off all the starboard boats as well as the main and mizen chains, which left the lee shrouds and backstays sagging in bights. The immediate consequences were naturally less serious than if the chains had been demolished on the opposite or weather side, in which event the masts would in all probability have gone overboard with a wind of the force blowing at the time. But the situation was urgent enough nevertheless, as the ship was a deep roller, and the masts had no support now in a weather lurch except from such ropes in the running rigging as happened to be belayed to the lee bulwarks, of which the principal were the topsail halliards.

Though accidents to lee standing rigging were so unusual as to fall outside the experience of most officers, they called for a prompt decision based on a seaman's accurate grasp of the particular conditions he had to deal with. In a light wind and heavy swell it might be advisable to let go all halliards at once, to reduce the straining leverage of the weight of yards and gear in a windward roll; but on the other hand, in a strong breeze, the pressure of the sails held the ship on a lee heel preventing any strain on the mast from the unsupported side, and under such conditions the natural impulse of a seaman to shorten sail on a partially unstayed mast might be an error. So decided the captain of the *Royal Oak,* who, on running on deck when the collision occurred, gave orders to keep the topsails hoisted and steer well full. This afforded the masts their best chance, though even so it was risky to send men aloft. But the risk had to be accepted and enough hands sent up in the first place to get a stout hawser round each masthead, which was set up on deck to leeward by luffs, and then reinforced by runners and tackles lashed to the mastheads' pendants. That having been done the masts were sufficiently held to starboard to permit of furling sails and striking topgallant masts; all of which being completed by the morning the ship was sent to Plymouth under steam for the necessary hull repairs and new boats, though not a spar was lost or sprung. It was a piece of work which can be appreciated by seamen brought up under sail.

Three months later the *Royal Oak* was sent 'up the Straits' once more, where Lord Clarence Paget was now Commander-in-Chief, with his flag flying in the *Caledonia.*

H.M.S. *Royal Oak*, at Malta about 1869 now rigged as a ship. (NMM. HAN/2)

Next year (1869) a second mishap befell her through no fault of her own. In November the Suez Canal was opened by the Empress of the French amid a lavish display of ceremony on an imperial scale; in which representative squadrons from several maritime Powers, including Great Britain, participated. The Egyptian authorities arranged that the larger vessels were to anchor in the open roads outside Port Said, and Paget was ordered to attend with four of his battleships and berth them accordingly. But owing to the entire absence of any heavy draught traffic on the desert shores of what had been a dead corner of the Mediterranean before the cutting of the canal, Egyptian waters were badly surveyed, and no pilot service existed beyond Alexandria. Except in the deep water fairway leading to the canal from the Mediterranean side, as dredged by French engineers, the chart soundings proved unreliable and an unknown shoal happened to lie in the line of access to Paget's intended position. As the British column approached in line abreast therefore three of the four heavy draught ships, including the *Royal Oak,* ran violently and almost simultaneously on the mud. Fortunately the bottom was soft and the sea smooth, but being the month of November, when northerly gales on the Syrian coast are not uncommon, it was an urgent matter to get the ships afloat again; and as they could neither be budged by reversing their engines nor laying out sheet anchors all hands worked day and night to lighten them. The *Royal Oak* was eventually towed off next day by the *Lord Warden* — the only battleship of the four to escape grounding — but the re-floating of the others was a longer operation. No serious damage was suffered in the end, and no Court martial was ordered, but the spectacular share in this very historic event was hardly a success in so far as the principal flag to be seen in the Suez 'ditch' in after years was concerned, though the blame for that lay elsewhere.

By the end of 1871 the *Royal Oak*'s second commission had lasted for almost four years, and as she had by then completed more than eight years of continuous fully manned service she was relieved by the *Lord Clyde* and paid off at Portsmouth under orders for extensive refit. As that happened to coincide with one of the national moods of so-called economy, however, her repairs — which included

H.M.S. *Royal Oak*, depicting the uniforms of officers and ratings in 1869 during the period when Captain Henry S. Hillyar C.B., was in command (1867 to 1871). (NMM. 3550)

H.M.S. *Royal Oak*, being launched at Chatham. (NMM. 58/2092)

H.M.S. *Royal Oak*, portrayed by an artist with her original double topsail rig. (NMM. A.527/29)

the expensive item of new boilers — were so indefinitely postponed that she fell into a condition not worth repairing, and after lying for fourteen years in the fourth-class reserve was sold to Castle for breaking up on September 30th, 1885.

The *Royal Alfred* was one of the three ships laid down in 1859 of which the conversion was postponed as described on a former page; the other two being the *Zealous* and *Repulse*. These vessels were five or six years on the stocks in consequence, but advantage was taken of the delay to make their conversion more efficient and up to date when finished. The *Royal Alfred* herself was launched in 1864, and by that time the 9-inch gun and 6-inch plate had come into existence, both of which were adopted for her final design; but with the experience of the *Prince Consort* class as a guide no attempt was made to increase the size of her engines beyond the dimensions originally planned for her use as an unarmoured ship-of-the-line. Externally her measurements and lines corresponded exactly to those of the *Prince Consort,* and internally her arrangement bore a very close resemblance to that of the *Royal Oak* (except in magazine and shell room details) as high above the keel line as the main deck, with only one group of boilers. But the plan of her main deck differed considerably from that of the other four lengthened ships. Instead of being open fore-and-aft it was divided transversely by two bulkheads enclosing the 9-inch ordnance; and instead of having all the ports of uniform size and equidistant from each other, the gun ports were larger than the remainder and more widely spaced. Moreover, unlike the others, which were built with flush decks to which poops and topgallant forecastles were only added as later improvements, the *Royal Alfred* from the first carried both. In general sea equipment she was on the same plan as the *Royal Oak* throughout, with the same arrangements of steering gear, messenger capstan and wooden quarter davits, and the same external scupper shoots fore and aft. Having neither bridge nor conning tower of any kind she was always handled from the poop, but it carried no chart house. Officers and men were berthed as in the *Royal Oak* along the whole length of the lower deck; but being a pooped ship her accommodation included quarters for an admiral and staff, for which reason she always served as a flagship except for a very brief spell at the conclusion of her commissioned career. For a good gun platform her rolling was excessive, but thanks to her light trim by the bows she rose well to a head sea. Like all the converted ships she was extremely easy to handle under steam or sail.

By restricting her machinery to the 800 h.p. standard about 200 tons weight was saved amidships, which made it possible without unduly straining her lengthened centre section to effect a measure of concentration in her fighting design approaching to the central battery ideal, and allow moreover some addition to the burden it imposed. Such a course had not been practicable in the case of the *Royal Oak* in spite of similarly lightened machinery, because it resulted from the introduction of heavier guns and armour after she had been launched. But of these such full advantage was taken for the *Royal Alfred* that she was not only the most formidable wooden built vessel ever to fly British colours, but the most formidable in proportion to displacement tonnage in the whole battlefleet of 1870, whether of timber or iron hull. Her armament comprised ten 9-inch guns mounted on the main deck broadsides, and eight 7-inch, of which four were mounted in pairs on the main deck at bow and stern, two under the topgallant forecastle and two on the quarter deck. That represented a heavier total weight of metal than was carried in any British armoured ship up to a displacement of 8600 tons, though the *Royal Alfred* only displaced 6500. In its composition and arrangement, and in its arcs of fire, it very much resembled the armament of the 7550 ton *Bellerophon,* though more powerful by three 7-inch weapons. Alone among British armoured capital ships completed for sea before 1866, the *Royal Alfred* never underwent any change even on paper in the number and calibre of her guns from the date ordering her conversion till she was sold to the shipbreakers. For saluting purposes she was one of the vessels supplied with six brass smooth-bore howitzers (carried in her case on the poop) which had only blank ammunition. Her complement was assessed on the Quarter Bill at 606 ranks and ratings, exclusive of admiral's staff and retinue when serving as a flagship.

In the matter of protection the *Royal Alfred* differed from all previously launched wooden capital ships. The others — with the sole exception of the *Zealous* — were completely armoured over every part of their external side surface from 6 feet under water to the upper deck level; but in the *Royal Alfred* the principle was adopted of leaving parts wholly uncovered in order to thicken the covering of the more vital areas without appreciably increasing the burden of plating to be carried, and of adding transverse internal armoured bulkheads. A complete waterline belt was retained, 6 inches thick on the beam and tapering to 4 inches at the vessel's extremities, from which a battery protection of 6-inch armour rose as high as the upper deck for a length of 115 feet amidships, and screens of 4 ½-inch armour at bow and stern to cover the chase guns; while across the two ends of the 9-inch battery stood 4 ½ inch armour bulkheads. Under this disposition of protective weight the main deck guns undoubtedly received a better degree of shelter than those in such ships as the *Lord*

Sail plan of H.M.S. *Royal Oak*.

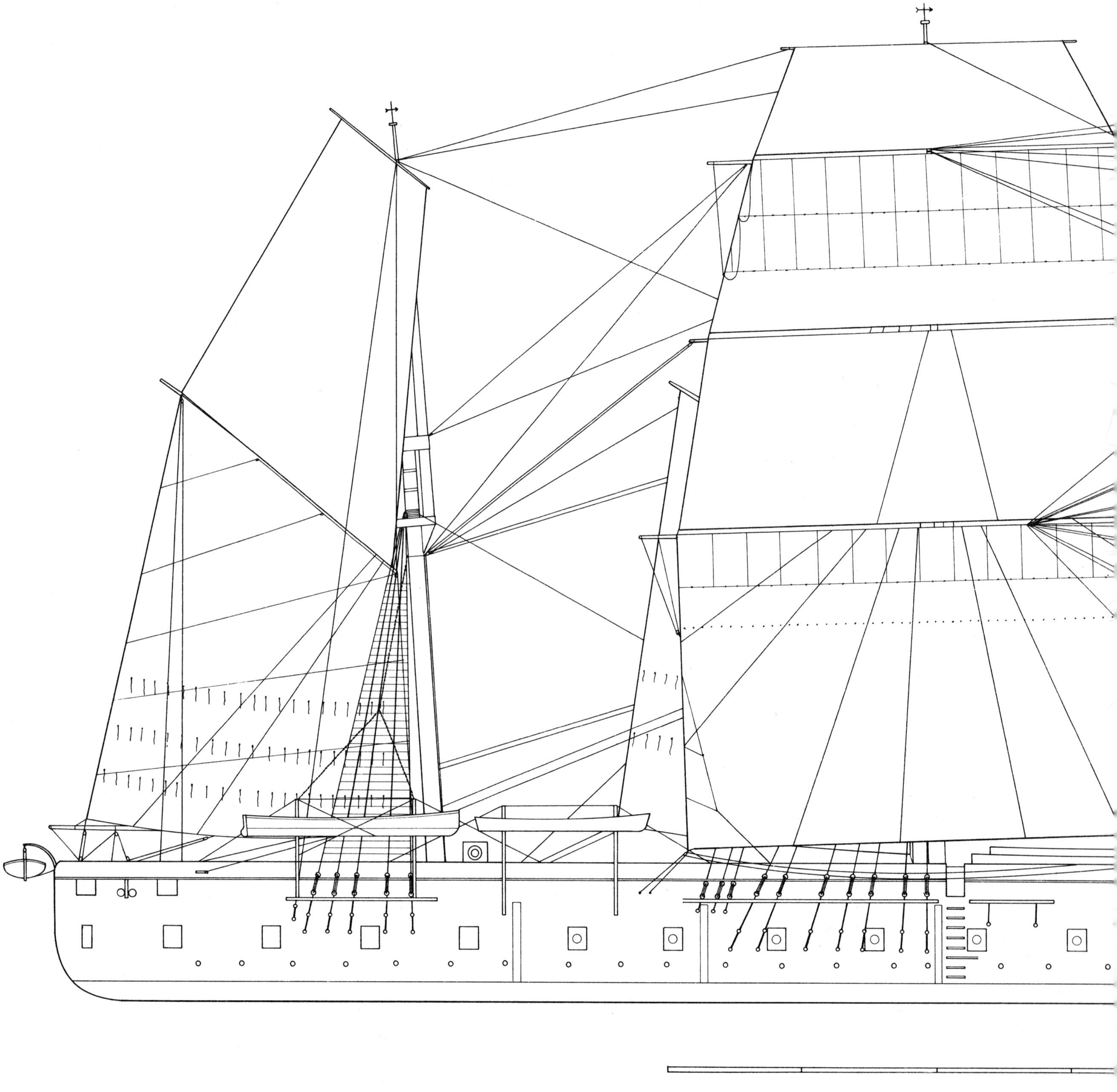

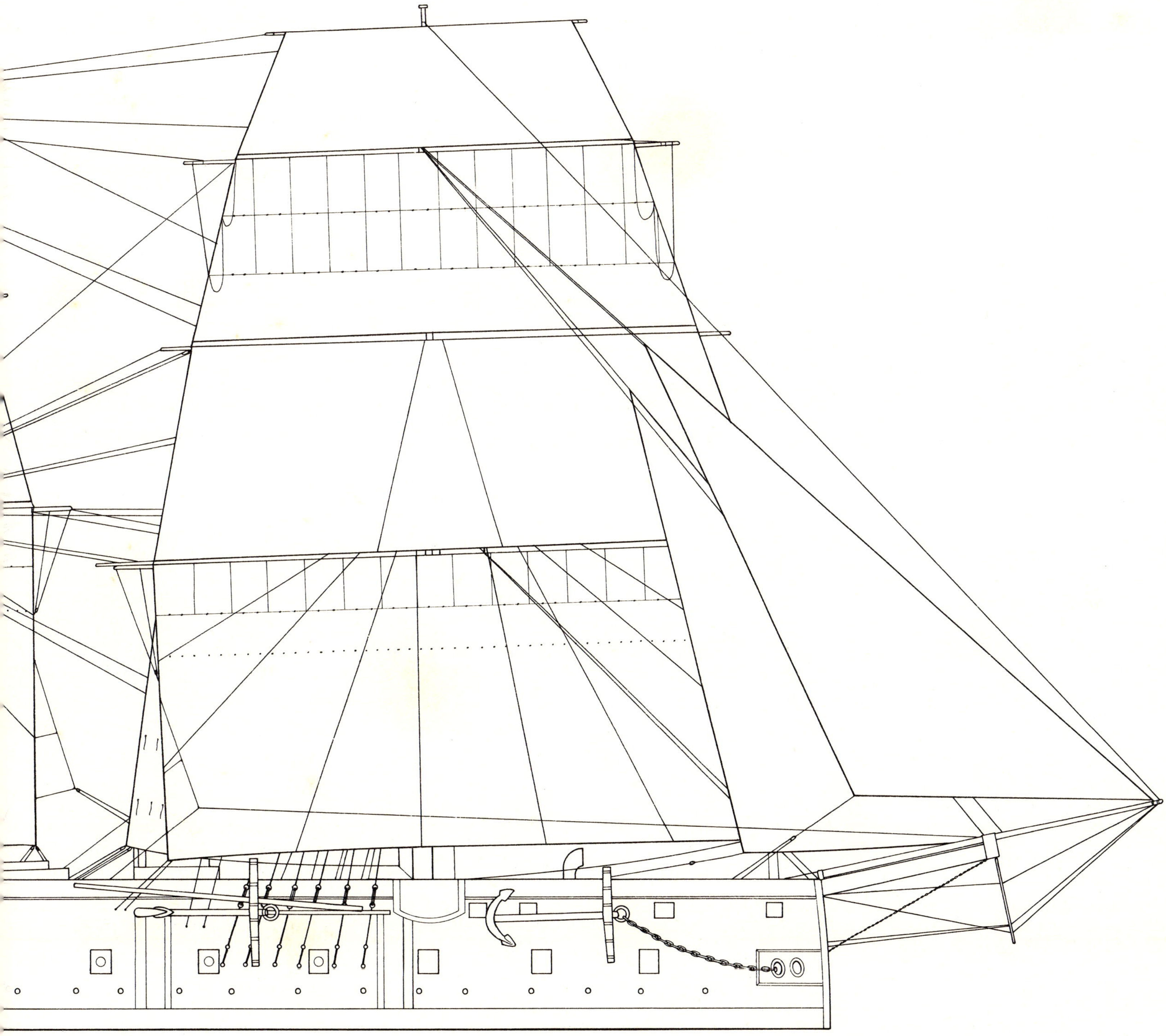

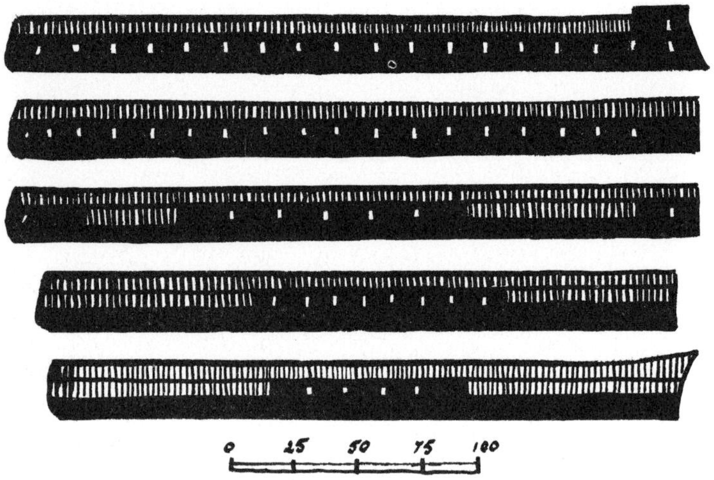

Diagram showing the area of side armour in each of the British wooden-hulled battleships of 1870, and giving the total weight of protection in each. From top to bottom: *Lord Clyde* and *Lord Warden*, 1380 tons, max. 5 ½ in. *Prince Consort, Caledonia, Ocean* and *Royal Oak*, 935 tons, 4 ½ in. *Royal Alfred*, 950 tons, including transverse bulkheads, max. 6 in. *Zealous*, 790 tons, including transverse bulkheads, max. 4 ½ in. *Repulse*, 1040 tons, including transverse bulkheads, max. 6 in.

Warden or *Prince Consort;* but on the other hand it was always a disputed point as to whether a target so inflammable to shell fire as a wooden side could be left exposed over any considerable area without undue risk to the ship as a whole; though in the *Royal Alfred* the proportion of the hull thus left naked was much less than in the iron central battery ships. For example, her ten 9-inch guns were spread over a battery 115 feet long, whereas in the *Bellerophon* the same number were crowded into a space of 98 feet only, though the latter was by 27 feet the longer vessel of the two. This debatable question, however, never received a solution under the ordeal of war, and as the advantages of the change were believed to outweigh the dangers its application was carried even further in the *Repulse,* which came after the *Royal Alfred* as the last British battleship to be put in the water with a timber hull.

The total weight of guns, mountings, ammunition and armour in the *Royal Alfred* was about 45 tons in excess of that in the *Royal Oak,* and thus in her case a minor portion of the weight saved by the lighter engines was re-imposed in another form. Moreover the centre of gravity of the armour lay about 20 feet abaft the midship section, instead of being almost exactly equidistant from bow and stern as in the wholly covered ships. These factors, in conjunction with the weight of a poop — which the *Royal Oak* did not carry — gave the *Royal Alfred* a less even keel, with a load draught of 23 ½ feet forward and 27 feet aft to the 24 feet and 25 feet of the *Royal Oak.* But from these figures it will be noticed that the mean draught was only a few inches greater, and the difference of trim seemed to suit the water, as the speeds of both ships were almost identical for any given number of revolutions with the same size of propeller, though the displacement of the *Royal Alfred* at 6500 tons was about 140 tons the heavier. As her machinery and boilers were practically a duplicate of the *Royal Oak*'s they need no further mention here, except to observe the detail that the funnel stood about 15 feet farther aft in order to keep the uptakes and casing behind armour.

Under canvas she was a very good performer for an armoured capital ship. Mention has been made on a former page of the fact that the *Royal Oak* was the only armoured vessel ever on any occasion to travel faster under sail than her own fastest under steam; and in that line of comparison the *Royal Alfred* stood next as the only one in which a rate of sailing ever equalled her own fastest under steam, both being 12·5 knots, which speed she once attained in a full gale under close reefed fore and main topsails and single reefed foresail. Being stiff and weatherly when close-hauled and very reliable in stays she carried out most of her station cruising while flagship in North America under sail, whether in open Atlantic or the intricate pilotage waters of the West Indies and Caribbean; and as she was masted after the abolition of double topsails her sail plan always remained a single topsail ship rig on the same scale as that shown in a former illustration of the *Ocean.* Her one peculiarity was a bowsprit running in under the topgallant forecastle instead of above, which allowed a better steeve, but reduced the training arcs of the upper pair of bow guns.

The *Royal Alfred* was built at Portsmouth, and begun as a 90-gun ship in 1859. Hers was the last timber keel ever laid at that yard for a capital ship and closed a chapter in its long history. By 1861 her frames were up when the decree went forth that all constructional work on ships-of-the-line was to take a new direction with armour as an essential qualification for capital rank; though in the cases of certain vessels conversion was to proceed slowly pending final plans. In her case it was slowed down for two years, whereby her launch was delayed till October 1864. It was then again slowed down to wait for guns, so that she was not ready for sea until the end of 1866.

Few mid-Victorian battleships went much beyond European waters, but it was the destiny of a minority to serve as lone battleships in perpetual and distant exile almost from beginning to end, and of these the *Royal Alfred* was one. Except for a few months when joined by the *Defence* on the other side of the Atlantic she was never at sea in line with a first-class consort. To Southsea beach or Plymouth Hoe she was quite unknown by sight, and amongst all the British battleships of her day she was the only one never in Gibraltar Bay.

By the time she was ready for sea it had been decided to send ironclads as flagships to the more important overseas stations. Among the vessels to be replaced was the two-decker *Duncan* in North American waters, and the *Royal Alfred* was selected to succeed her.

She was commissioned at Portsmouth in January 1867 as flagship to Sir Rodney Mundy, and took her departure thence when ready; arriving at Halifax after a very tempestuous passage of twenty-five days — of which about half were under steam — with only coal enough left for one more day at 5 knots speed, having lost two men killed by falls from aloft, carried away her main yard, buckled the starboard fore chains and split five sails. In spite of the fresh gale that continued throughout the operation the heavy broken halves of her 91 foot main yard were sent down, fished, and swayed up again, with the reefed topsail sheeted home on them in 48 hours, which was good work for a new ship's company. The fished spar lasted to the end of the voyage.

The North American station extended from Labrador to the Spanish Main, and there the *Royal Alfred* spent seven years with only one very short absence, though her maiden commission was devoid of any notable events. It passed in carrying out the usual annual round of visits to Nova Scotia, the St Lawrence, Newfoundland, Bermuda, Jamaica, Trinidad and all the principal West Indian and Central American anchorages, with a yearly official call and salute at a port in the United States. As no dock existed on that side in those days which could take a ship of her lines, she was ordered home to pay off and be docked and re-commissioned towards the end of 1869; the *Defence* being sent out in the interval in order that British interests in the West Atlantic should not lack the near comfort of an armoured white ensign. On re-commissioning she hoisted the flag of Vice-Admiral Wellesley and returned to the same duty; and an incident which occurred a few months later is worth a passing mention as an instance of bygone methods in the Navy. Excessive propeller vibration necessitated an examination of the cheese coupling and seat of the banjo frame, but the Bermuda floating dock — which had now arrived out — was occupied by another ship undergoing extensive underwater refit. Accordingly after the *Royal Alfred's* propeller had been hoisted on deck and her rudder unshipped, a wooden cofferdam was built in the yard, launched and fitted round her stern; within which when it had been pumped out the artificers executed the necessary repairs.

About two years later, at the end of her annual West Indian cruise, she was working through the Straits of Florida under sail in light winds on her way from Nassau to Chesapeake Bay when she grounded on one of the maze of reefs on the Bahama Bank. With the rise of the tide she was re-floated 7 hours later, but not before losing her false keel and the sheathing of her garboard strakes, which necessitated proceeding to Bermuda for docking. The ensuing Court Martial reprimanded her captain and admonished him to be more careful in future, but the Admiralty did not take so severe a view of the accident as in the groundings of the *Agincourt* and *Lord Clyde* and his career was not terminated.

A ball held aboard the *Royal Alfred* at Barbados whilst she was serving as Flagship of Admiral Mundy. (NMM. 58/2920)

After a four years second commission the *Royal Alfred* was relieved by the *Bellerophon* and returned home to pay off in January 1874. She was then commissioned once again as guardship of the First Reserve for the Portland district; which was her only service either in home waters or under a pennant instead of an admiral's flag, and a brief one at that. Even then she remained without company at sea, for at that period the annual cruise of each coastguard ship was carried out independently instead of in combination. But she was suffering badly by this time from the boiler erosion which was a very common source of trouble in mid-Victorian machinery; pressure having been reduced to 10 lb. which limited her best speed to 7½ knots. That was no worse than in other vessels with old boilers, but it had to be taken in hand before going further and she was paid off at the end of March 1875 for a thorough survey of hull and engines. Her hull proved perfectly sound from bow to stern, with a promise of lasting for many years, but the expensive item of new boilers caused an indefinite postponement of any effective refit until she had become obsolete for fighting purposes, and after lying for 7 years in dockyard reserve she was sold for breaking-up in 1885. No other ship in the British Navy ever bore the name *Royal Alfred.*

CHAPTER ELEVEN

THE 'ZEALOUS' AND 'REPULSE'

H.M.S. *Zealous* on passage through the Straits of Magellan, the first British armoured vessel to make such a passage, and showing her damaged main yard. (NMM. 58/2862)

Of the seven converted wooden ironclads of the Black Battlefleet, five had been lengthened in the process of conversion, but two, the *Zealous* and *Repulse,* retained their original mould. Like the *Royal Alfred,* they were a year ahead of their sisters when the decision was taken to convert them to ironclads, and their reconstruction was delayed pending experience with their sisters.

The retention of the original submerged lines and dimensions in these two vessels lent a special interest to their completed forms, since their new form was carried on the same hull as the old, and could be directly compared with it. It affords a clearer illustration than can be found anywhere else of the unmistakable superiority of the new type of fighting equipment over the old. Both types were carried on an identical floating support, but the difference in combatant power was so great that it is as difficult to estimate how many vessels of their originally intended class would have been a match for the *Zealous* and *Repulse* in their actually finished form as to estimate how many sheep are a match for a wolf.

Though remaining unaltered up to their line of load immersion, their changes of form above it were carried further in the direction of what were then modern ideals than in any other wooden-hulled battleships except the *Royal Alfred,* through improvements in naval architecture which were introduced during the interval while their conversion was delayed and could be adopted in their final design. But, nevertheless, for reasons to be explained hereafter they were not finished as sister ships, and the modernising of the *Repulse* went much beyond that of the *Zealous.*

The latter was the first to be launched by as much as four years. Being built with a poop and topgallant forecastle she always furnished a flagship's accommodation. In other respects her berthing and messing quarters for officers and men accorded with the usual custom in wooden ships; as did her disposition of magazines, shell rooms, provision holds, tiers and stores, with the necessary special details for an armament not exactly duplicated in any other vessel. In general system of structure her hull was the same as those of the other oak-built armoured ships, with its limited use of ironwork for beams and stanchions. Bow and stern were precisely similar to those of her lengthened consorts; scupper pipes and deadeyes for the lower rigging carried externally like theirs. As in the case of several contemporary battleships her cutters were hoisted at wooden davits and her single-banked boats at iron; though in her case alone the iron pairs stood abaft the wooden instead of before. For working cables she was provided with the 'patent' capstan with chain lockers on the fore side of the boilers, but no steam drive. Three treble hand wheels stood on as many different decks having connection to an upper and lower iron tiller on an iron norman head, stepped in the bronze neck socket of an unbalanced oak rudder. She was a good seaboat and a steadier gun platform than any wooden armoured vessel except the *Repulse.* Her manoeuvring qualities were excellent.

The armament of the *Zealous* from the time of her completion for sea till the end of her service was never changed. She was the only battleship with 7-inch guns throughout; and except the *Repulse* the only one in the

H.M.S. *Zealous* as completed in 1866, before leaving for her first commission in the Pacific. (NMM. 9075)

British fleet of 1870 to carry and retain through her whole career an absolutely unmixed armament of any calibre. Sixteen of her twenty 7-inch rifled muzzle-loaders stood on the main deck, two under the topgallant forecastle, and two on the quarter deck, all on wooden carriages and slides. Her complement was 510 ranks and ratings exclusive of Admiral and staff.

During the period when the conversion of the *Zealous* was subject to delay the earlier practices in regard to armouring had undergone a modification in the direction of reducing the covering of the sides to the more immediately vital parts; and thus instead of the complete hull casing provided for the *Prince Consort* and *Lord Warden* types, the sides of the *Zealous* were only protected on the battery and waterline; but the battery was screened against raking fire by internal athwartship bulkheads. All the side and bulkhead armour was 4 ½ inches thick, but the waterline belt tapered to 2 ½ inches at bow and stern. Partly through her shorter model, though more through the restriction of protected area, her total weight of armour was only 790 tons to the 935 tons of the *Prince Consort.*

Engines and boilers in the *Zealous* remained unaffected by the change in hull. These were identical to the 800 h.p. return connecting-rod sets in the *Royal Oak* and *Royal Alfred* and made by Maudslay, who engined all the converted ships except the *Repulse;* but she had a four-bladed instead of a two-bladed screw, a difference adding nothing to her speed under steam while reducing it under sail.[1] Thanks, however, to a saving in her weight of armour she carried more coal than any of them, her bunker capacity being 660 tons. The engines were situated exactly amidships with the engine room and boiler room all in one. She was single funnelled like all the British battleships of less than 6800 tons in 1870.

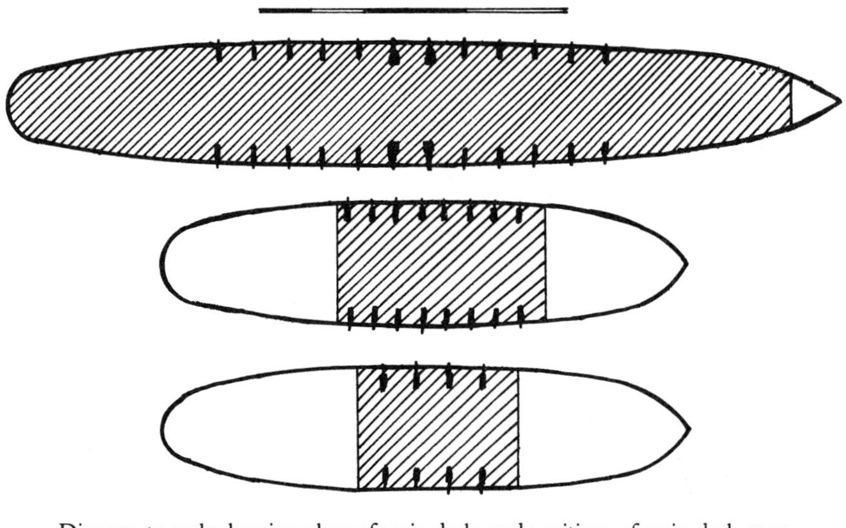

Diagram to scale showing plans of main decks and positions of main-deck armaments in the two longest and the two shortest vessels in the British battlefleet of 1870. Shaded portions represent areas sheltered by side armour. Top: *Minotaur* and *Agincourt*. Length of main deck 410 feet. Main deck guns, four 9-inch, twenty 7-inch. Centre: *Zealous*. Length of main deck 252 feet. Main deck guns, sixteen 7-inch. Bottom: *Repulse*. Length of main deck 258 feet. Main deck guns, eight 8-inch.

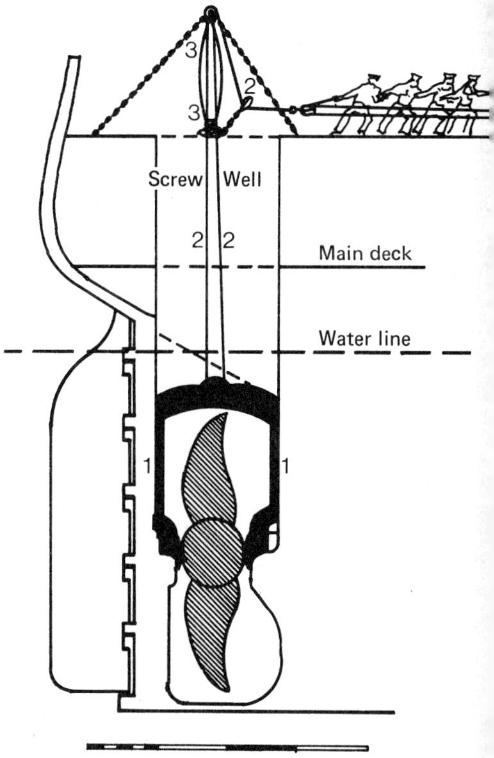

Diagram to scale of a hoisting screw as fitted in nine British battleships of 1870. 1-1: Banjo frame. When hoisted this was held up by two pawls at the corners of its arch, working in ratchet strips on the rudder post and stern post. The sheers and purchases were then unrigged till required for lowering. 2-2: Pennants and purchases. These were rove in separate twin sets side by side, of which only the further is shown in the diagram. Each pennant passed over a roller sheave at the sheerheads, thence down and round another in the arch of the banjo frame, and up again to make fast at the sheerhead. The purchases led forward on opposite sides of the quarter deck. 3-3: The sheers. These straddled the circular top of the screw well, with their heels stepped in socket hinges on the deck and the heads joined by a stout crosspin carrying the roller sheaves for the pennants. Chain guys leading fore-and-aft held them upright. When not in use they were stowed on chocks on each side of the quarter deck.

The *Zealous* was always a full-rigged ship with the same sail plan as that of the *Ocean,* but although very handy, stiff and weatherly under canvas, she moved very slowly owing to heavy propeller drag, and never logged more than 10 ½ knots. In spite of this, however, no contemporary armoured vessel covered so great a total mileage with cold boilers, thanks to six years in the Pacific, by far the widest in extent of all the foreign stations of the British Navy, and yet the one in which steam was least in use during the mid-Victorian era. She was one of the vessels with a main mast stepped on the lower deck on account of the position of the after cylinder of the engines.

This battleship was laid down at Pembroke as a 90-gun two-decker in 1859, but was not ready for sea till 1866. In September of that year she hoisted the flag of Rear-Admiral Hastings, and took her departure from Plymouth to relieve the frigate *Sutlej* in the Pacific, in which ocean she remained for two full commissions of three years each. It was a part of the world where political conditions were at that time almost chronically unstable from Mexico southward, and where British blue-water tonnage had more than equalled the shipping under all other flags put together since the decay of the United States mercantile marine. No armoured vessel of any European nationality had appeared there before 1866, but the French were sending out the new *Belliqueuse* and the British government followed suit on principle, the *Zealous* being the first British armoured ship to be sent farther from home than the Mediterranean. On passage out she carried away her mainyard after clearing the Magellan Straits, and as it was too badly splintered for fishing or repair a jury affair was rigged up by using the spare topsail yard in its place with the spare mizen topsail set above it and a reefed main topsail bent as a course, a new mainyard being made by her carpenters ashore at Valparaiso when she reached that port.

The *Zealous* reached the station headquarters at Esquimalt in July 1867, ten months after commissioning, where the Admiral found cabled instructions from home to remain till further orders. Telegraphic communication with other ports in the Pacific was unreliable in those days owing to the absence of any direct cables combined with the inefficiency of the land lines from the Atlantic side, and no sort of telegraphic connection with any Pacific islands existed. Once a Pacific flagship had left Esquimalt, therefore, on her usual long yearly round of the station, it was a very difficult matter for the Admiralty to get into quick touch with her after she was south of Panama or westbound for Honolulu. And it happened that Pacific affairs just at that time were in such a condition of chaos and unrest from north to south as to make the display of a large White Ensign possibly advisable at any place or moment as a reminder of British interests. Four years of domestic bloodshed in the United States had only just concluded, leaving a vast wake of disturbance; Mexico was in a condition of anarchy following the withdrawal of the French army of occupation and the execution of Emperor Maximilian; Chile and Peru had provoked Spain into bombardment and blockade by a Spanish fleet; Presidents were being assassinated in punctual rotation in the lesser republics of South America.

The British government therefore deemed it prudent to keep the Admiral at the end of the wires in case of emergency, from the time of his arrival at Esquimalt — a British harbour in telegraphic reach — in July 1867 till

The officers and men from the *Revenge* seen on the left of the picture, ready to be transported across the isthmus of Panama in order to join the *Zealous* for her second commission in the Pacific. (NMM. 58/570)

Rear-Admiral Farquhar, with Officers and ratings of the *Zealous* photographed during her second commission in the Pacific, leaning on one of the quarter deck 7-inch slide mounted guns. (NMM. 58/604)

April 1869. For those twenty-one months in consequence the *Zealous* swang round her mooring swivel in Esquimalt harbour except to go out for two days in each quarter for heavy gun firing. This was a record stay in one port without a break for any British battleship in full sea-going commission in Victorian days, and by the end of it her third year since leaving England was more than half through. She was then released to proceed on a nine months' station cruise with orders to be at Panama in January of the year following to pay off and re-commission, the new ship's company being sent out in the two-decker *Revenge.*

The exchange of officers and men between the *Revenge* at Colon and the *Zealous* at Panama was effected by the railway across the isthmus; an unusual proceeding, requiring prior diplomatic negotiation.

The flag of Rear-Admiral Farquhar was hoisted as successor to Hastings, and during this second commission the ship was continually cruising, political conditions no longer requiring her prolonged presence within telegraphic reach. As nearly all the principal American Pacific seaports south of San Francisco are open anchorages easily entered from seaward, and as coal was extremely expensive throughout the station, the *Zealous* seldom moved under steam.

After nearly three years of further station service she was relieved by the *Repulse,* which ship met her at Valparaiso in November 1872. Thence she started for home, but having been more than six years without docking — a record for a British armoured ship — her sheathing had become so extremely foul in spite of frequent hoggings that her best speed under steam or sail was little better than 7 knots, and she took five months to reach Plymouth. In steaming through the Magellan Straits at 5 knots she was caught in one of the violent local tide eddies and thrown against an uncharted rock, fortunately without sustaining serious damage to her stout and sound timbers. As an instance of the standard of efficiency of fuelling arrangements where traffic was rare a hundred years ago it may be mentioned that four days were required to put 500 tons of coal on board at Sandy Point.[2]

After paying off at Plymouth in April 1873 the *Zealous* was refitted and commissioned for two years as guardship of the Southampton district Reserve, during which service she was twice employed as Sovereign's Naval escort in Cowes Roads while Queen Victoria was in residence at Osborne. Then she was finally paid off and laid up at Portsmouth in 1875, and eleven years later sold to the shipbreakers in September 1886. It has been mentioned that the *Royal Alfred* was never at sea in company with any other battleship except the *Defence,* but the career of the *Zealous* was even more isolated from her own kind than that, for she was the only battleship in the British Navy of 1870 that never once found herself in line with another.

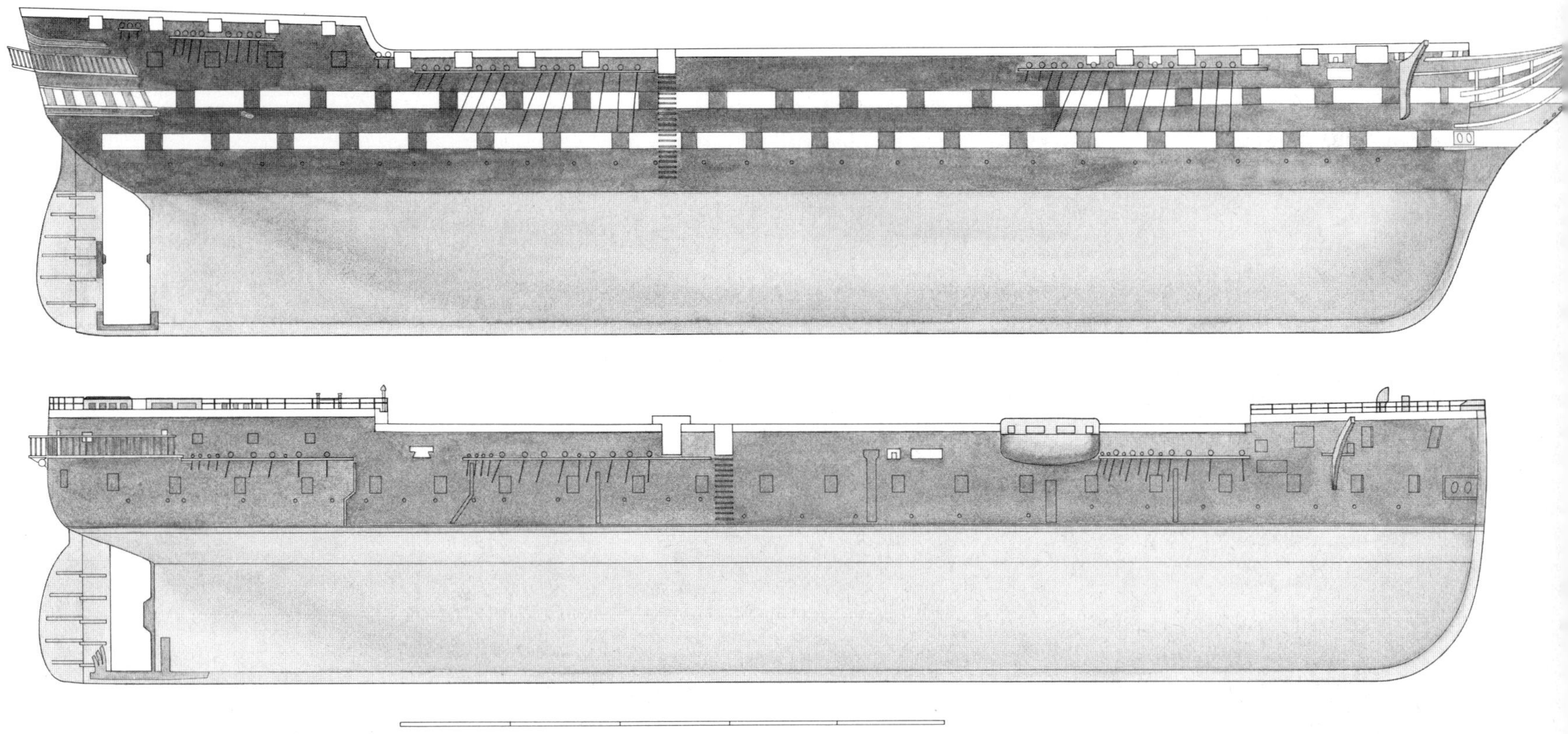

Hull profile of H.M.S. *Zealous.* Above: As a wooden-hulled ship of the line. Below: After conversion to an ironclad.

The *Repulse* has an enduring historical interest, not only as the very last built British capital ship of all the hundreds with timber hulls, but as the last of the long succession of capital ships — reaching back to Tudor times — whose frames rose into shape in the once famous Royal Dockyard of Woolwich, the original birthplace of their race, where the black and stinking foreshore mud is hallowed by the memory of such masterpieces in English oak as the *Henry Grace à Dieu*, the *Sovereign of the Seas*, the earlier *Britannias* and many others. Her launch indeed ended full-time work in the yard, which had been already marked for abolition as one of the political advertisements of the Gladstonian administration of the late 'sixties and closed its gates for ever a few months after the Thames received the *Repulse* from the launching ways on a flood tide. No iron hull was ever banged into form at Woolwich, where the fine redolence of timber work held its ancient monopoly to the end.

Like the *Zealous* at Pembroke, the *Repulse* was originally laid down in 1859; but in her case, when the question of conversion came up for Admiralty decision two years afterwards, operations were entirely suspended for five years instead of being merely slowed down for three; and as great advances in armouring occurred during that interval her eventual transformation followed much more advanced design than that of any other wooden vessel in its finished form above water, though less changed than any of them except the *Zealous* below. Reed had introduced the central battery type in the interval, which was adopted in its extremest development, for a single gun-deck in the *Repulse;* and that, together with certain re-distribution of weights as compared to the *Zealous,* altered her line of immersion; so that although her length, beam and mould remained the same as in the latter-named vessel, her draught of water was 22 ½ feet forward and 26 aft, giving a maximum of 3 inches more than the *Zealous* but a mean of 9 inches less. On that flotation the displacement of the *Repulse* was calculated at 6190 tons.

Apart also from its battle design her hull embodied some special features in its upper parts. Alone among British capital ships of her period she displayed a noticeable rise of sheer plan at both ends. Experience had shown that short models with little or no rise were apt to dip both forward and aft when end-on to a sea, and this improvement therefore had a practical purpose. But she had another peculiarity for an armoured vessel that had no such recommendation — and was shared only with the *Warrior* and *Black Prince* — in the possession of an obsolete knee bow, which in her case had doubtful merits even from the standpoint of appearances which was its sole inspiration. Such a bow was properly associated in naval architecture with the light shape of stern of its own earlier period, as in the *Warrior* and *Black Prince,* having high counter and projecting taffrail. It looked incongruous when combined with a stern of the heavy low countered type introduced with square noses, of which the stern of the *Repulse* was a pronounced example. Moreover, hers was the only knee bow in the service surmounted by a horizontal running-in bowsprit and this broke the upward continuity of line possessed by such a bow when carrying a highly steeved fixed bowsprit as in the old ships-of-the-line and frigates.

By the date of the *Repulse*'s completion the British custom of employing armoured flagships on certain distant stations had become permanently established, and as the many flush-decked battleships were unsuitable for such service on account of insufficient accommodation, the *Repulse* — like the *Zealous* — was among the very few provided with poop and topgallant forecastle from the first, in order that they should be available for flagship duties. In her case, however, the usual arrangement whereby the Admiral's quarters were under the poop and the Captain's on the main deck was inverted, mainly because of the space absorbed by two stern guns. And instead, therefore, of being enclosed underneath by a transverse bulkhead below the break, the poop was open to the quarter deck with cabins ranged fore-and-aft on either hand.

Even for a period in which all sizes of vessels under the White Ensign carried bulwarks at least 5 feet high those of the *Repulse* were so exceptionally lofty that the gangway entrance was through an arch. This added weight to the main structure of the hull with no marked advantages; but, taken in conjunction with the fact that she was the steadiest gun platform among the seven converted 90-gun ships, it served to keep out green water and obviated the necessity of the external scupper pipes carried by the rest. In addition to her own individual excellence as a seaboat she shared their admirable steering qualities under steam and sail.

Internally, her berthing arrangements for the men presented a compromise between those of the iron-hulled and the wooden-hulled battleships of 1870. In the former all the men messed on the main deck (except some stokers) and in the latter all on the lower. But in the *Repulse* they were divided between the two. As the only representative of the central battery type among those with wooden hulls she alone among them had sufficient main-deck length forward of armament to berth an appreciable proportion of her ship's company without putting mess tables between the guns; and at the same time had an open lower deck like the rest, free from the bulkheads and wing passages of the iron vessels. Thus her total space available for berthing was on an exceptionally roomy scale. About one-third of her men, including all the petty officers, messed on the main deck forward of the battery, leaving the whole length

H.M.S. *Repulse* during her last commission in the 1880s entering Portsmouth with the aid of the Admiralty paddle tug *Camel.* (NMM. HAL/3)

of the lower deck on the fore side of the officers' quarters available for the remainder. Her accommodation for officers was also better than that in any other British capital ship afloat at the time of the *Repulse*'s launch; for although the wardroom was down on the lower deck as in all of them, the proportion of space required for guns on the main deck was so much less than in the others that room was left available for putting most of the cabins up there instead of down below, and she was the first British battleship, therefore, in which the majority of the officers could sleep with an open port at sea in the tropics.

In certain details of construction and equipment also this battleship had special features. Of the twenty-eight British armoured vessels in 1870 that carried side armaments she alone resembled the two contemporary turret ships in stowing boomboats abaft as well as before the main mast, her steam pinnace and Admiral's barge resting in crutches over the quarter deck. Her wooden davits hoisted two boats to each pair as in several of her consorts, but stood much higher. She was the first capital ship with the innovation of a charthouse on a bridge forward of the funnel; and on returning from her first service abroad was fitted with a unique structural peculiarity in the shape of a pair of large iron boat crutches on the crest of the waist bulwarks each side, to receive two second-class torpedo boats. These resembled miniature destroyers 60 feet long, steel built, with a speed of about 12 knots, and an armament of two 14-inch torpedoes carried in boat's dropping-gear. They were intended for night attacks on harbours and anchorages after being brought to the offing in the parent ship under cover of darkness, and were supplied to several battleships and a few cruisers in the late 'seventies; but in all vessels except the *Repulse* their crutches were amidships. They were found to roll so violently, however, as to be practically useless in any seaway, and discarded after a time as having little war value. The *Repulse* in fact never received hers, though the fittings for their stowage were left in place. In other details her structural sea equipment was in accordance with the Admiralty standards of her date, cables being worked by a 'patent' hand capstan which was steam geared when the ship was eleven years old, and steering being carried out by threefold hand wheels under the break of the poop and in the central battery, which were never changed for steam. Rudder, tillers and norman head were the same pattern as in the other converted ships.

Like the *Zealous,* the *Repulse* had the rarity for a capital ship built in the 'sixties of a uniform calibre armament. Four 8-inch rifled muzzle-loaders occupied each side of the central battery; two more stood under the topgallant forecastle as chase guns, and two under the poop. But as it was expensive and in other ways undesirable to fire salutes with 8-inch ordnance, a pair of small 20-pounder Armstrongs were put on her quarter deck on truck mountings as saluting guns. In 1880, when twelve years old, she received the addition of a 16-inch Whitehead armament, consisting of four carriages placed in pairs on the main deck forward and aft of the central battery and twelve torpedoes, being the only timber-hulled British ship ever to carry White-

Sail plan of H.M.S. *Repulse*.

heads except the *Lord Warden.* Her fighting complement was 515 all told, exclusive of Admiral and staff.

As her qualification for battleship rank, under the revised standards of valuation which were forced on all navies by the introduction of armour, this vessel carried 6-inch waterline protection tapering to 4 ½ inches at extremities, and a 6-inch side plating over the central battery, which was screened also against raking fire by the 4 ½-inch internal bulkheads usual at that period in vessels where the external armouring was incomplete. But all the rest of her hull above the waterline belt was unprotected, as well as the four chase guns forming one-third of her attacking power, and no vessel in the British battlefleet of 1870 left so much wooden structure exposed to shell fire. Among the timber hulls she presented the extreme application of the principle of reducing the armoured area in order to increase the thickness of the armour where it remained without adding to the total weight. Less important defensive details consisted of a pair of rifle-proof towers built into the bulwarks abreast of the poop, with voice-pipes to the fighting wheel in the battery, and a torpedo net defence, added with the installation of her Whitehead armament, and like it a feature of equipment unprovided for any other wooden-built vessel except the *Lord Warden.*

Of the seven converted 90-gun ships the *Repulse* alone was driven by trunk engines, of which a set of the 800 h.p. medium weight standard was installed for her machinery that had originally been made by Penn for the three-decker *Prince of Wales,* a vessel hulked as soon as her steaming trials were finished, owing to the abolition of her type. More powerful trunks had proved unsuitable to a timber hull from the experience of the *Lord Clyde;* but on the lighter scale used in the *Repulse* they ran well enough. Six boilers supplied steam at 20 lb., standing in the same compartment as the engines, all forward of the main mast, with the cylinder beds just midway between stem and stern. As the very large four-bladed propellers tried in the older vessels had been found to give no better results than smaller patterns the *Repulse* was furnished with a two-bladed hoisting screw of 19 feet diameter Although the last of the converted ships to be launched — and coming seven years after the first — she was no faster than the rest except the *Zealous,* and her coal stowage was the least of all. This last was due to the fact that on the same hull dimensions as the *Zealous* she carried 250 tons more weight of armour, and to avoid over-immersion her bunkerage had to be cut down to 460 tons, whereby her steaming radius was reduced by a full third.

From the above general account it will be seen that the *Repulse* was a ship of marked individuality of design in every factor of importance. Except that her submerged body had the same lines and dimensions as that of the *Zealous* she had nothing afloat that was anything approaching to a sister vessel, differing as she did completely from all others in internal construction, external hull profile, armament, arrangement and area of protected surfaces, engines, coal stowage and many details of lesser interest. As completed for service she was the farthest departure in the way of transformation from an original and partly built design ever undertaken in the British Navy to meet contemporary changes, and the final example of timber naval architecture as used for the purposes of war in British dockyards.

H.M.S. *Repulse* after her 1877 refit, and the addition of both light quick firing guns and torpedo net booms. (NMM. PX.21212)

Even her masting plan was not exactly the same as in any other vessel under the White Ensign. Her fore and main lowermasts were 3 feet shorter than in the others sparred on the second-class scale, but the loss of drop in the courses which that would have caused was met by raising the tops 3 feet higher, so that her sail plan was the same as in the *Ocean* class save for the minor details of a mizen topgallantmast 5 feet shorter and a jibboom 7 feet longer. Her rig and sail area were never altered as in so many other battleships. The main mast was stepped on the lower deck. Her best speed under sail was only 10 ½ knots, but she was very easy to handle in tacking or wearing; and next to the *Zealous* no armoured ship ever cut so many thousand miles of water with drawn fires. During her long commission of nearly five years in the Pacific her screw was far more up than down. She was stiff and heavy for the scale of her spars, however. On one occasion she carried away her foreyard; and on another in a heavy gale between Esquimalt and Magdalena Bay, with the ship battened down, she carried away mainyard and jibboom and lost her close-reefed fore topsail, fore topmast staysail and three davit boats.

The *Repulse* was laid down at Woolwich in 1859. Two years later her building was stopped pending instructions from the Admiralty, though work on the *Caledonia*

— begun a year after the *Repulse* on the next launching slip — was pressed forward. In 1866, however, full plans for the conversion of the *Repulse* were sent to the yard, and operations having been resumed she took the water in 1868; but before she was nearly ready for sea the 400 years of Woolwich dockyard history were abruptly and unceremoniously ended and the ship had to be towed to Sheerness for completion and armament. Having passed the naval trials she hoisted the pennant in March 1870 as guardship of the First Reserve for the east district of Scotland, with headquarters at Queensferry. It had been very unusual for a new ship to commence her service in the Coastguard squadron, and possibly the limited coal stowage of the *Repulse* was connected with the step in her instance as it certainly was in the case of one or two others.

But different work awaited her within two years. A sheathed battleship was required as relief for the *Zealous* in the Pacific, and as the facilities for repair and maintenance in that part of the world were still of the scantiest description it was expedient to send out the newest available. All the other armoured vessels with sheathed bottoms had already completed at least five years of sea service, and hence the choice of the Board fell on the *Repulse.* Her place in the Reserve squadron having been filled therefore by an older ship she was re-commissioned as flagship to Rear-Admiral Hillyar, the new Pacific Commander-in-Chief in 1872, and sailed for that station, where she took over duty from the *Zealous* at Valparaiso and exchanged salutes with the French and American station flag officers.

A very long commission followed — only falling short of five years by one month before she paid off — during which she flew in succession the flags of Rear-Admirals Hillyar, Cochrane and Hancock, being kept constantly on the move with screw up and funnel down between all the ports from British Columbia to Patagonia and westward to the then independent Hawaiian Archipelago, much of the Admirals' work being semi-diplomatic on behalf of the Foreign Office.

It was not till 1877 that the *Shah* was sent out as her relief. On handing over station command, Hancock proceeded to England by crossing from Panama to Colon and taking passage by West Indian mail steamer. This left the *Repulse* to be brought home by the Captain, and though the fact was not realised at the time, that officer's mind was beginning to fail. The ship's machinery was in serviceable condition — having in fact undergone comparatively little wear — and the normal and shortest route for a vessel with steam power was through the Magellan Straits. In spite of that, however, the Captain insisted on making a sailing passage right round Cape Horn, which was the first symptom of trouble, though his malady showed no signs of affecting his competence to handle the ship as a seaman. But the Horn was in bad mood and by her seventh week at sea from 'Vallipo' the *Repulse* was only abreast the coast of Brazil. By that time the Captain's disciplinary extravagances left no doubt as to mental derangement, and the senior medical officer having certified him insane to the Commander the ship called at Rio Janeiro and transferred him to a homeward-bound mail steamer under charge of the junior Surgeon. Forty-five days later the *Repulse* anchored at Spithead, having been the only armoured ship ever to round Cape Horn under canvas.

H.M.S. *Repulse* leaving Portsmouth during her final commission about 1880. On the right of the picture, the old three-decker *St. Vincent* can be seen giving her boys experience of setting sail. (NMM. A11618)

On paying off in 1877 she underwent extensive refit and received the torpedo armament with other improvements to war efficiency previously mentioned. Her next employment was guardship of the First Reserve for the Hull division with headquarters in the Humber, in which she passed four years of duty, taking part in all the annual combined cruises of the coastguard ships which by that time had become an established custom. In 1885 she was relieved there and paid off for the last time, remaining in the 4th Class Reserve till sold for breaking up in 1889 though still in sound condition.

With the disappearance of the *Repulse* from the Navy List the British battlefleet paid a last farewell to the mother oak which had formed its substance in whole or part through many centuries though in rapidly dwindling quantity towards the end. Thenceforward its entire material composition derived being from the womb of the earth, and one of the three main lines of its mid-Victorian transformation as a great war machine was concluded. Internally mounted armaments and square canvas still lingered on for a few years more, but the change from timber to metal in its architecture had reached completion. The name of the *Repulse,* however, retained such favour at Whitehall as to share with that of the *Royal Oak* the claim of having been twice[3] revived for capital ships since borne in the Black Battlefleet of 1870.

CHAPTER TWELVE

THE 'HECTOR' AND 'VALIANT'

H.M.S. *Hector*, as she was originally completed in 1864 with pole topmasts and double topsails. (NMM. HAS/1)

Of the nineteen vessels forming the iron-hulled portion of the battle fleet of 1870 the oldest ten belonged to the broadside type as distinct from the central battery. Six of the ten were the largest fighting ships in the world, and these have been described already. The other four were the *Hector, Valiant, Defence* and *Resistance,* of which the first two and the last two were sister pairs. All four had many points of similarity, and bore about the same relation to the larger six as the two-deckers of the past era to the First Rates. Having guns of the same calibre as the bigger group, and armour of the same thickness as three of them, their only definite inferiority lay in fewer weapons and lower speed; while in manoeuvring capacity they were actually better. Taken all round, therefore, these four when launched were useful second-rank additions to the battle line; and though neither quite so powerfully armed nor so extensively shielded as their wooden-hulled second-rank contemporaries, such as the *Royal Oak* and *Lord Clyde,* they had advantages over them in other directions through watertight internal partitioning and non-inflammable structure.

As all four were laid down within eighteen months of the *Warrior,* they were designed during the opening stage of iron battleship construction, when the Admiralty were feeling their way without experience but inclined to a forward policy. At that point in the history of naval architecture scientifically calculated forecasts of the displacement and line of flotation of ships on the stocks had not reached the precision of a later period, and considerable errors were not unknown. The *Hector* and *Valiant* were cases in point. Though only a little larger than the *Resistance* and *Defence,* they had 300 tons more armour to carry in addition to heavier machinery, with the result that they floated nearly a foot deeper than was intended at full-load draught; and although less heavily burdened in the matter of guns than the wooden-hulled battleships of the same length, their lesser beam (by nearly 3 feet) and finer lines reduced their power of flotation by more than the weight thereby saved. Moreover their finer extremities affected their behaviour in a head sea, which they had neither the length to cut through like the *Achilles* and *Minotaur,* nor the forward buoyancy to rise over like the bluff-bowed wooden-hulled *Prince Consort* Class. To save further weighting as far as possible, therefore, their coal stowage was limited to 450 tons, and thus restricted they were never employed on a foreign station. For the same reason neither ever had a poop or topgallant forecastle added, although all the other originally flush-decked battleships of the period received one or both as later improvements at some point of their career.

The *Hector* and *Valiant* had a proved full-load draught of 24 ½ feet forward and 26 feet aft, which was more than 10 inches in excess of calculations, and led to their exceeding their official displacement by about 300 tons. Alone among British iron-hulled battleships, except the *Achilles,* they carried a vertical bow profile as a compromise between a knee and a ram. Moreover, they were the first in which the original draught building plan (never altered) showed the new type of stern soon to be common, which rose straight upwards from the counter without overhang or quarter galleries. Internally they resembled the *Warrior* on a smaller scale as regards general system of construction and arrangement, the men berthing on the main deck as in all the iron-hulled battleships, but the officers on the lower deck aft as in every British armoured vessel launched before 1869 whether iron or wood. As details of general equipment, they were fitted with the unbalanced rudder, messenger capstan, wooden davits and three hand-wheels (one on each deck) which were customary in the original design of the broadside battleships of armoured types; and like all these, except the six longest, they were very handy in manoeuvring, but bad rollers owing to an extremely low centre of gravity.

For armament they were intended to carry eight guns on the upper deck and twenty-four below; but, as a consequence of their over-immersion, that full number was never put on board. During her first commission the *Hector* mounted four 7-inch Armstrong breech-loaders as upper-deck chase guns and twenty 68-pounder smoothbores on the maindeck broadsides; but in the general adoption of rifled muzzle-loaders for the fleet which took place in the middle sixties, this first armament was replaced (as in most of her contemporaries) by the new ordnance. As re-armed, she received a pair of the new 8-inch muzzle-

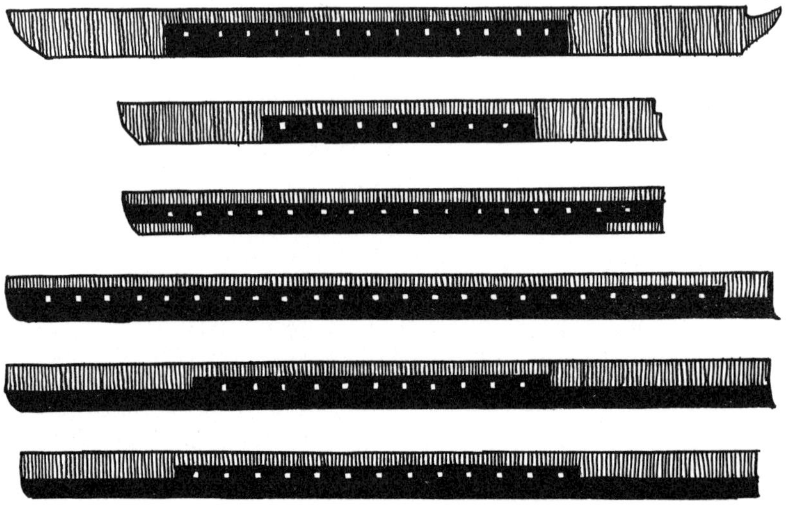

Scale diagram illustrating the arrangement of armour in the ten broadside battleships of 1870. From top to bottom: 1. *Warrior* and *Black Prince* – 975 tons – max. thickness 4 ½ in. 2. *Resistance* and *Defence* – 607 tons – max. thickness 4 ½ in. 3. *Hector* and *Valiant* – 912 tons – max. thickness 4 ½ in. 4. *Minotaur* and *Agincourt* – 1776 tons – max. thickness 5 ½ in. 5. *Northumberland* – 1549 tons – max. thickness 5 ½ in. 6. *Achilles* – 1200 tons – max. thickness 4 ½ in.

H.M.S. *Hector* about 1875 as rerigged with single topsails, topgallant and royal yards. (NMM. 9296)

loaders and sixteen 7-inch. It was a peculiarity of the *Hector* and *Valiant* that their two heaviest guns were on the quarter deck; for though many other broadside vessels of the period carried a pair of quarter deck guns, they were never heavier than those below, because on general principles it was considered desirable to put the chief weapons of a ship behind armour. But a quarter deck gun did have the advantage of standing much higher above the water, where it could be used under weather conditions that kept the main-deck ports closed; and for vessels so deep as these two such a consideration counted for a good deal. Four of the *Hector*'s 7-inch guns were also mounted on the upper deck as chase guns. The remaining twelve formed the main-deck battery. It was a feature of her first equipment that she carried a cupola furnace for red-heating the 68-pounder roundshot, but this was removed with the smoothbore guns.

Through being delayed for three years longer in building and completing for sea than her sister, the *Valiant* never shipped her originally intended armament at all, but was armed from the start with a combination of 8-inch and 7-inch rifled muzzle-loaders corresponding to the *Hector*'s, and never changed it to the end of her career. The complement of both ships was 530 officers and men.

In these two vessels the armour was disposed in a fashion unique in a way, as an illustration of the original conception of iron plating as a protection for the guns rather than for the vulnerable parts of the hull. With the first four British armoured capital ships the covering was confined to a rectangular patch amidships, screening little more than half their length either on the gun-deck or the waterline. Then followed the *Hector* and *Valiant,* in which it was extended fore and aft so as to from a complete belt at battery level, though on the waterline only carried amidships. In the *Minotaur* and *Agincourt,* coming next in order of designing, it was spread over the entire hull, waterline included; while in the *Northumberland* and *Achilles* it was disposed on an inversion of the plan in the *Hector* and *Valiant,* the waterline being completely belted but the hull extremities above it left uncovered. No less than four different systems of protective casing were represented, therefore, in these ten original iron battleships, all designed within two years of each other. Like all those in which the main deck was behind armour from bow to stern — whether iron- or timber-hulled — the *Hector* and *Valiant* had no athwartship armoured bulkheads. Externally they carried the 4 ½-inch plating of most of the broadside vessels, tapering to 2 ½ inches at their extremities.

Though engined by different makers, the machinery of this pair was practically identical. Each ship was driven by return connecting-rod engines turning a two-bladed screw. These propellers were constructed to hoist; but as it was the policy of the Admiralty to retain the ships permanently on the home station on account of their small coal stowage, the sheers and hoisting purchases were never put on board. Steam was provided by six boilers at 20 lb. pressure when new. For the *Valiant,* the machinery was made by the Maudslays; but for the *Hector* by the Napiers, who built her, and she was the first British armoured ship in which hull and engines came from the same firm.

By original intention both ships were to be double topsail barques in rig, with pole topmasts and no royals, on exactly the same plan as that of the *Royal Oak,* except that stunsails were omitted altogether, and that as their jibbooms measured 58 feet (the longest in the Navy) their jibs were large in proportion. But although the *Hector* was so rigged accordingly when first sent to sea, she was altered only a matter of a few months later to a single topsail barque with fidded topgallantmasts and royal yards, under the general order of 1864 abolishing double topsails in the British service; the change being carried out by her own ship's company. The *Valiant,* on the other hand, was not completed for sea till after the abolition order had been issued, and carried a single topsail rig identical to the second sail plan of the *Hector* from the beginning to the end of her sea-going career. Alone among British battleships in 1870 they never carried stunsails.

When iron-battleship construction commenced the Admiralty divided their contracts for the first six vessels equally between the Thames and the northern private dockyards, which latter were rapidly forging ahead under the advantage of closer proximity than their southern rivals to the main iron-producing centres. These six ships were in three pairs, and one of each pair was allotted to the Clyde or the Tyne. Thus the *Hector* came to be built at the Govan yard of Napier and Sons, who laid her keel in March 1861, a month after launching the *Black Prince,* the northern sister of the *Warrior.* The *Hector* slid into the Clyde in her turn in September 1862, as the second of the long line of armoured battleships that entered the water from Scottish soil. When completed for sea she hoisted the

H.M.S. *Hector*, despite the fact that she was by 1900 an obsolete hulk, is shown in this picture rigged with a light main topmast and a wireless telegraphy gaff at its truck for the initial trials in this method of communication within the service. (NMM. 59/1708)

pennant at Portsmouth in January 1864, under Captain G.W. Preedy, for service in the Channel Squadron, taking the place of one of the many unarmoured ships-of-the-line whose fate was sealed under the new standard of qualifications for battlefleet status. For three years she remained in the Squadron as one of its most important units in that early stage of armoured designs.

At the end of her three years in the Channel command, the *Hector* was paid off for re-armament in 1867, on relief by the *Warrior.* When again ready for sea in 1868, she was commissioned as guardship of the First Reserve in the southern district, on which service she remained continuously employed for the next eighteen years, incidentally doing duty as Queen Victoria's guard in Cowes Roads almost every summer when the Court was in residence at Osborne. No vessel in the Navy past or present, therefore, except the Royal Yachts, had a more extensive record of attendance on the Sovereign than the *Hector.* In 1886 she was paid off into the dockyard reserve after twenty-two years on the sea-going list, and for another fourteen remained lying dismantled up Portsmouth harbour. Then in 1900 she was selected as a store hulk for the *Vernon* torpedo school and used as such till sold for breaking up at Portsmouth in 1905. She was the first ship in the service to be fitted with wireless, albeit only when a hulk. Her name has not since been revived in the Navy.

The *Valiant* was the Thames' representative of this pair, being laid down at Poplar in the yard of Westwood and Baillie a month before the *Hector* at Govan. But the first signs of difficulty in regard to the future of shipbuilding on the London river were already beginning to appear; and the industry was handicapped in situation from the very beginning of the change from timber to iron in naval architecture. Before the *Valiant*'s shell was half riveted together, Westwood and Baillie went into liquidation and their works were taken over by the Thames Shipbuilding Company. As the transfer entailed considerable temporary suspension of activity, however, the ship was not launched till October 1863, more than a year after her Clyde sister. And then another long delay followed, while she waited, in company with several other new vessels, for the Ordnance authorities to provide guns of the muzzle-loading rifled description with which the whole Navy was being rearmed. When eventually completed for service a period of more than six years had elapsed since the laying of her keel.

In December 1868 she was commissioned at Plymouth under Captain J. Kennedy as guardship of the First Reserve in Southern Ireland, and thereby commenced a record of service unique in one respect, for she was kept to that same duty for seventeen years with only one short interval for re-boilering, and never knew any other. Some vessels — such as the *Minotaur* — passed an equal period in the Channel Squadron, but varied their service on the home station by long spells in the Mediterranean when the Squadron was sent out there as a reinforcement. Some again — such as the *Warrior* and *Hector* — spent their whole career in home-station commands, but varied their time between the Channel and Reserve Squadrons. The *Valiant* alone remained not only in the Reserve Squadron always, but in the same coastal district for divisional duty from beginning to end of her career on the effective list. Her district headquarters were first on the Shannon and afterwards at Bantry; and her routine work lay in carrying out the annual drills of the Fleet Reserve men in the Southern Irish coastguard, besides the annual cruise of the First

H.M.S. *Valiant*. Unlike her sister she was completed as shown in this picture as a barque with single topsails. The crossing of spare yards on the mizen was commonplace on many barque rigged vessels in the Royal Navy. To many it made for seemliness aloft, and at the same time resulted in a little less clutter on deck. (NMM. HAT/1)

Reserve Squadron and the regulation quarterly gun practices, with a yearly docking and overhaul at Plymouth. From 1868 to 1885, therefore, she was familiar to all the southern coast of Ireland from Cork to Killary Bay, and her men were in those days popular and welcome with the local Irish wherever she went, not only on their own account but because their monthly rations absorbed many thousand pounds of Irish beef, and no small part of their monthly pay went in Irish whiskey and the entertainment of Irish colleens.

Seventeen years of peace duty in the old Coastguard Squadron produced very little to record in the way of noteworthy incident. Except that the *Valiant* was present with the other ships when the Squadron formed part of the force specially concentrated in 1878 in anticipation of serious trouble with Russia over Constantinople, her service followed the same routine from year to year, only varied by the list of places visited in each annual drill cruise. Two minor incidents, however, are worthy of a passing reference. One of these occurred when the Squadron visited Cronstadt under command of the Duke of Edinburgh. On the forenoon of the last day of their stay, they were hoisting in boomboats a few hours before departure when the *Valiant*'s foreyard snapped in the quarter as it took the weight of the launch. It was a 91-foot spar of 4 ½ tons, but the broken halves were sent down, fished together, re-rigged, swayed up again and ready for use with the sail bent in sufficient time to hoist in boats and weigh for sea with the Squadron at three o'clock the same afternoon. Old logs may be searched in vain for any mention of a smarter bit of work than that in repairing a fractured lower yard,

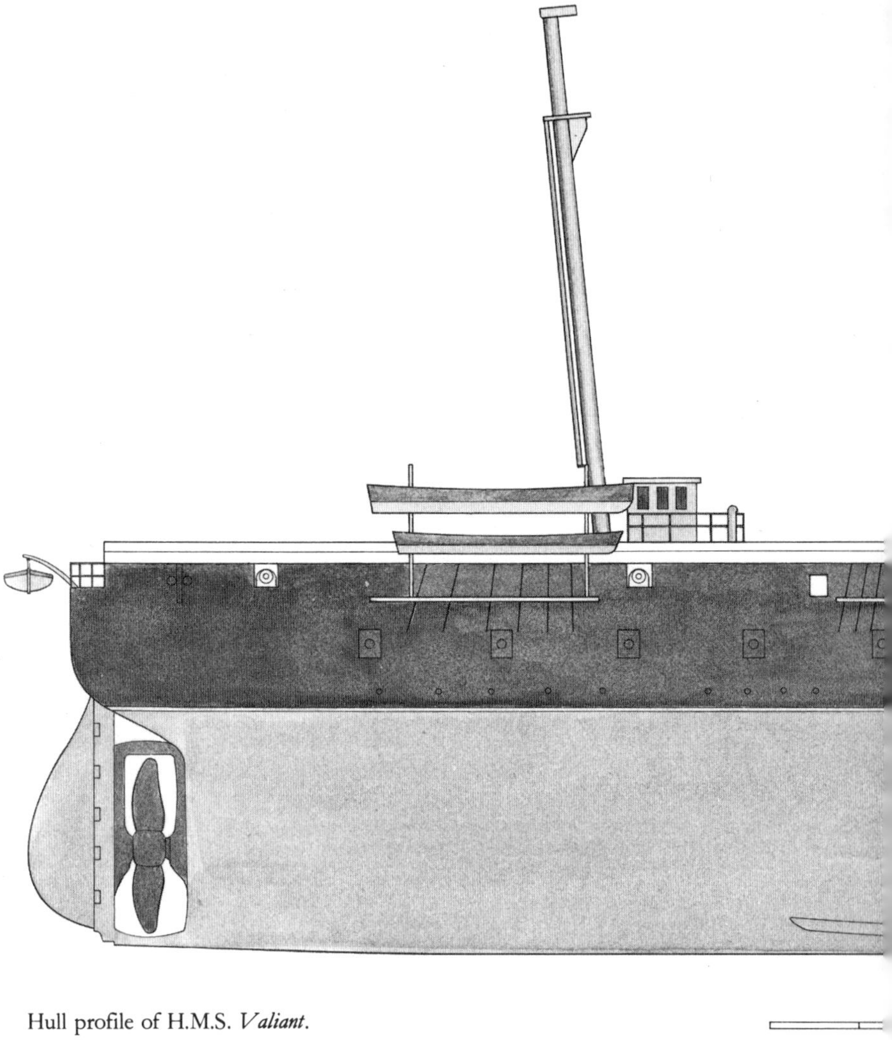

Hull profile of H.M.S. *Valiant*.

H.M.S. *Valiant*, now as *Valiant III*, a floating oil tank at Devonport in the early 1930s. (NMM. AAC/1)

although of course it was carried out under the advantage of harbour conditions. The other occurrence was a very unusual form of emergency. All large ships in those days had telescopic funnels lowered and raised by a powerfully geared winch, and kept down in harbour, or when on a sailing passage, till fires were lit to get up steam. On one occasion the *Valiant*'s winch gave out when the funnel happened to be down and the ship under orders for moving. The funnel therefore was hoisted by using the two large forestay tackles supplied for getting boomboats in and out, with their tail pendants adjusted at the lowermast heads to give an accurate plumb to the falls. In mid-Victorian days the pipe *Clear lower deck, Up screw* was very familiar; but uncommonly few ships' companies ever heard the pipe *Clear lower deck, Up funnel.*

At the end of 1885 the *Valiant*'s second set of boilers had become worn out; and being by that time too obsolete to be worth further expenditure, she was paid off into the dockyard reserve and removed from the sea-going list. But her career of usefulness was by no means ended, and like others of her generation her dismantled hull served national requirements in war and peace for a much longer period than it did as a battlefleet unit.

For thirteen years after hauling down the pennant she lay moored in the Hamoaze. In 1897 she joined the *Indus* establishment, briefly losing her name. In 1904 she was again re-named, as *Indus IV,* and became a depot-ship for destroyers. During the Great War she served as the Kite Balloon storcship *Valiant III.* In March 1922 she appeared in the sales list, but in 1926 was converted into a floating oil tank, and as such lay in the Hamoaze throughout the Second World War. At last she was sold to Belgian shipbreakers, and on December 8th, 1956 was towed away from Devonport bound for Bruges.

CHAPTER THIRTEEN

THE 'RESISTANCE' AND 'DEFENCE'

H.M.S. *Defence* portrayed with her double topsail rig, and roller reefing to her upper topsails. (NMM. 58/2057)

In the efforts of the naval architects in all maritime countries during the mid-Victorian age to augment the offensive power of the ships they laboured to design they fell back on early history to borrow the idea of using the ship herself as a projectile. It had certainly achieved practical results in the day of the archer, the pikeman and the galley slave; but it demanded a surety and impetus of tactical movement which was impossible under sail, and had been abandoned for centuries with the sweeps of the trireme, the longship and the galley till the invention of steam machinery permitted its disinterment from oblivion.

A half-step in that direction was first taken by the French, who, when building the *Gloire* discarded the knee bow of long tradition and launched her with a square stem. This example was not copied when the British built the *Warrior* as a reply to the *Gloire;* but the hesitation was brief nevertheless, and when Whitehall did adopt the idea the Board took the lead themselves in their plans for two new ships, which were designed with not merely square stems but with a very pronounced under-water projection at the bow, shaped in the rounded outline of a ploughshare. These were the *Resistance* and *Defence,* iron-hulled and armoured sister vessels of medium tonnage which were both laid down only a few months after the *Warrior;* and to such a degree did the innovation stamp their combatant qualifications in professional and public eyes alike that they were invariably referred to under the special designation of 'steam rams' until ram bows became the rule in all new battleships. After these two had been launched every maritime Power followed suit sooner or later; but the day of the ram was marked as much by disaster as success before it came to an end. In its very nature it was like a weapon always loaded even among friends, and hundreds of skulls lie on the sea floors which were sent down by the rams of their own countries. The *Vanguard* and *Victoria,* the *Grosser Kurfürst* and the *Yoshino* — each a prime unit of her class at the time — all went under to a stab in the vitals from a comrade ship. Others escaped a like fate only by the narrowest of margins. But the ram had one day of brilliant success at Lissa; and although that was largely, if not mainly, due to the inefficient use of the gun as a deterrent, the spectacular decisiveness of the victory achieved by the use of this instrument in Tegetthoff's determined hands created such an impression as to sweep aside any doubt as to its value for twenty years or more.

H.M.S. *Resistance*, an early picture of her still wearing the red ensign, with double topsails, but with her chequered sides painted out. Contrary to other comments it clearly shows that she was originally given iron quarter davits at the beginning of her career. (NMM. A.527/30)

With a moderately fine entry and run the *Resistance* and *Defence* were shaped on the general lines of a frigate of the period on an extended scale, except for a rather squarer bilge and less rise of floors. Forward they carried the 'plough bow' described above; but aft a very handsome wide quartered stern, precisely corresponding in design to that of the frigates of the day enlarged in all its proportions, with which the whole mould of the hull above water was in keeping. During their first commissions they were painted with the chequered sides of a Victorian frigate; and as all the main-deck ports were alike in size and spacing, a view of these ships from any point abaft the beam (in which the ram bow was obscured) at that stage of their career, presented a hull aspect indistinguishable from that of a contemporary frigate of unusual size, although they were somewhat different aloft. In general details of hull equipment they were fitted in the old style with wooden messenger capstans, threefold hand-wheels on the quarter deck and main deck, and a light flying bridge just forward of the mizen mast carried on the high bulwarks. Only in one minor point of external detail did the two ships differ from each other; for whereas in the *Resistance* the quarter davits were wooden, the *Defence* carried iron davits throughout. As seaboats they behaved on the whole rather better than other battleships of about their own size in a head or a following sea; but when it came abeam they were amongst the worst rollers in a fleet where heavy rolling was a general fault. Both handled easily under steam or sail except in a beam wind, when they were slow to answer helm — a defect which nearly caused the loss of the *Defence* on a critical occasion.

Internally they were built with the three decks of all British battleships of 1870 (except the *Captain*) and the same general arrangement in regard to vertical watertight bulkheading, but only a partial double bottom. Next to the *Warrior* and *Black Prince* they were the first British vessels of single gun-deck design in which the men were berthed on the gun-deck instead of below it, though most of the accommodation for the officers, including the wardroom, remained below in the old fashion. Having no poops they had no Admiral's quarters; nor as first built had they topgallant forecastles, though these were added later.

Like all British capital ships completed for sea during the period of transition from smoothbore to rifled ordnance, they were armed at first with a combination of both, though originally designed for smoothbores alone. The only rifled guns then in existence were the newly-introduced but short-lasting Armstrong breechloaders, of which the supply was very limited at the time. Six 7-inch Armstrongs were allotted to the *Resistance* and eight to the *Defence;* each vessel carrying two on the upper deck as chase guns forward and aft, with the remainder divided between the main deck broadsides. In addition each mounted ten 68-pounder smoothbores on the main deck. On the quarter deck the *Defence* had four 5-inch Armstrongs, and the *Resistance* a pair of 32-pounder smoothbores, which were the last two weapons of that truly historic denomination in the British battlefleet and the only two ever floated in a British armoured hull. All the armament was wooden mounted in each ship.

After their maiden commissions they were re-armed throughout with rifled muzzle-loading guns on slides, each receiving two 8-inch and fourteen 7-inch. But a slight difference existed in their disposition in the two ships: for whereas the *Resistance* carried six of her 7-inch on the upper deck, the *Defence* only carried four. Each alternative had an advantage of its own. Since neither vessel had any shellfire protection for upper-decks guns, the *Defence* had the smaller number left exposed, but on the other hand the *Resistance* had the smaller number carried too near the water for use in heavy weather. Neither vessel ever carried torpedoes or net defence. Both had complements of 460 officers and men. The growth of ramming tactics greatly increased the importance of end-on fire, and the early ironclads, unlike their unarmoured predecessors, mostly carried stern chasers on the upper deck. Where only one was mounted it usually had at least three alternative ports, and sometimes four, to enable it to come into action not only astern but on either beam. In the first four British armoured capital ships — the *Warrior, Black Prince, Resistance* and *Defence* — only one was carried, but two different systems of transferring it from port to port were tried. For the *Warrior* and *Black Prince* a central turntable was devised, flush with the deck, on which the whole gun and mounting were swung round in shifting ports. In the *Resistance* and *Defence,* on the other hand, an arrangement of traversing racers laid on the deck was combined with the training racers on each port, somewhat complicated in design, but as quick for working with as the other. Being the less expensive of the two, this principle was adopted in preference to the turntable for ships built afterwards, with such modifications in each case as might be required, and only one vessel thereafter was fitted on the turntable system.

The armouring of these two vessels was distributed on the earliest and simplest plan — which they alone shared with the *Warrior* and *Black Prince* — consisting of merely a rectangular side patch 4 ½ inches thick, extending for just half the ship's length over her central body from deck-level to a fathom below the waterline; the ends being joined athwartships by transverse bulkheads of the same plating. Within this enclosure were seven gun ports a-side. The rest of the hull, waterline and armament had no protection.

Both ships were trunk engined by Penn on an identical design. Four rectangular boilers generated steam at 20 lb. pressure. Their machinery was the lightest and lowest powered among the British battleships of 1870, being no heavier than in several of the larger contemporary frigates.

At the period when the *Resistance* and *Defence* were built, all screw steamers in the British Navy were fitted with hoisting propellers, to which rule these two were no exception. Their prime object, of course, was to diminish drag

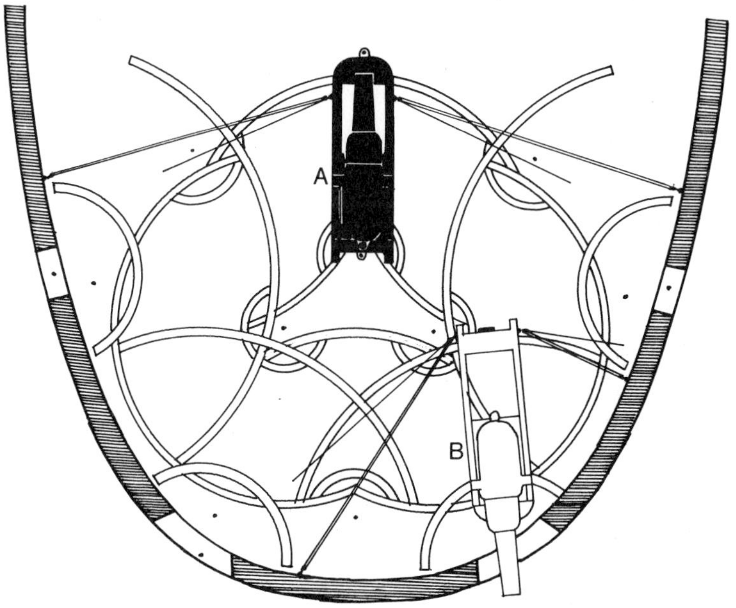

Plan of the racer arcs of the stern gun in the *Resistance* and *Defence*, showing at *A* the gun in its securing position for sea, and at *B* in firing position at one of its four ports. When secured for sea it was held on the centre line of the ship by dropping hinged flaps at the ends of the slide over the two 'securing' pivots. To bring it into action at any port ordered the gun tackles were hooked to the front end of the slide and the front flap raised, so that the slide could be swung round pivoting on its rear till the front was over the nearest 'traversing' pivot. The front flap was then put down on this, the tackles shifted to the rear end of the slide, the rear flap raised in its turn, and the swing round repeated but with the slide turning on its front till the rear was over the next traversing pivot. By a third lateral movement in this fashion the gun was finally brought opposite the selected port with its muzzle pointed for running out. The front of the slide was then connected by a pivoting bar to the 'firing' pivot on the port sill, but this bar is obscured in the diagram by the chase. When so placed the tackles were shifted to the rear and both flaps raised, so as to leave the slide free to turn on the firing pivot. The gun was now ready for action, a few minutes having sufficed for the whole operation, but care being necessary in traversing if the ship was rolling.

H.M.S. *Defence*, photographed about 1862, after her chequered sides had been painted out, but still retaining her double topsails, and with the so-called topgallant yards, which, as Admiral Ballard makes the point she did not cross them, were perhaps only used for a very brief interval of time. Like her sister photographed at this period she is wearing the red ensign. (NMM. A.527/31)

on the ship when under sail alone. Disconnecting and feathering screws had the same purpose, but neither answered it quite so well. And apart from this a hoisting screw relieved the lining of the stern tube of a good deal of friction and wear, since its considerable weight never bore on the tail shaft even when down, but on the brackets carrying the banjo frame. On the other hand it involved increased outlay not only in the fittings of the propeller itself but in the stern structure of the hull. And in addition to that it had two special drawbacks. As the cheese coupling only received the drive of the blades when the engines were turning ahead their backward pull, when the engines were reversed, fell entirely against the rudder post instead of the thrustblock on the shafting. Moreover, the connection of the tongue and slot in the coupling had to be a good fit; and if the screw was by any chance kept raised for too long, barnacles might get into the slot and prevent the tongue from sliding home on lowering, resulting in so tight a jamb that the tongue stuck half in and the screw could be neither properly lowered into steaming position nor rehoisted. Cases of that sort actually occurred. In spite of those drawbacks, however, hoisting propellers remained in favour with the British service for the majority of unarmoured classes of all sizes from iron-hulled frigates to composite gunboats for quite twenty years after the *Resistance* and *Defence* were laid down, and only disappeared finally with the universal adoption of twin screws.

As regards rig, the *Resistance* and *Defence* were the first two vessels in the British Navy in which double topsails were tried, and in their cases these were attended by other special innovations at the outset. As originally fitted, the upper topsails were reefed on the Cunningham system, where the yard rolled the upper part of the sail round itself as it was lowered, on the principle of a window blind. In theory this should have worked all right, but in practice it was found that the unequally shrunken canvas of a wet sail caused irregular rolling and great trouble. After a year's test therefore they were changed for ordinary sails. The *Defence* carried no topgallantsails, but in the *Resistance* an arrangement (unique in the Navy) was rigged for a time on the Gunter plan, whereby topgallantsails were hoisted on spars lying close up and down abaft the topmasts with their heels resting as far down as the lower caps, and their upper ends pointed through reversed topmast caps with 27 feet of length above them. Relatively therefore to the topmasts the position of these officially termed 'topgallantmasts' corresponded exactly to that of a flying jibboom alongside the other jibboom, with the lighter spar lying close along the whole length of the heavier beyond the latter's housing, and then projecting past its extremity far enough to set the added sail. This peculiar contrivance, however, represented the maximum of extra top weight for the minimum of extra canvas — the so-called topgallantmasts being actually 7 feet longer than the topmasts in extreme measurement — and was discarded soon after its introduction.

When the order abolishing double topsails in the British Navy was issued both of these ships were converted to single topsail rig with fidded topgallantmasts by their own men; but whereas yards were added on the mizen in the *Defence,* provision for a similar completion of the *Resistance* was countermanded to reduce her reported tendency to carry excessive weather helm; and in 1870 she remained as one of the only three barque-rigged vessels in the Black Battlefleet. The *Defence* received a running-in bowsprit on her change to single topsails, but the *Resistance* kept a fixed bowsprit till ten years later. They were handy under sail but slow, neither ever doing better than 10 ½ knots under the most favourable conditions.

Both vessels were built by contract in private yards. The keel of the *Resistance* was laid in December 1859 at Poplar, by Westwood and Baillie. She was launched in 1861, and being ready for sea by the middle of 1862, was commissioned at Portsmouth for the Channel Squadron by Captain W. Chamberlain, where — together with the *Warrior* and *Defence* — she was subject to continuous and critical interest and trial as one of the earliest representatives of a radical departure from long tradition. Experiments with their novel types of guns and mountings were among other matters of report, and the conditions under which these were sometimes carried out sound strange today. For instance, it was an understood thing that the ancient political connection between Britain and Portugal gave the Channel Squadron an unwritten privilege to use Lisbon as its headquarters for a few months in each winter, to the great

Hull profile of
H.M.S. *Resistance*.

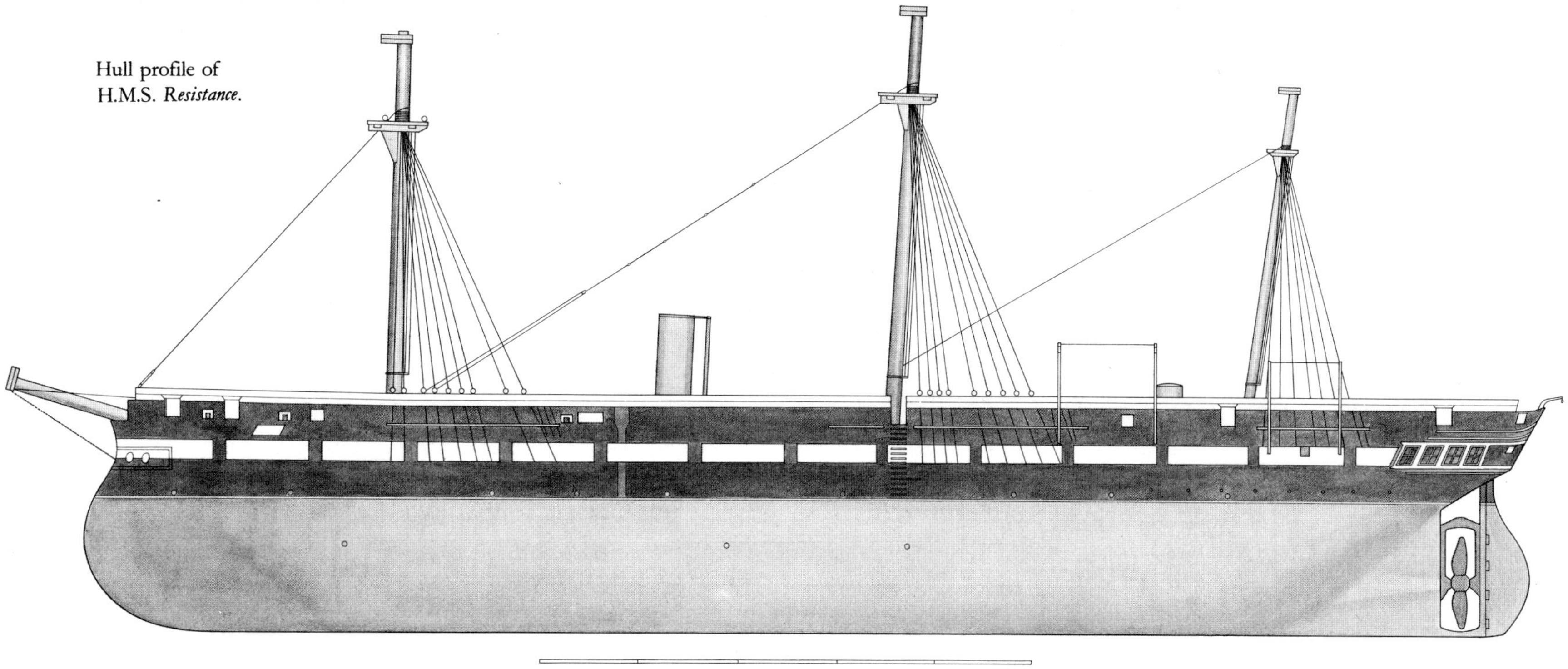

pecuniary advantage of its citizens. The Portuguese, therefore, were accommodating in regard to the conduct of squadron training in their territorial waters; and so far was that carried that when on one occasion the *Resistance* rigged a target on their beach near the mouth of the Tagus for certain ranging tests, and then proceeded to fire 7-inch projectiles at it from seaward, the matter occasioned no particular comment from anyone.

After eighteen months in the Channel Squadron she was ordered to join the flag of Lord Clarence Paget in the Mediterranean, on which station she was the first British armoured ship to appear. There she remained till relieved by the *Lord Clyde* in 1867, paying off at Portsmouth at the end of a maiden commission lasting just under five years. A long refit and change of armament came next, followed from 1869 by four years as Guardship of the Mersey district First Reserve. Paid off a second time in 1873, she recommissioned for another four years of service, this time with the Channel Squadron once more; at the expiration of which she reverted to her former duty as Guardship in the Mersey till paid off finally at Devonport and laid up in 1880, after eighteen years under the pennant.

In 1885 her dismantled hull was selected after special preparation as the subject of a long series of experiments to ascertain the comparative efficacy of various forms of protection against gunfire and torpedoes; in the course of which she was battered on occasions into a sinking condition, but patched up repeatedly till the tests were concluded, after lasting intermittently over years; by which time she had been half reduced to scrap metal. In that condition her mangled shell was laid aside pending further treatment

H.M.S. *Resistance*, now rerigged after 1870, with a topgallant forecastle, running in bowsprit, and wooden quarter davits. (NMM. HAR/3)

H.M.S. *Resistance* shown as she glides down the ways. With her sister, the *Defence* they were the only two vessels of the armoured Navy to be painted in the traditional frigate manner, and that was only during their first commission. (NMM. 58/1913)

on the same lines; but other obsolete craft were substituted and what was left of her tough old iron carcass went to the breakers in November 1898, twenty-nine years after the laying of its keel plates. On March 4th, 1899 she foundered under tow in Holyhead Bay, and was later raised and finally broken up at Garston. As the first ram-bowed British capital ship the *Resistance* retained the service nickname of 'old rammo' to the end of her days.

The laying of the keel of the *Defence* at Palmer's celebrated Jarrow yard in December 1859 initiated the vast industry of Tyne fighting-ship construction. It was a sign of the times and a direct result of the gradual abandonment of wooden hulls then in process. So long as oak forests provided the raw material for fleets, the southern estuaries were the nearest at hand for putting it afloat in the finished form; but when iron ore took its place, with the needful coal for smelting conveniently adjacent, then the Tyne, the Clyde and the Mersey all held an advantage in situation over the older slipways, and were not long in utilising it.

Though the *Defence* was not launched till a few days after her Thames sister, she was so much more expeditiously completed for sea when once afloat as to be ready first by six months, and hoisted the pennant at Chatham in December 1861 under Captain R. Powell only two years after the rivetting of her keel. Then followed five years' continuous hard service in the Channel Squadron, a force which during the early 'sixties had become more of an experimental and severely testing assembly of entirely new and untried types of naval architecture than a strategic combination of homogeneous units as in the past. In 1866 she was paid off at Plymouth for a much needed refit and re-armament, and in 1868 again commissioned for the Channel Squadron. In 1869 an armoured battleship was temporarily required in North American waters during the absence of the *Royal Alfred* in England for docking and recommissioning, and the *Defence* was ordered out. Leaving the Channel Squadron in the Tagus she arrived at Barbadoes after a good run across in the north-east trades of twenty-five days, of which twenty-two were under sail and three under easy steam with two boilers. Remaining for a year on that station she was then transferred eastward to the Mediterranean to relieve the *Prince Consort,* and spent the remainder of her commission 'up the Straits', under Captain Nowell Salmon, who had received the Victoria Cross for conspicuous bravery in the Indian Mutiny.

It was during this stage of the *Defence*'s history that the *Lord Clyde* was nearly lost on Pantellaria, and the *Defence* was the ship despatched from Malta to recover the anchors, cables, guns, etc., which had been either lost or hoisted into local small craft in the operation of lightening the stranded battleship. While so employed the *Defence* in her own turn narrowly escaped leaving her bones in that treacherous roadstead. All the work had to be carried out in an open anchorage with bad holding ground, directly exposed to the locally prevalent north-westerly gales of the March equinox, and steam had to be kept in constant readiness for immediate use. Caught at midnight suddenly by one of these, with her job almost completed but both anchors still down, the *Defence* first parted one cable and then, when a sheet anchor was dropped in place of the lost bower, proceeded to drag both that and the other bower rapidly towards a dead-lee shore with a heavy surf breaking. As there was no time for weighing with her old-fashioned messenger capstan, Salmon was compelled to slip both cables and endeavour to head out under steam; but before the chain had finished tearing out of the hawsepipes her stern bumped violently in the 'scend between two big seas, and the instant the cables were gone the bow fell off the wind before the ship could gather headway, and brought her broadside on to the rollers. If she had struck when in that position, total wreck would probably have resulted; and for a few minutes the danger of this was extreme, since her usual slowness in answering helm in a beam wind was aggravated now by a partly-twisted rudder. Fortunately she just headed up in time, and having gained a clear offing was able to shape course for Malta to be docked for examination of the stern. But it was a near thing, and her rudder and screw were both damaged. A sheet anchor was seldom used in the mid-Victorian Navy, except in cases of

Sketch illustrating the usual method of transporting a sheet anchor of a mid-Victorian ship to its stowing position after weighing and catting. When catted the cable was unshackled to reduce weight on the tackles, but a hawser was hitched to the ring as a precaution for recovering the anchor in case of accidents.

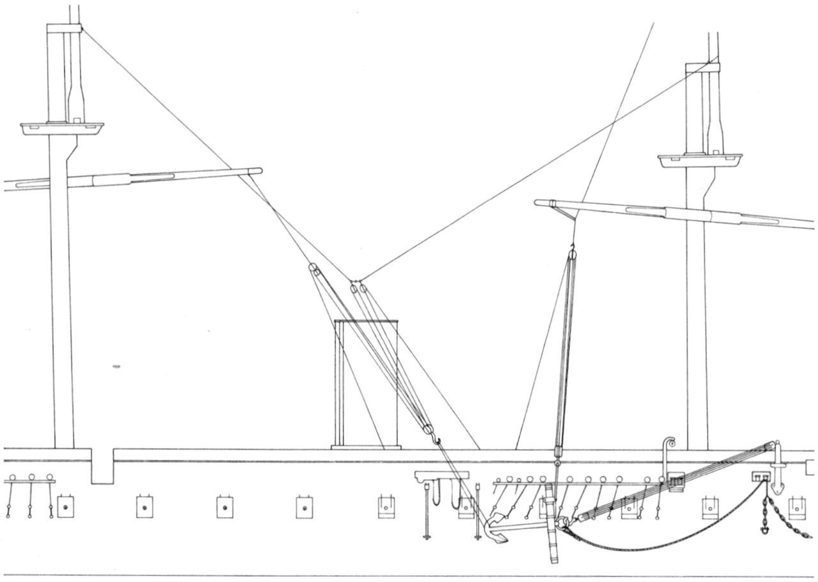

H.M.S. *Defence*, after 1870, now ship-rigged with a topgallant forecastle and a running in bowsprit. (NMM. 58/3005)

that kind, on account of the labour of restowing it again when weighed. The berthing and handling arrangements for these anchors had undergone no appreciable improvement for at least a hundred years, and gave a good example of the unreasoning adherence to ancient convention which has so often been noticeable in the Controller's department in details of technical equipment. The anchors were stowed against the ship's side with the stocks vertical, at some distance from the bows, and as no appliances were fitted or furnished in the way of catheads, davits or purchases for transporting them from the bows on weighing, an extemporised arrangement had to be rigged with heavy tackles supplied to the ship for other purposes whenever these anchors were used.

After completing her second commission of four and a half years, the *Defence* was paid off at Plymouth in 1872 for the regulation survey and overhaul. Then came two years as First Reserve Guardship for Southern Ireland with headquarters in the Shannon, and then (beginning in 1876) a second spell in the Channel Squadron, commanded at that time by Lord John Hay. This lasted till the end of 1879 and included the period in which that force did a year's duty in the Mediterranean while the proper station squadron was up the Dardanelles in support of British interests during the closing stages of the Russo-Turkish war. While in those waters the *Defence* was present at the transfer of Cyprus to the British flag, and assisted in disembarking the occupying troops. Paid off for the fourth time in 1879 she was next commissioned to take the place of the *Resistance* as First Reserve Guardship in the Mersey, this being her final service as a seagoing ship and lasting for five years. In 1885 she hauled down the pennant as such and was laid up at Plymouth, whence in the ordinary course she would sooner or later have passed into the hands of the shipbreakers, but it happened that in 1890 the Admiralty were establishing the floating factory system of mechanical training and the Board's selection fell on the *Defence* as the first of the obsolete vessels to be converted to that use. She was permanently moored in the Hamoaze accordingly and rechristened the *Indus,* as a successor to the old two-decker of that name which had long flown the flag of the Admiral Superintendent of Devonport Dockyard in the days when it was compulsory for every admiral to fly his flag afloat even when his work lay ashore. There the *Defence* — to use her original name — was to be seen until she was sold out of the service in August 1935, thirteen years after she had last paid off but four years before her name finally vanished from the Navy list at the end of 1939.

CHAPTER FOURTEEN

THE 'AUDACIOUS'

H.M.S. *Audacious* as she appeared in the early 1870s during her first commission as Flagship on the China Station. (NMM. A.2236)

In the 1860's, when ships and ordnance alike were passing into obsolescence so rapidly the British Admiralty found it advisable to relinquish the old and generally sound custom of making good periodical wastage by building groups of homogeneous model; and to substitute a wide degree of experimental variety among the larger vessels of each year's programme of fresh construction. As a result thirteen of the armoured ocean-going hulls built in the 'sixties were absolutely individual in design, and ten more were only designed in pairs. Even to the very end of the 'sixties only one great change had been established as final and comprehensive for large tonnage under the White Ensign. No more timber hulls were to be built. But even in a time of such revolutionary advance, the shipbuilding policy of the British Admiralty was determined in the main by the constructive activity of foreign Powers. The French were launching battleships of medium displacement avowedly for despatch to distant waters where British commercial and colonial interests predominated.

In framing the building programme for 1867, therefore, it was decided that ships of moderate size and cost would best meet immediate national needs; but that whatever their actual features might be, at least four would be necessary to keep abreast of other countries. And if a thoroughly acceptable design was forthcoming, all four might conform to it as a homogeneous class; thereby reverting in some degree to previous custom, with its unquestionable advantages in economy and simplification of maintenance as well as in familiarising officers and men with a whole group. Under those instructions, therefore, the Controller's Department took as a starting point the displacement tonnage of the medium-sized five years old pair *Resistance* and *Defence;* upon which Chief Constructor Reed based the plan of a capital ship much superior in power, with excellent qualities for sea, though in certain directions not free from defects. In part, the improvements were due to progress in the manufacture of ordnance and armour since the older pair of ships had been built; and, in part, to practical experience gained in the same interval as regards the specification and distribution of structural weights in iron hulls.

Except for some criticism in details, this design found favour among the naval officers who happened to see it; but it came before the Board just at a time when they were suffering under political pressure to alter their unenthusiastic attitude to the *Captain,* then building, and so they temporised. Four ships were to be built; but only two, the *Audacious* and *Invincible,* were laid down at first to Reed's drawings. Then the Admiralty followed the unprecedented course of addressing a letter to each of the seven leading private shipbuilding firms in the kingdom, promising that any that could submit the plan of an obviously better fighting ship within the same limits as to size and cost should have the contract for the remaining pair.

Not altogether unnaturally, this caused some feeling among the Board's own subordinates; but they had no voice in the decision, and all the seven firms to whom the invitation was sent — as well as others who had not been included — accepted immediately, each submitting a plan representing their own particular notion as to the most formidable capital ship producible under the conditions imposed. The result, however, only served to disclose the marked divergencies of theory prevailing among civilian designers. One firm did indeed tender a design that was scarcely distinguishable from Reed's own; but the other six each had some palpable objection or demonstrably extravagant claim, one in fact suggesting a 9-inch armament to be run about the decks on a sort of tramway system. In the end therefore, the Admiralty decided to adhere to Reed's plans for the whole quartette of battleships to be built. The *Iron Duke* and *Vanguard* were thereupon laid down as sisters to the *Audacious* and *Invincible;* these four constituting the only additions to the British battlefleet begun and completed as a group from one original design between the first adoption of armour in 1859 and the building of the *Camperdown* class twenty-seven years afterwards.

Although this particular account only professes to deal with the history of the *Audacious,* one general description of their common features will answer for all four vessels; since their only differences, except in comparatively minor details, lay in an experimental underwater zinc sheathing for the *Audacious* which was not adopted in her sisters, and a hinged instead of a balanced rudder for the *Iron Duke.* In dimensions, they measured 280 feet between perpendiculars by 54 feet maximum beam like the *Resistance,* but as originally completed for sea only drew 22 ½ feet of water to the older ship's 26, though carrying 500 tons more weight in equipment and stores. This pronounced superiority was largely due to a skilfully calculated adoption of a lighter hull framework; but largely also to much fuller lines and squarer cross section, which would have been detrimental to speed and manoeuvring power had it not been for the reduction in draught they helped to procure. But the great novelty in design, distinguishing them from all predecessors and dominating their entire conception, lay in a double tiering of the central battery; whereby on the one hand a great saving was obtained in the longitudinal space devoted to armament, and on the other an advantageous heightening of the centre of gravity as compared with its position in earlier armoured ships. Thus, although the main armament consisted of ten 9-inch guns well protected (as in the larger *Bellerophon* and *Royal Alfred*), only six of these were mounted on the main deck, while the other four were carried immediately over them on the deck above instead of all being ranged on the same level after the practice in older ships.

In cross-section, they were very full with a high freeboard; giving rather a top-heavy appearance to the eye in so comparatively short a model. To permit fore-and-aft fire from the angles of the upper battery, the sides had more tumble home than in any earlier armoured types. Forward they carried a short pointed ram — which was accidentally used with fatal results by one of the group on another — but the visible portion of the stem rose almost vertically, giving no indication of the spur beneath the surface. In his first sketch, Reed had given his design the bow and stern profile of a frigate above the waterline, in order to meet the First Lord's expressed partiality for a handsome vessel according to the ideas of the period. But the practical difficulty of combining a frigate's projecting knee with a ram below it led to a revision of the bow outline in the final draught, whereby the knee disappeared except for a slight prominence at the top of the stem to allow extra space for the latrines under the forecastle. At the other end of the ship, however, the frigate stern remained, though imposing more weight than the contracted underhung sterns of contemporary vessels. But the *Audacious* had ample reserve buoyancy, and the wide counter and quarters of this form of stern afforded room for flag officers' accommodation without a poop.

The upper deck was cramped and restricted as compared with that of any other British capital ship afloat or building in 1870. It was narrowed by the unusually extreme convergence in the upward slope of the sides and bulwarks, and divided athwartships by the enclosure forming the upper battery; so that instead of presenting a clear run from quarter deck to forecastle, it was broken transversely (as in a cargo steamer) into two short sections with a raised section between them, through which the only fore-and-aft passages led through narrow armoured battery doors. Though not actually detrimental to fighting efficiency, this reduced the space for manning the ropes of a large sail area, or hoisting boats, catting anchors, etc.; besides leaving little room for company and squad drills. It was more in the nature of an inconvenience, therefore, than an absolute defect. A topgallant forecastle was included in the design, but poops were not added till these vessels were nearly obsolete.

Internally the design presented several novelties rendered possible by the small requirements of armament space consequent on mounting only three guns a-side on the main deck. This left such ample room for berthing allowance on that deck that the officers' messes and cabins could all be brought up from what might have been correctly described as the basement position which they occupied in all the older armoured ships. The period had at last arrived when officers lived in the enjoyment of full daylight and ventilation; with a wardroom hatch opening direct on to the quarter deck instead of through a deck intermediate, and cabins having square ports high enough above water to remain open sometimes even at sea, instead of six-inch plug scuttles only three feet above the surface. Cabin candles were now no longer needed in the daytime. As for the men, they lived forward of the guns as in all central battery ships, but with a better mess deck than in older vessels, and enough slinging room for all hammocks without using the flats. Naturally therefore the *Audacious* class were popular in the service.

They suffered, however, from a badly planned internal system of watertight partitioning, both horizontally and vertically; never recognised till it failed to save the *Vanguard* after being rammed, nor remedied even then in the other three, though in part it was not difficult to amend. In the first place, the vertical bulkheading was deficient through the absence of any wing passages. Some constructional weight was saved thereby, but only at the price of an immediate flooding of any compartment in the event of a penetration of the ship's side anywhere between the sea surface and the double bottom, which only reached up to within four feet of it. And no ships could less afford to risk an inflow of that kind than the *Audacious* class, because their horizontal partitioning was defective in having the main deck placed so low that the upper edges of the bulkheads dividing up the hull beneath it barely rose three feet above the waterline. Any loss of trim through the swamping of a large compartment might reduce that narrow margin to nothing at the depressed end of the ship, and thereby permit an overflow by the main deck hatch of the flooded compartment to all parts. This defect might have

H.M.S. *Audacious* after her refit which included the addition of a poop and extra quarter deck davits, ready in 1883 to return to the Far East. (NMM. HBC.1)

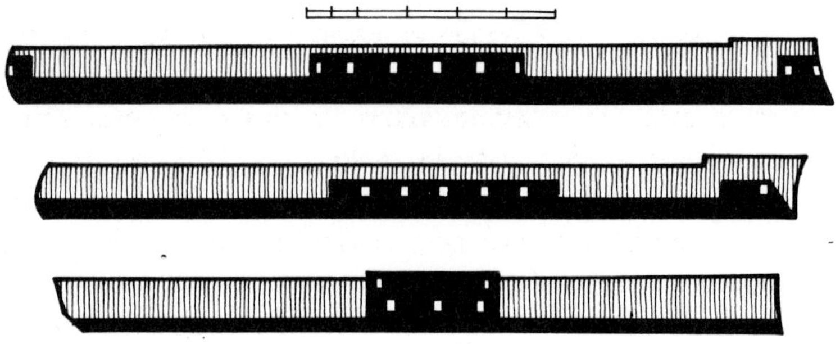

Diagram illustrating the disposition of the side armour above the waterline in the six iron-hulled British central battery capital ships of 1870, with the total weight and maximum thickness in each. All these vessels had transverse internal armour in addition. Top: *Hercules* – 1481 tons – max. 9 in. Centre: *Bellerophon* – 1090 tons – max. 6 in. Bottom: *Audacious, Invincible, Iron Duke, Vanguard* – 924 tons – max. 8 in.

been easily and greatly lessened by enclosing the main deck hatches with cofferdams up to any height deemed advisable (as in ships built twenty years after), but such an expedient was never thought of when the *Audacious* was designed, or even when the *Vanguard* was sunk.

As a compensation for internal faults of that kind, however, they possessed admirable qualities in other directions, having no equals in the supremely important matter of steadiness as gun platforms, in spite of being in point of tonnage the smallest of the twenty-seven British capital ships of 1870. Nor was that excellence the result of chance. In the days before the adoption of armour, when naval architects worked largely by conventional rules, the reason as to why the ships that are most susceptible to the motion of the waves are the least easily capsized by the wind had never been sufficiently investigated to prove that it is due to the position of the centre of gravity relative to the metacentre. Hence, in the design of most of the earlier armoured hulls, the centre of gravity had been placed so low (to allow a large margin of safety against turning turtle) that though no amount of sail pressure could jeopardise their stability, they 'rolled their guts out', in seaman's parlance, which made accurate shooting very difficult in a beam sea.

When designing the *Audacious,* Reed devoted special attention to this subject, by various adjustment of weights; such as the placing of an armoured upper half to the central battery and a diminution in the quantity of framing for the ship's bottom. As an insurance of steadiness, however, these measures were so successful that in the end it was found he had gone too far. When first sent on her trials, the *Audacious* scarcely rolled enough to spill a glass of rum in a gale; but heeled far too readily to the wind even under bare poles; and when empty of coal and ammunition, actually sat on a bilge. A step in the reverse direction was taken, therefore, by putting 360 tons of cement in the double bottom as ballast; which on trial proved a satisfactory mean, for she was now much stiffer under canvas without rolling appreciably more than before. The extra dead weight would have taken a preferable form had it been worked into her original construction as additional watertight bulkheading or stouter plates, or some other useful extension of design, since it increased her draught of water by almost a foot and decreased her speed by nearly a knot. But that was impracticable in a finished ship; and at least the main point was attained of securing her safety under sail without interfering with a steadiness so remarkable that she could make first-class practice under conditions of weather when other and larger battleships could only fire wildly, even if they dared open their ports.

As the *Iron Duke* carried a hinged rudder, while those of the three sister vessels were balanced, exceptionally good conditions arose for ascertaining the relative merits of each. Under steam, the balanced proved much the better. Six men could put the helm hard over at full speed, while the other required twelve; and the average diameters of the respective turning circles were 370 yards and 500. But the balanced rudder was against good sailing, as its great hold on the water checked headway so much that with three turns of helm on her the ship became a log, unless the engines were moving. Moreover, as a square rig on a broad-beamed hull always forces a lot of weather helm when the wind is before the quarter, the drag of a balanced rudder necessarily prevents fast sailing unless with the wind nearly dead aft. Only the *Iron Duke,* therefore, proved manageable without steam; though in all four vessels their large sail plan saved coal appreciably when used in conjunction with the screws. Each carried a triple wheel on the quarter deck for ordinary use, and another as a fighting wheel behind armour in the battery. None had steam steering gear when first built, which had only just been invented, but all except the *Vanguard* received it later. In shallow water they yawed erratically — because, on account of their very full runs, the turning effect of the rudder was largely dependent on the pressure of the uprush from below — and when passing through the Suez Canal they either used a tug ahead or dragged a chain cable.

Several features in their general equipment were novel to the British Navy. The bower cables worked on the upper deck — in which departure from former usage, the *Hercules* alone among capital ships had preceded them, for a special reason — but owing to lack of space under the topgallant forecastle, the sheet cables worked on the main deck as of old. It was in the *Audacious* class also that the subsequently prevalent custom of carrying the bower anchors at topgallant forecastle height was first introduced in British battleships, and with it the swinging davit type of cathead. The sheet anchors were stowed on the top of the bulwarks close abaft the forecastle, instead of outside the bulwarks abaft the fore rigging. And when new, this quartette were the smallest British vessels with steam capstans; which stood between the fore mast and the funnel with upper and main deck barrels and the engine below the waterline. The stowage of the boomboats also marked a conspicuous change from long tradition. In the older vessels — except the two turret ships then building — they were carried before the main mast on crutches so low that only the outline of their gunwales was visible from outside the ship; whereas in the *Audacious* and her sisters they were at first placed abaft the main mast, but in order to avoid congesting the narrow quarter deck cradled on stanchions so high that the whole boats were completely in view. As it was found, however, that this obscured an important arc of bearings from the bridge they were all transferred at later stages to crutches on the top of the upper battery, but still remained high and visible.

The question was then being debated as to any practicable means of obtaining a good end-on fire otherwise than by burdening the extremities of these new vessels with heavy chase guns behind armour, liable to increase pitching; a method carried rather too far in some of their predecessors. Reed solved the problem by introducing his upper central battery projecting outboard on both sides in sponson fashion, so as to allow arcs of fire from right ahead to the beam for two 9-inch guns, and from right astern to the beam for two more. Six others were divided between the broadsides immediately underneath the first four. For a ship of only 6000 tons these ten weapons provided an unusually powerful attacking equipment; disposed on an excellent plan for giving a complete all-round fire from a well-covered central position unobstructed by masts or rigging; and moreover presenting the further solid advantage that four of them were carried so high that they could engage an enemy under weather conditions which were too bad for the use of any other guns behind armour in the whole British fleet except those of the *Monarch,* and the bow pairs of the *Lord Warden* and *Lord Clyde.* In fact, although the *Audacious* class were the smallest units of capital rank under the British pennant in 1870, they would have been more formidable antagonists in a heavy gale than any of the remainder except the much larger *Monarch.*

As a minor addition to the 9-inch ordnance, four 6-inch truck mounted guns — officially denominated 64-pounders — were placed on the upper deck outside the armoured enclosure, two forward and two aft; each of these guns having three alternative ports; and six 20-pounder Armstrongs were supplied for saluting. When the ships were about twenty years old the 64-pounders and the Armstrongs were replaced by either six or eight 4-inch breechloaders as an anti-torpedo defence, but the main 9-inch armament remained unchanged throughout their seagoing service. After their second active commissions these ships received the addition of four main-deck 14-inch torpedo carriages, two placed forward of the central battery and two abaft. Their full complement stood at 450 ranks and ratings, being the smallest for any British capital ship in 1870.

In the disposition of her 920 tons of armour, the *Audacious* marked a notable advance in the practice of increasing the thickness in new designs by reducing the area, so as to improve the protection of vital parts without adding to the total weight. As in the older and rather larger *Bellerophon,* the central battery carried 6-inch plating on the ship's sides and 4-inch on the internal bulkheads; but the waterline was covered by 8-inch amidships, which gave the stoutest belt in the fleet except the 9-inch of the far larger *Hercules.* Above that belt, however — which only reached three feet up from the water at full draught — the *Audacious* and her sisters were bare of any side armour except the central patch 52 feet long covering the 9-inch guns. Towards the end of their sea-going service their protective equipment received the addition of a torpedo net defence.

It might fairly be claimed for the *Audacious* that she was the first armoured capital ship afloat in all the world with twin screws, as the only armoured twin-screw vessels laid down before her happened to be the ill-fated *Captain,* which was not launched till later, and the much smaller *Penelope,* only classed officially as a corvette.

No foreign navy, and no mercantile marine foreign or British' had up till then adopted twin screws for heavy tonnage; and even in the British Navy they had many strong opponents except for the small gunvessels in which they were already installed. Private owners and shipbuilders viewed double propellers as yet with almost unanimous disfavour; but the Admiralty began to recognise their advantages at this period, which enabled Reed to adopt them for the *Audacious* and thereby reduce propeller size sufficiently to design her with three feet less draught aft than the *Resistance* of the same displacement. Her twin-screw machinery was made by the Clyde firm of Ravenhill, and comprised two sets of double-cylindered horizontal and outward turning return connecting-rod engines; with cylinders of each set placed between its shaft and the angle of the bilge, and the condensers amidships. Each drove a bronze Mangin propeller with four blades, not arranged to form a radial cross but placed in two pairs one behind the other in tandem on an elongated cylindrical boss. Steam came from six rectangular boilers at 30 lb. pressure all in one compartment. It was an admitted defect in her design that she only stowed 460 tons of coal. Later, the Mangin screws were replaced by two-bladed Griffiths.

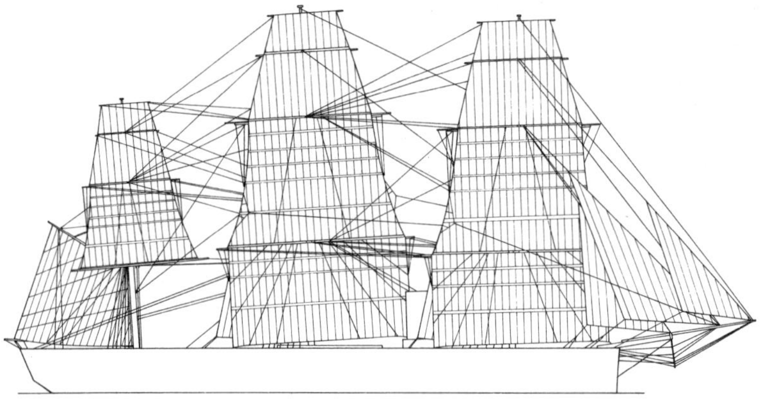

Sail plan of H.M.S. *Audacious* before change to barque rig.

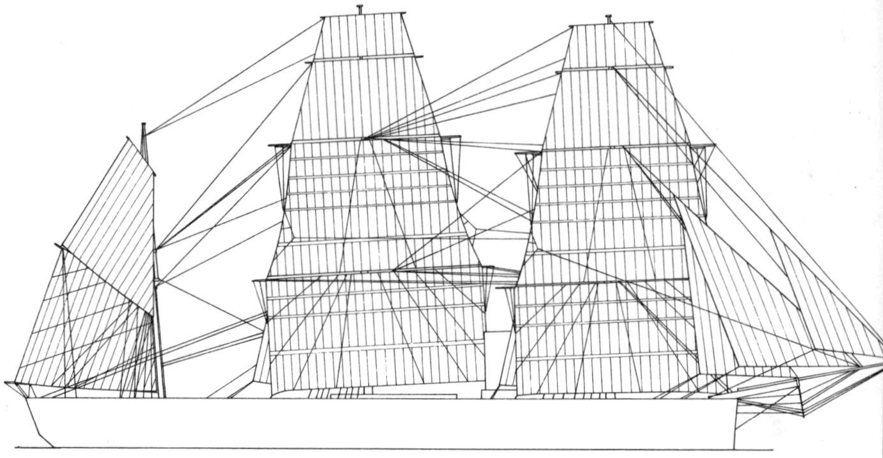

Sail plan of H.M.S. *Audacious* after change to barque rig.

Among British armoured capital ships the sail area of the *Audacious* quartette was the most extensive in proportion to hull dimensions ever carried, even when somewhat cut down on conversion to barques. In their original ship rig the total hoist from trucks to deck equalled that of the five specially heavily sparred and much larger vessels in the fleet. The upper yards were as square also, and the only reduction in spar lengths was in topsail yards and lower yards. After cement ballast had been put into their double bottoms, the ships possessed quite enough stability to stand up to this great spread; but the loss of the *Captain* at this period led to an Admiralty policy of increasing the margin of safety in the masting of new types. While still in their first commission, therefore, the yards were removed from the mizen in the *Audacious* class and the fore and main topmasts shortened by six feet, and in the *Vanguard* by ten. In the last-named ship the topgallantmasts and upper yards were shortened also, but not in the other three.

Even as thus cut down, however, their canvas remained on a larger plan relatively to tonnage than in any other British battleships of the period; and, owing to the narrow angle of spread of the lower rigging in the absence of external chains, they could brace their yards up sharper and point correspondingly higher. But in spite of these advantages the *Audacious* and her sisters proved slow and leewardly under sail. Slow, partly on account of the double drag of twin screws — and partly on account of the interference with the free flow of water in the run caused by the protuberant casings of the tail shafts. Leewardly, because their flat bottoms and comparatively shallow draught diminished lateral hold on the water when sailing close hauled. On the other hand their large sail area economised coal very appreciably when steaming with the wind abaft the beam; and as their bunkerage was small, the rig was retained until they were about twenty years old and past any chance of service on distant stations. They were therefore then reduced to pole masts.

When the government reached the decision to begin four new iron ships, the only Royal dockyard with the plant necessary for iron construction was Chatham; which happened to be working at full capacity already on the *Monarch* and *Hercules*. The orders for the first new couple, therefore — the *Audacious* and *Invincible* — were entrusted unavoidably to private industry and the contract for both went to the Clyde firm of Napier and Son, at that time the leading yard in the north, having been amongst the very earliest in the kingdom to undertake iron hulls for the Admiralty when they built the *Black Prince* and *Hector*. This was the first occasion in history on which one private establishment was utilised for the simultaneous building of a pair of capital ships for the nation.

The *Audacious* was laid down on the Govan ways accordingly in 1867, whence she was sent afloat in 1869. A strong gale was blowing right up the river on her launching day in sufficient force to cant her when half waterborne, whereby some bottom plates and framing were damaged; an accident attributed by the builders to weakness in the Admiralty design, and by the Admiralty to bad placing of the builders' launching cradle. There, however, the argument ended, as in any case the responsibility for delivering the ship in proper condition for sea rested with the Napiers, who had to bear the whole cost of any repairs while she was still in their charge. She was the first iron-hulled ship in the British service to float on a lighter draught than designed, and in a varying degree her three sisters did the same. All former errors in calculation for iron-built tonnage had been the other way, and this reversal of previous experience went to enhance Reed's reputation as a designer.

She was taken to Plymouth for arming and trials, and when ready for sea hoisted the pennant there in October 1870 under Captain Spain as Guardship of the First Reserve in the Kingston district of Ireland. Next year, however, she was transferred for similar duties to the Hull district in England, and remained so employed till paid off at Chatham three years later, for preparation to relieve the *Iron Duke* as flagship of the Commander-in-Chief in China; a station with which her name had a longer connection ultimately than that of any other British capital ship, extending to ten years before the end, though with an interregnum. She was commissioned accordingly in September 1874 and sailed eastward flying the flag of Vice-Admiral Ryder. Through her tendency to erratic steering in very shoal water, she grounded twice in the Suez Canal in spite of escorting tugs, and took four days to get through. Moreover, steam passages were slow in the early Victorian period, as the economical speed of a battleship rarely exceeded 5 knots. Thus, for instance, the *Audacious,* with extra coal in the battery to help, was three weeks at sea between Aden and Ceylon, averaging only 100 miles a day against the north-east monsoon.

After taking over from the *Iron Duke* at Singapore, she passed the three following years in the normal routine duties of station flagship in the far east. It was the custom, for the Commander-in-Chief to appear at all the Treaty Ports in turn every year, besides making occasional visits to Singapore, Saigon, Vladivostock and Manila; the annual refit of the flagship taking place while wintering at Hongkong or in Japan.

By an arrangement with the Japanese, the British Admiralty were allowed to maintain a store and repairing depot at Yokoska, where the neighbouring anchorage of Yokohama Bay was frequently in use accordingly. Here it was that the *Audacious* was caught in a typhoon, dragging both anchors till she collided with a nearby steamer and smashed all her own starboard boats. The steamer was dragging also and shouting for help, and in clearing away the stream anchor to send to her, the carpenter was killed; but the death of that officer was the only fatality in the *Audacious* during a long commission, in which respect she was extraordinarily fortunate as compared with most ships, for her men had their full share of falls from aloft and falls overboard. On one occasion at night the circumstances were so dangerous, that an officer — Lieutenant Sandilands — who jumped to the rescue received the Stanhope medal. On another a man was jerked off a lee yardarm when reefing topsails, to drop a full hundred feet into the sea but to be picked up little the worse, although striking the water from that height is apt to be fatal.

In 1878 Vicc-Admiral Hilyar relieved Ryder as station Commander-in-Chief, but towards the end of the same year the *Audacious* returned home on the arrival out of the *Iron Duke* once more as relieving flagship. Paying off at Chatham after four years in commission, she reverted a few months later to her old duty as Guardship of the First

H.M.S. *Audacious*, stripped of her square rig, with light quick-firing guns along her topsides, ready to take up her rôle as Guardship of the Hull District about 1890. (NMM. HBC.3)

Reserve in the Hull district, where after a year and a half she was again paid off in 1880 to undergo a long refit including new boilers and the addition of a poop. Being ready again by March 1883, she was commissioned and dispatched a second time to relieve the *Iron Duke,* and now remained on the China Station for two successive commissions of three years each, flying in turn the flags of Vice-Admirals Willes, Dowell, Vesey Hamilton and Salmon. The new ship's company for the second commission went out in the *Invincible,* which brought back the old. When the second was nearing its end, in its turn the *Audacious* was ordered home and paid off at Chatham in 1889, having worked 'watch and watch' alternately with the *Iron Duke* as British flagship in the far east for eighteen years between them. She was then deprived of her square rig, reduced to pole masts and supplied with eight 4-inch breechloaders in place of her four 64-pounders; and after a general overhaul was sent back for a third and last time as Guardship of the Hull district; at the end of which, in 1894, her seagoing service came to a close twenty-three and a half years after it began. On removal from the effective list, she lay up in the 4th class Reserve for seven years, and was then appropriated as a harbour service hulk with engines removed; first as a torpedo depot and then as one of the group of obsolete vessels fitted as training ships at Plymouth for boy artificers under the name *Fisgard.* Being still thus in use on the outbreak of the Great War, she was towed to Scapa Flow as a general receiving hulk under her third name of *Imperieuse*, and after the war taken to Rosyth where she remained till sold to Ward of Inverkeithing for breaking up in 1927. Her service therefore, though long, was devoid of dramatic episodes and terminated through sheer old age.

CHAPTER FIFTEEN

THE 'INVINCIBLE', 'VANGUARD' AND 'IRON DUKE'

H.M.S. *Vanguard* as completed for sea at Plymouth in October 1870. (NMM. HBH/1)

The *Invincible* was the second battleship of the *Audacious* class to be launched, and formed with that ship the original pair ordered to Reed's design, of which the *Vanguard* and *Iron Duke* were reproductions. She was laid down in Napier's yard on the Clyde at the same time as the *Audacious,* of which ship she was an exact duplicate in hull, armament, masting and general equipment, except for the absence of zinc sheathing. Her engines, however, were made by her builders instead of by an outside firm, and differed in minor details, the cylinders being 72 inches in diameter and the propellers slightly finer in pitch. As regards design, they seemed an improvement on those made by Ravenhill for her twin sister, as she reached a speed of 13 ½ on the measured mile to the latter's 13 ¼. But the material or the workmanship must have been of poor quality for they very quickly wore out.

She left the ways in June 1869 and after completing for sea commissioned at Plymouth in October 1870 under Captain C.W. Hope as First Reserve Guardship of the Hull district; but only remained till it was taken over a year later by the *Audacious.* Her next service was in the Mediterranean, a station where much of her future was to lie, and for which she sailed with a new ship's company in August 1872 for a four years' commission. Most of it was passed near the centres of contemporary political unrest on Mediterranean shores; that is to say, the coast of Catalonia during the brief rule of King Amadeo of Spain and the turbulent Spanish Republic which followed; and the coast of the Levant, at the other end of the station, during the Balkan troubles leading up by degrees to the Russo-Turkish war of 1877-8. She also acted as escort to the *Serapis* on the stage from Egypt to Malta, when that ship was bringing home the Prince of Wales — afterwards Edward VII — from the first visit to India of an heir-apparent to the throne. Towards the end of that commission, however, her machinery had lasted so badly that on paying off at Plymouth in 1876 all her boilers and cylinders needed renewal, though only six years old. For ten years, thereafter, the makers were awarded no more contracts for battleship machinery.

Being re-engined and re-boilered she was commissioned, for the Mediterranean by Captain Brine in March 1878. Directly on arrival from home, she was sent to join the *Swiftsure* watching the Russian armies in the Gulf of Xeros. Next she was sent to assist in taking over the island of Cyprus, newly ceded to Britain by the Congress of Berlin. After that she was sent to join the Mediterranean battle squadron still up in the Sea of Marmora under Admiral Hornby; whereby arose an incident provocative of much comment among naval officers. Hornby was the ablest seaman of his day at handling a fleet but apt to be hasty in measures of discipline. On a certain morning during tactical exercises he considered that the steering of the *Invincible* risked the safety of his flagship, which produced forthwith a signal placing Captain Brine under arrest and ordering his second-in-command to take charge of the ship; thereby censuring Brine publicly before the whole squadron, including his own officers and men. In the general opinion of the Service such an extreme step was prejudicial to discipline rather than helpful, and in the sequel it recoiled on the Admiral's own head. He brought Brine to trial by Court Martial on a charge of hazarding the two ships by danger of collision; but after sitting for seven days the Court found the charges were not proved, and acquitted the accused. By endorsing this verdict the Admiralty administered a severe indirect rebuff to Hornby, and trans-

H.M.S. *Invincible* at Plymouth in August 1872 before her departure for service in the Mediterranean. (NMM. 7377)

H.M.S. *Invincible* in dry dock about 1876 when undergoing repairs to her machinery and boilers. (NMM. PX.21219)

H.M.S. *Invincible* at Malta during the early 1870s with funnel lowered and awnings rigged. (NMM. 9556)

ferred the *Invincible* to the Channel Squadron till Brine received another appointment.

When Captain Fremantle succeeded him, the ship was sent back to complete her four years' commission in the Mediterranean, where she was re-commissioned with a new crew in March 1882. Two months later she was cruising with the rest of the station battle squadron in the Levant, when the Mediterranean powers were disturbed by the rebellion of the Egyptian army and the troubles which followed. Sir Beauchamp Seymour was ordered to Alexandria, in readiness for emergency measures. It happened that the *Invincible* was the only armoured ship at his disposal of light enough draught to enter the inner harbour (except the *Monarch* already there) so he transferred his flag to her and arrived on June 5th.

By July 11th Seymour's ultimatum, warning the Egyptian rebel commander that he would be compelled to attack the sea-front batteries unless they disarmed, was due to expire. The last foreign vessel had left on the 10th and shortly before midnight the *Invincible, Monarch* and *Penelope* slipped out and anchored off the western approaches in readiness to bombard the forts on that side next morning, while the five heavy draught battleships took up station opposite the powerful defences to the eastward.

The *Invincible,* with Seymour on board, brought to in a position about 1500 yards from Fort Mex, having a spring on her cable so as to keep her port broadside of five 9-inch guns to bear on it. This work was armed with one 10-inch, one 9-inch and three 8-inch rifled guns, besides nine heavy smoothbores; and at dawn it was seen that these were all manned. No reply to the British ultimatum having been received from the Egyptians by 7 a.m. (the last hour of grace), the attack began from all eight ships. The batteries replied with a fire well-sustained, but so badly laid that in the case of the *Invincible* — a ship at anchor within less than 2000 yards — of about a hundred shot and shell that fell around during the next three hours and a half only thirteen scored a hit. Moreover, as not one penetrated her armour, her sole casualties lay among the few hands outside its shelter, of whom six were wounded. By half-past ten the western defences had been silenced, and the three vessels at that end ceased firing, the *Invincible* having expended 140 rounds of 9-inch shell, 131 of 6-inch and 5000 of machine-gun ammunition. Two days afterwards the *Invincible* re-entered the inner harbour — still deserted — and landed 200 men to participate in the occupation of the city.

The remainder of the Commission was uneventful and in 1885 she returned to Portsmouth to pay off. Next year she was temporarily commissioned to carry a new crew out to the *Audacious* in China and bring back the old. She was then deprived of her square rig and exchanged her 64-pounder guns for six 4-inch breechloaders in preparation for commissioning in November 1886 as First Reserve Guardship in the Southampton district. There she continued to serve till 1893, carrying out the annual routine duty of attendance at Cowes when Queen Victoria was in residence at Osborne. In 1893 she was removed from the effective list, and in 1901, after her engines had been taken out, was converted into a harbour depot for the local destroyer flotilla at Sheerness.

Five years later she was transferred thence to join her sister the *Audacious* at Plymouth as one of the assemblage of hulks forming the newly established training school for engine-room artificers under the name *Fisgard II.*

When the Great War broke out, these hulks were requisitioned for duty as general receiving ships at the northern fleet bases, and despatched thither in tow of tugs, having no motive power of their own. But the removal of their main engines and other weights low down, in conjunction with the erection of a mass of factory machinery on their decks, had reduced their stability so greatly that when the *Invincible* ran into a gale on the passage her workshop plant broke adrift, causing her to founder on September 17th, 1914, in the rough water of Portland Race and drown twenty-one of her scratch crew of sixty-four hands. Had this happened during peace it must have created a public outcry against the Admiralty; but as it chanced to occur when the country was engrossed in a struggle sacrificing lives by the hundred day after day, the incident passed almost unnoticed among the burden of greater anxieties.

The keel of the third of the *Audacious* class was laid in Laird's yard at Birkenhead in 1867, where the turret ship *Captain* happened to be already under construction. And as the *Captain*'s name specially recalled Nelson's exploits at the battle of St Vincent, so the Admiralty decreed that her cradle mate should bear the name *Vanguard* to commemorate his still greater fame at the Nile. Yet, by a perverse destiny, both these fine ships after brief careers came to untimely, disastrous ends in a period of profound peace.

A 9-inch, 12½ ton muzzle-loading rifled gun in the battery of the *Iron Duke.* A typical mounting for such a weapon, shown in the 'run in' position, ready for loading. Within the framework of the slide beneath the muzzle of the gun can be seen the long strips of metal comprising part of the 'compressor', which through the friction it creates governs the recoil of the gun after firing. The lever on the gun carriage, below the gun's trunnion regulates the degree of compression applied. Nearer the after end of the carriage can be seen the 'elevating drum', with part of the elevating arc showing above the carriage.
With guns having a bore larger than 7 inches in diameter, the training races were grooved, and training was effected by means of geared training winches at the rear end of the slide, one of which is shown on the extreme left of the picture. Even 'Jack' had his fun with the photographer by placing a very small model of a truck mounted gun on the carriage of the larger weapon almost below the cascable. (NMM. B.7679)

Sketch to scale of a British 9-inch 12½ ton muzzle loader of 1870, of the type forming the main armament of H.M.S. *Invincible* when engaging Fort Mex at Alexandria; with the gun's crew at elevating stations as follows: 1. Captain of the gun, laying for elevation. 2. Standing by to prick the charge, insert the friction tube, and hook the firing lanyard on the order 'ready'. 3, 4. Loading numbers—waiting to reload. 5, 6. Clamp and compressor numbers, attending clamps of elevating drums. 7, 8. Lever numbers, elevating the gun as ordered by No. 1. 9. Winch number, attending brake and pawl of training winch. 10, 11, 12. Winch numbers, manning training winch and standing by to train as ordered by No. 1. 13, 14, 15. Rear numbers bringing up shell, not shown. At General Quarters gun's crews wore their waist belts in case of being called as boarders. Arms were racked near each gun.

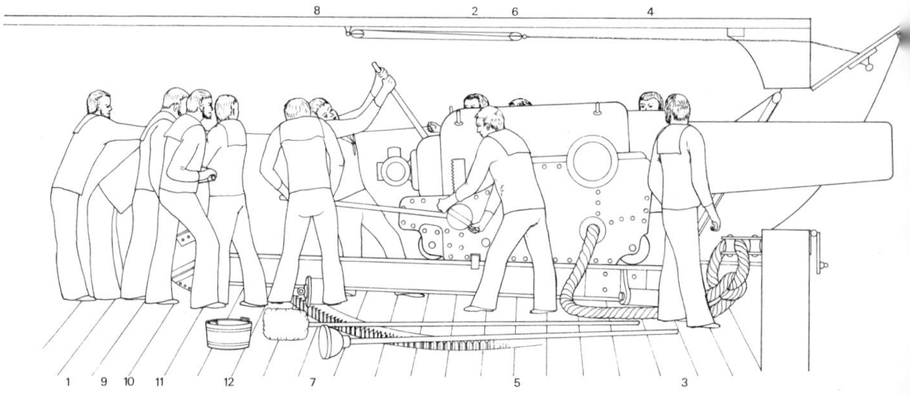

H.M.S. *Invincible* leaving Portsmouth about 1890, whilst employed as Guardship at Southampton. She is shown with light weapons mounted along her topsides, searchlights on both her forecastle and poop, and torpedo netting stowed on her beam over her gun ports. (NMM. AAJ/1)

The *Vanguard* was afloat and ready for service, after highly successful trials, by September 1870. In design she was identical with her sisters, though the fastest of the quartette under steam by as much as three-quarters of a knot. She was engined by her own builders, with 72-inch cylinders like the *Invincible,* and machinery of the same general type and dimensions as the others; but, by reaching a speed of 14 ½ knots on the Plymouth mile, proved herself the best steamer of the nineteen British capital ships of medium tonnage afloat in 1870. Like her sisters, she was ship rigged at first and changed to barque soon after. But in her case the reduction of canvas was carried further; for though the lowermasts and fore and main yards retained their full lengths, all the higher spars of her barque rig were shortened to the second-class scale of the *Resistance, Caledonia,* etc.; with 55 foot topmasts, 68 foot topsail yards, 42 foot topgallantmasts, and upper yards in proportion. Hence the cut of her topsails was unique and non-interchangeable with any other ship's in the service, having a 68 foot yard for the head and a 96 foot yard for the foot.

Though the four ships of her class were specially built for distant service, they were ready before being required on any foreign station, and started their active careers in home waters as Guardships in the various First Reserve districts; the *Vanguard* being allotted to the east coast of Ireland with headquarters at Kingstown. From home employment, three of the four passed on sooner or later to long spells abroad, and remained afloat for forty odd years. But the *Vanguard*'s short day ended where her first duty began and she was never sighted from an alien shore. She hoisted the pennant at Plymouth in October 1870 and served continuously, thenceforward, as First Reserve Guardship of the Eastern Irish district. Little more than four years had passed in this way, however, when her career was brought to an abrupt close by a disaster of historical interest as the first occasion of the loss of a British capital ship by collision.

It was towards the end of July 1875 that seven out of the nine First Reserve Guardships had assembled at Portland to carry out their annual cruise under Admiral Tarleton, whose flag was in the *Warrior;* and as certain Irish Members of Parliament were wont to complain that Ireland did not see enough of the Navy, Tarleton was ordered to take his squadron round the whole island, calling at half a dozen ports and finishing in Dublin Bay. On arriving there a month later, he received further instructions to take four ships on to Queenstown — which had been omitted from the original programme — and send the others back to their districts. His choice fell on the *Warrior, Hector, Iron Duke* and *Vanguard;* the last-mentioned being, by his own account, the smartest ship under his orders. She was now commanded by Captain Richard Dawkins, whose last previous service had been as Flag Captain in the Pacific. The *Iron Duke* had only joined the squadron a few weeks before this cruise, on returning from her first commission as flagship in China. Her new captain was Henry Hickley, an officer of long experience and a close friend of Dawkins.

Sailing from Dublin Bay with these four battleships at half-past ten on the morning of September 1st, Tarleton elected to follow the route for Queenstown which passes outside the twelve-mile ridge of shoals known as the Kish Bank, lying parallel to the Irish coast at about six miles distance with its north end, opposite Dublin Bay, marked by the Kish lightship. He began, therefore, by steering right out to the east till a couple of miles past the lightvessel, so as to get well beyond the bank, and at noon turned to the south. Up till then his ships had been in single line ahead in the order *Warrior, Hector, Vanguard, Iron Duke;* but as soon as he altered course he made the signal to form in divisions line ahead, which meant that the two rear ships were to move up to positions on the port beam of the two ahead. This particular change of formation can be effected either by the leading pair slowing down till the others have moved up, or making a turn first to starboard and then to port while the others hold straight on; or by the leading pair standing straight on at the same speed while the two rear ships sheer out to port and increase speed. Tarleton elected to carry out the manoeuvre by the last-mentioned method; although as the squadron had only been ordered to have steam available for 8 knots and was already going 7, he must have known that a full half hour would elapse before the two rear vessels — *Vanguard* and *Iron Duke* — were up on his beam. The captains of those two ships knew that also, and both went below to snatch some lunch, leaving orders with the officers on watch to keep on at full speed; these orders being emphasised in the *Iron Duke* as she was a cable astern of her proper distance from the *Vanguard.* But only ten minutes after they had gone down, the squadron ran into one of the dense fogs which sometimes form very suddenly at sea on a day of light and variable winds. Every ship then lost sight of every other.

Fog is the seaman's most dangerous enemy, and never more so than when it falls on a whole squadron. Worst of all if it catches that assembly in the actual process of changing formation as happened now. Here we find, in consequence, the first item of unavoidable adversity in the sequence of events that ensued. On striking this fog, the officers on watch immediately sent down to their captains, who returned on deck and assumed responsibility.

But the lieutenant on watch in the *Iron Duke* had done something more, and here the element of avoidable blunders entered in. He was an officer of junior standing in his rank with little experience as a watchkeeper in charge, and now — no doubt with the best intentions — committed a flagrant breach of Admiralty regulations which proved to be the prime cause of the catastrophe close at hand. When Hickley glanced at the compass on returning to the bridge of the *Iron Duke,* he saw to his surprise that the ship's head was a couple of points to port of her proper course. On demanding to know why, he received the answer from this officer that on losing sight of the *Vanguard* ahead he had deemed it safer to steer a little outwards instead of following exactly in her wake. Such an action could only arise from ignorance combined with thoughtlessness; for besides disregarding the regulations whereby, for the general safety of a fleet in a fog, every vessel therein is directed to do her utmost to keep in correct station by her next ahead, it obviously forgot that the latter herself might be forced out of line by a strange vessel suddenly sighted on either bow, which was exactly what happened. Moreover, his deviation from the *Vanguard*'s wake had caused the *Iron Duke* (already astern of station) to lose distance still more. Hickley, therefore, immediately ordered a return to proper station (in so far as that could be judged) by a gradual sheer back to starboard and any increase of speed of which the engines were capable with the steam available. This action was perfectly correct, although possibly rather precipitate, as delay was occurring in getting steam on the whistle to warn the *Vanguard* of his position, and his advance was made in silence.

Unhappily, other events occurred just then to compel the *Vanguard* to steer to port. No sooner had Dawkins arrived on deck in response to the fog report than the forecastle look-out hailed the bridge that a ship under sail was nearly right ahead, and Dawkins, seeing something slowly crossing his bows to starboard, altered course to pass astern of her, reduced speed, and blew his whistle. He might, perhaps, have avoided the stranger by reducing speed without altering course, or vice versa, either of which alternatives would have kept him out of the path of the oncoming *Iron Duke;* but in the absence of any knowledge that the latter was on his port quarter instead of right astern as she ought to have been, he acted as most captains would have done, and so all-unwittingly threw the *Vanguard* across the ram of this approaching, though still invisible and inaudible, danger.

Barely three minutes later came the crash like a thunderclap, as the *Iron Duke* struck her sister ship a tremendous blow nearly amidships at an angle of about 46°; the fog being so dense that the colliding battleships saw nothing of each other till less than forty yards apart, when it was too late to do anything. In the *Vanguard*'s length of 300 feet from ram to taffrail lay one small section of about ten feet where a blow under water was more serious than anywhere else. The odds were thirty to one, therefore, against that narrow target receiving the stroke; and yet precisely there it went home. Again the odds reached incalculable figures against the sailing vessel passing ahead of the *Vanguard* just at the moment when an alteration of course was most fatal to the battleship. If the stranger had been sighted a few minutes sooner, or later, the colliding pair would not have reached the point of intersection of their respective tracks simultaneously as they did.

As her headway had been checked abruptly by the bump and her engines were at full speed astern almost before the collision, the *Iron Duke* backed clear very quickly with trifling damage to her bows, although her bowsprit and head gear were a wreck. But she left a gaping vertical wound through the *Vanguard*'s ribs deep down; which, as afterwards measured by divers, had a height of nine feet and width of three at its centre; and lay so close abaft the transverse bulkhead separating the engine room from the boilers, that in the absence of any support from the wing passages in the design of her class, this most important of all her internal safety partitions was so damaged that the compartments on both sides of it were flooded, which chanced to be the two largest in the ship. In the diagram the disastrous outcome of discarding wing bulkheads can be traced by comparing the cross sections of the *Vanguard* and *Resistance,* and noting the penetration of the *Iron Duke*'s ram. Its actual point never pierced the inner skin of the double bottom; but that availed nothing, as its upper edge cut through not only the single skin above the double bottom, but through the horizontal plating which formed the shelf or lid whereby the double bottom was capped. In omitting wings from the design, Reed seems to have trusted to the 8-inch waterline armour as a collision safeguard, though the divers reported it as bent inwards as if it had been tin.

Dawkins now found himself suddenly in the difficult

Crippled by collision off the Kish Bank in the Irish Sea, the *Vanguard* rapidly settles down watched helplessly by the crew of the *Iron Duke.* (NMM. 58/4288)

position of sole responsibility for making a hard decision at once, under conditions which can never fall to the lot of a landsman, wherein an error of judgment may have most tragic consequences. Like any captain so placed, he was desperately anxious to save his ship but had good reason to fear her injury as mortal, in which case the lives of 350 officers and men rested on his shoulders. Within five minutes of the collision the engine room was flooded 6 feet deep, and within ten the rapidly rising water in the boiler compartment had extinguished the furnaces and deprived him of the use of the steam pumps. As the two officers whose duty it was to advise on constructional questions, the Chief Engineer and the Carpenter both reported their opinion that the ship had only a short time left to float. Dawkins had to lose no time, therefore, in making up his mind whether life saving was a matter of instant emergency or not. Fortunately he was very soon helped to decide by the sight of the *Iron Duke*'s boats approaching through the fog, in which that vessel herself had again been lost to view on backing clear, though Hickley kept as close as reasonable care permitted. Thus assured of ready aid in the event of a hurried abandonment becoming necessary, Dawkins retained all hands on board except a few sick; and though short of nearly half his complement of executive

Outline of cross section of Vanguard (left) and *Resistance* (right), showing their respective systems of waterlight wing passages and double bottoms, with the shape of the transverse waterlight bulkhead between the two largest compartments in each in block silhouette. Also the penetration of the ram of *Iron Duke* (far left.). From official drawings.

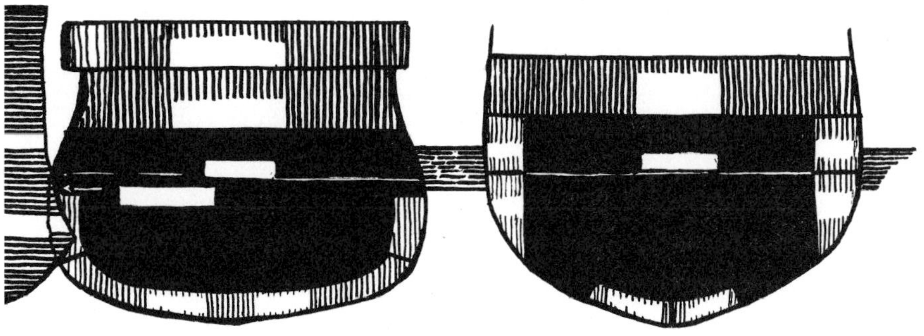

All that remained to be seen of the *Vanguard* at low water after the disaster. (NMM. 58/4290)

Diving operations on the wreck of the *Vanguard*, as viewed by those on board of the despatch vessel *Amelia*. (NMM. 58/4306)

officers as well as about a hundred men, sent one watch to hoist out the boomboats and the other to man the hand pumps; asking at the same time that one of the *Iron Duke*'s boats should take a message requesting a tow to the shoaler water about a mile distant.

Possibly the last mentioned of these measures might have stood a remote chance of success if tried; though it would have been a difficult job to pass a heavy hawser between ships in a fog quickly enough to be of any real use, and a risky undertaking for the *Iron Duke* to feel her way towards the sandbanks at any speed, especially as the tide was falling. But in the haste of transferring sick men, pay accounts, government money, etc., from ship to ship, a misunderstanding as to which boat was to convey the message prevented it from ever being delivered.

All other steps were futile, as the water continued to gain fast; and when about half an hour after the collision the Chief Engineer reported it as overflowing the engine-room hatch along the main deck to other compartments, the last hope vanished. By the time all hands were in the boats (after having been called back to the pumps once) and Dawkins himself stepped down as the last, the quarter deck was a bare 5 feet above the surface. As if to lift the curtain for the final scene, the fog cleared when the sinking vessel was abandoned and the boats pulled for the *Iron Duke*. Watched from the crowded decks and ports of the latter, the *Vanguard* settled fast and about seventy minutes after the collision gave a sudden lurch by the stern and vanished in 19 fathoms, striking the bottom so violently as to snap off her topgallantmasts, but leaving the topmast heads still showing. She had drifted about three miles to the southward since being rammed. The flagship and the *Hector* were visible about six miles ahead; but being totally ignorant of the accident and anxious to reach Queenstown on next morning's tide, Tarleton pushed along, and only heard of it upon his arrival there by a telegram from Dublin, whither the *Iron Duke* had returned to report as it was distant only a few hours' steaming.

Great public concern was aroused by the disaster; and the ensuing Court Martial sat at Plymouth for sixteen days. Legally every officer and man on the books of the *Vanguard* was on trial for losing his ship, but only a few were required to attend. On five of these fell a wholesale condemnation; which no impartial historian with any practical knowledge of handling ships will consider as justified by the evidence. The investigations of the Court were directed first to the causes of the collision, and secondly to the steps taken after it. As to the former, any blame due naturally fell on those officers who were on duty on deck; but the loss of the ship as a result of the damage arose from a defective design for which they could in no way be held responsible.

As regards the causes the verdict declared that these were (1) the high rate of speed of the squadron in a fog; (2) the absence of Dawkins below while the change of formation was still incomplete; (3) the unnecessary reduction of speed by the *Vanguard;* (4), (5) and (6) the increase of speed, sheering out of the line, and absence of fog signals on the part of the *Iron Duke*. Of these (1) and (3) seem contradictory; but even if the *Vanguard*'s reduction of speed did contribute to the collision, that was only because the *Iron Duke* was out of station as well as out of sight. By what process of reasoning the Court arrived at (2), there is absolutely nothing in the evidence to show. But in the case of (4), (5) and (6) the evidence was abundant and clear. Yet nobody in the *Iron Duke* was brought to trial.

From a modern seaman's standpoint it is hard to avoid the conviction that in the conclusions at which the Court arrived in their second line of investigation — that is to say, the steps taken after the collision — the members were influenced by their early training and service in wooden ships, all being old or middle-aged officers, some of whom had never so far served in an iron-hulled vessel. In the fleets in which they had passed their careers, collisions never led to founderings, because the point of impact was always above water, and submerged injuries were confined to small perforations easily plugged, such as shot holes, or to strained timbers consequent on grounding, of which the leakage could often be kept under by hand pumps.

Brought up under these experiences of a vanished past, it would seem that the professional outlook of at least a majority of the nine members of the Court had failed to move with the times when they placed on official record their opinion that '*the foundering of H.M.S.* Vanguard *might have been delayed if not averted by Captain Dawkins giving orders for immediate action being taken to get all available pumps worked instead of employing his crew in hoisting out boats. And if Captain Dawkins, Commander Tandy, Navigating Lieutenant Thomas, and Mr. Tiddy, Carpenter, had shown more resource and energy in trying to stop the breach from outside with the means at their command, such as hammocks and sails.*'

The first of these accusations was directly opposed to hard facts. The Court ascertained from Constructor's evidence that the five hand pumps of the *Vanguard* working together could eject water at the rate of a ton and a half a minute. But they never attempted to find out how fast the water was coming in, or they would have discovered it was at a rate of about 50 tons a minute, since the ship's displacement increased by quite 3500 tons in the seventy minutes between the collision and the foundering. It was very far indeed from the truth, therefore, to suggest that the loss of the ship could have been 'averted' by the use of the hand pumps, or even 'delayed' by more than three minutes. Moreover, to later generations of naval officers, with all their modern knowledge regarding submerged injuries to steel hulls, it sounds almost incredible that the

The last page of the log of the *Vanguard*, recording her loss. '12.50 Observed HMS *Iron Duke* close to on Port beam. Ditto ran into the ship'. '2 H.M.S. *Vanguard* sank stern first heeling over to Starboard in 16 fathoms. All hands saved'. It is signed by Dawkins and the officer of the watch. This log was saved with the ship 's papers. (Public Record Office, ADM. 53/10409)

Court Martial proceedings concerning the loss of the *Vanguard* in progress. (NMM. 58/4302)

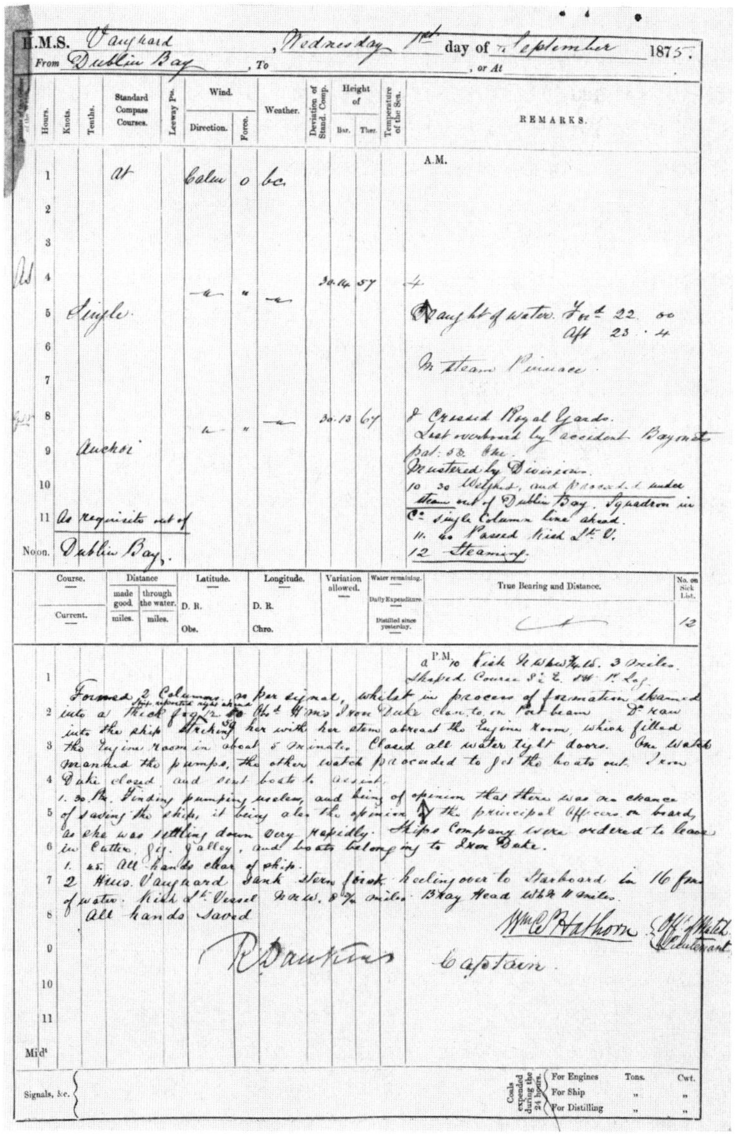

H.M.S. Vanguard, Wednesday 1st day of September 1875
From Dublin Bay, To , or At

A.M.

Noon. Dublin Bay. 12 Steaming.

P.M.

Formed 2 Columns as per signal, whilst in process of formation steamed into a thick fog. 12.50 Obs^d H.M.S. Iron Duke close to on Port beam. D^o ran into the ship striking her with her stem abreast the Engine Room, which filled the Engine room in about 5 minutes. Closed all water tight doors. One Watch manned the pumps, the other watch proceeded to get the boats out. Iron Duke closed and sent boats to assist.
1. 30 P.M. Finding pumping useless, and being of opinion that there was no chance of saving the ship, it being also the opinion of the principal Officers on board, as she was settling down very rapidly. Ships Company were ordered to leave in Cutter, Gig, Galley, and boats belonging to Iron Duke.
1. 45. All hands clear of ship.
2 H.M.S. Vanguard sank stern first heeling over to Starboard in 16 fms of water. Kish Lt Vessel N.N.W. 8½ miles. Bray Head W.b.S. 11 miles.
All hands saved

W^m C Hathorn, Off^r of Watch, Lieutenant
R Dawkins, Captain

Court should have propounded the theory that the pressure of such a terrific inrush could have been effectively stopped by hammocks and sails; and the fact that such a notion never entered the heads of Dawkins and his officers shows that their understanding was far nearer to the realities of the emergency than that of their judges.

Nevertheless, on the decisions of the Court as to the causes of the collision and the action taken thereafter, five officers were convicted of blame contributing to the loss of the *Vanguard* and professionally ruined. Dawkins was severely reprimanded and dismissed his ship, never being employed again; while the Commander, Chief Engineer, Navigating Lieutenant and Carpenter were all reprimanded in varying degree and never promoted to any higher rank. The verdicts met with very little approval among seamen, and much openly voiced suspicion of a miscarriage of justice among the general public, whose views found reflection in a music-hall song greatly in vogue at the time. But the Admiralty endorsed the findings — except as regards their reference to the speed of the squadron — and confirmed the sentences imposed. They went further with more evident justification. For some never-explained reason they brought nobody in the *Iron Duke* before a Court Martial, but though expressly exonerating Hickley from all censure they dealt summarily with the lieutenant on watch by dismissing him from his ship as being mainly responsible for the collision, as indeed he was. The end of this unfortunate young officer was not long in following. After a spell on half pay he was appointed abroad and died of fever. In his final delirium he imagined himself once more on deck in a fog and the very last words that passed his lips were '*Vanguard* ahead!'

H.M.S. *Iron Duke* dry docked for a refit during her first commission on the China Station during the early 1870s. Looking down on to the ship it can clearly be seen that the picture was taken before she was fitted with a poop. (NMM. AAK.3)

To pass now to the record of service of the *Iron Duke,* the last of the quartette to be built and the only one turned out by a Royal Dockyard. The keel of this battleship was laid in April 1868 at Pembroke, which came next after Chatham among the national yards in receiving the plant required for iron-hull construction. She was launched in March 1870 and except for a hinged rudder differed in no way from the others in design and armament. Her engines, by Ravenhill, were an exact duplicate of those made by the same firm for the *Audacious,* but she was half a knot the faster vessel.

This *Iron Duke* was the first of her name, in which she was unlike her sisters. Her career, though long, varied and useful, included no service before an enemy; but she was fated to be rather frequently in trouble, for in addition to sinking the *Vanguard* she was twice in some risk of being lost herself through grounding. She began like the others as a First Reserve Guardship, though in her case this duty only lasted for the summer of 1871; and in September of that year she was selected as the new flagship in China under Vice-Admiral Shadwell, of 'Star Table' celebrity.[1] In sailing for the east it fell to her lot to be the first capital ship ever to use the Suez Canal, recently opened and in those days only 24 feet deep. As all her coal except 50 tons was first discharged, she was taken through by three tugs in three days, yawing less in shallow water than the others with their balanced rudders. But on the comparatively long stage of the voyage which lies between Aden and Ceylon, the deficient bunkerage of her class was demonstrated for the first time with practical emphasis; for in spite of filling the whole main-deck battery with extra coal she could only just get across at four knots, and having too little left towards the end to reach her intended destination of Trincomalee was forced to put into Galle instead, with less than a day's fuel left after being three weeks at sea.

More than three years later she returned home on relief by the *Audacious* and paid off at Plymouth in May 1875. To save tug hire she went through the Canal under her own steam on this occasion, scraping both banks liberally and grounding four times in a four-day transit. Two months after paying off, she was re-commissioned as First Reserve Guardship of the Hull division and joined the squadron for the cruise which ended in her collision with the *Vanguard*; one of the results of which was her own transfer to the East Ireland district to take the *Vanguard*'s place with that ship's former crew. On that duty she remained till July 1877, when she was paid off again for a long overhaul in preparation for a second term as flagship in the Far East.

H.M.S. *Iron Duke* at Malta, outward bound to begin her second commission on the China Station. (NMM. HBE/2)

She sailed for China again in August 1878 as flagship of Admiral Coote, and remained in commission till relieved a second time by the *Audacious* more than four years later; during which period Admiral Coote was succeeded by Admiral Willes, and among her midshipmen was Prince Arisugawa of Japan, who in later years served with distinction in the naval wars of his own country.

The unvarying routine of a very long foreign peace-time commission was broken in the *Iron Duke*'s case by three incidents arising from the ordinary risks of navigation. Passing down the Red Sea on her way out — incompletely surveyed in the 'seventies — she found the P. and O. steamer *Bengal* on a reef and after two days' effort pulled her off little the worse. Next year the *Iron Duke* herself, when entering the Shanghai River, was put hard on the dangerous Woosung bar by the local European pilot taking her in, and stuck fast for five days. Very luckily the weather remained calm, but the rushing tides made the boat-work of laying out anchors extremely difficult, and the shallow soundings kept large ships at too safe a distance to pass a hawser, though three British and three foreign pennants were trying to help. Her own engines were useless through sand choking the condenser inlets. Eventually shc was towed off with no serious damage by the *Monocacy,* one of the very light draught river paddlers of the United States navy then serving on the China station.

Nor was that her only 'trip on shore'. A year later again the squadron were making what in mid-Victorian days was an annual visit to the Russian fleet at Vladivostock, and in those fog-laden regions the *Iron Duke* struck a rock near the west coast of Hokkaido. Once again the universal brotherhood of the sea was manifested by the speedy arrival of three French, one Russian and one Japanese ship from Vladivostock and Hakodate, all offering assistance, with the French and Russian station Admirals; but a British corvette pulled off first the *Iron Duke* and then one of the Frenchmen, which had struck also in trying to aid the British flagship. This time the *Iron Duke* was considerably damaged and spent a month in dock at Hongkong. Moreover, a Court Martial followed (as the ship was not in charge of a pilot) by which the Captain, Staff Commander and officer of the watch were all reprimanded and admonished to be more careful in future. Incidentally the occasion had an interest in being apparently the very last on which a British capital ship hoisted out and in any of her heaviest guns with her own resources.

On returning home she paid off in March 1883 and after a very long refit (including new boilers) commissioned in 1885 for the Channel Squadron. In that she remained till 1890, when her square rig was removed and she reverted in her old age — like the *Audacious* and *Invincible* — to a First Reserve Guardship, the duty for which her maiden pennant had been hoisted nineteen years before. Never being appropriated for use a harbour hulk, however, after the fashion of those other two, the *Iron Duke* was sold in May 1906 to Galbraith of Glasgow, and, although the last of her sisterhood to be launched, preceded the *Audacious* — which had been the first — to the shipbreaker's yard by no less than twenty-one years.

H.M.S. *Iron Duke* as she appeared after her refit of 1885. Apart from a new funnel, she had now been fitted with a poop, together with light armament added to her topsides, and torpedo net defence. (NMM. AAK/2)

CHAPTER SIXTEEN

THE THREE BRITISH ARMOURED CORVETTES

H.M.S. *Pallas*, an artist's impression of her appearance as originally completed. (NMM. 58/2644)

In describing ships forming the Black Battlefleet of 1870 mention was made more than once of the classification problem confronting the Admiralty when a fresh scale of grades became necessary, through the unsuitability of the old system of rates for the new types of naval architecture being built in the mid-Victorian era. So completely did these differ from their predecessors that the defining qualifications of each rate had ceased to exist as a standard of values, and bore no sort of descriptive relation to the changed order. Nevertheless, in the absence of any new suggestions for nomenclature, several of the long-established general terms — such as frigate, corvette and sloop — were retained for convenience, in spite of the fact that they were now applied to ships having no resemblance whatever to the vessels formerly classed under the same denomination, nor any similarity of duties in war. Ignoring the special tactical functions of the novel designs as a useful guide to a decision (though these differed in no respect from the special functions of the ships they now rendered obsolete) and finding no satisfactory grouping by armament practicable in the heterogeneous assembly of models to be sorted out, the Admiralty fell back on a scale of tonnages as the only method of division. The term ship-of-the-line dropped out entirely for no other reason than because the much more powerful armoured vessels now forming the line of battle had only one gun-deck and therefore must be called frigates. And as by convention the next thing below a frigate in size must be a corvette, that designation was conferred on three armoured ships of comparatively small displacement, the *Favorite, Pallas* and *Penelope*; although these had practically nothing in common with each other except a kind of third-grade tonnage, and not even that in common with their very numerous unarmoured contemporaries classified as corvettes, which were far smaller again. But except in their case the term 'armoured corvette' was never used in the British service, nor presumably ever will be again.

In the revolutionary transformation of the 'sixties this little group[1] came into existence holding a position analogous to that of the 50-gun ship of the Napoleonic age, midway between the true capital ship and the true frigate. Though only corvettes in name they were more than a match for any unarmoured vessel, even if a First Rate by the old rank. But they were less than a match for the heavier-armoured types representing the true capital ships of the new order. Having ill-defined tactical or strategic functions they might never have been built but for a temporary phase of belief at Whitehall that all future fighting ships would be armoured irrespective of size, whether intended as squadron units or cruisers. In the details of their respective building processes all three differed; for whereas the *Favorite* was one of the wooden hulls converted to armour from an unarmoured earlier beginning, the *Pallas* was a wooden-armoured hull designed as such from the first and the *Penelope* an iron-armoured hull designed as such from the first.

The term 'corvette' had a French origin to denote a vessel corresponding to a large ship-rigged British sloop; and having been retained to distinguish some of these which fell as British prizes was adopted in later years as suitable for the increasingly numerous British type coming between the Sixth Rate frigates and the medium tonnage sloops. As used in the beginning it was never applied to a vessel with a main deck armament; but that rule became disregarded in the middle of the century, and thenceforward the scale of armament next to a small frigate classified a vessel as a cor-

H.M.S. *Favorite*, at anchor in Plymouth Sound, the only known photograph of this vessel. The 'bulwark gates' which permitted a degree of end on fire from her battery can be seen just forward of her funnel, and the after one just abaft her mainmast shrouds. (NMM. HGF/1)

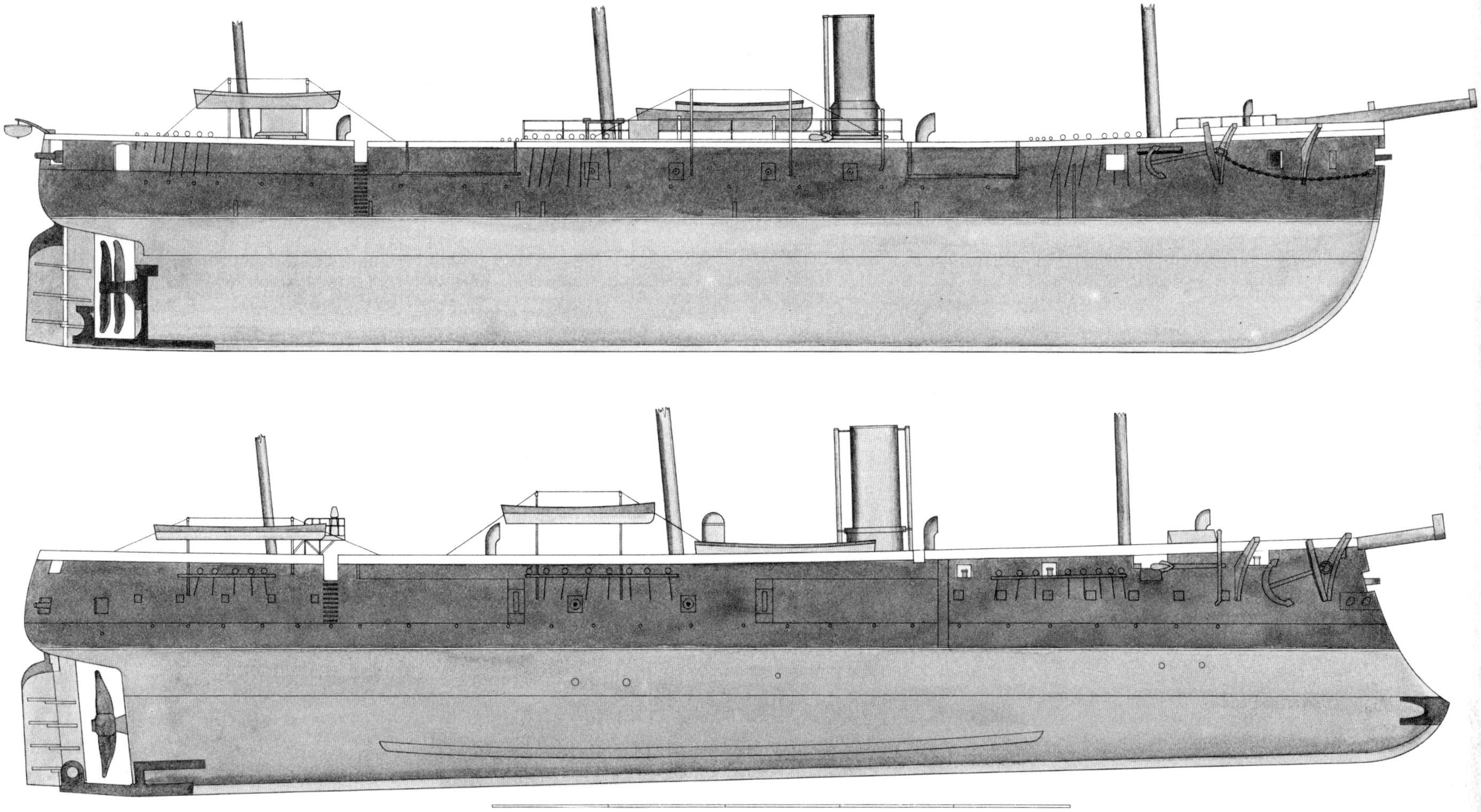

Hull profiles of H.M.Ships *Favorite* (top) and *Pallas*. (Diagrams to scale based on official plans.)

Diagram illustrating one of the bulwark gates outside the battery corners of the *Favorite*, as opened to allow a corner gun to fire through a battery armoured door.

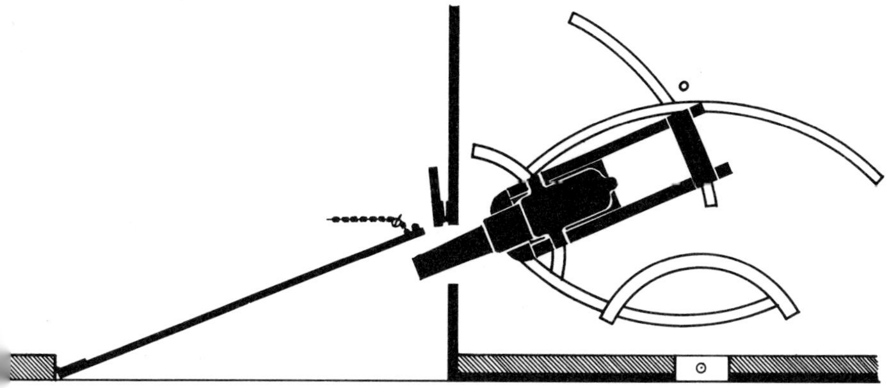

vette whether the guns were carried above or below. The *Favorite* was first designed as one of a class of large 21-gun corvettes which did answer to the old description of a vessel with only two decks, an upper for armament and a lower for berthing quarters; in which respect she differed from the *Pallas* and *Penelope*, which had three like the armoured capital ships. Her keel was already laid (in 1860) when the French programme for an entire new armoured fleet, issued in 1861, caused a wholesale and immediate change in British construction whereby nearly all ships building or projected at the moment — including the *Favorite* — were converted to armoured design.

As converted the *Favorite* was almost exactly half the tonnage of the smallest sized capital ships of her day. She steered extremely well under steam or sail, and rode easily when lying-to, but rolled heavily and shipped a lot of water. Structurally she was a central battery ship on a reduced scale (though that was not yet officially recognized as a special type of vessel when she was launched), but differed from the larger examples in mounting all guns on the upper deck. Her stem was without either ram or knee, and her stern semicircular in plan and nearly vertical in rise from a very low counter, with no exposure of the rudder or screw aperture. She always carried a topgallant forecastle, but never a poop. All ranks and ratings, including the captain, berthed and messed on her one internal deck; but for only 250 the accommodation was very ample.

The *Favorite*'s armament always consisted of eight 7-inch rifled muzzle-loaders and two 6-inch as chase guns (officially designated 64-pounders), the whole ten weapons being wooden mounted. The 7-inch ordnance was carried on the upper deck broadsides in a central 'box' battery covered by a spar deck and enclosed at the ends by transverse armoured bulkheads, through which a fore-and-aft passage was provided each side by armoured doors at both ends. By means of a contrivance unknown in any other British ship each of these four doors was made available as a second port for a corner gun in the battery, so as to increase its angle of fire towards the fore-and-aft line. For that purpose the sections of the lofty bulwarks immediately contiguous outside the four corners of the battery were made movable by swinging on hinges like gates 20 feet long and 6 feet high. To fire a gun through a door the masking bulwark at that corner was swung in and formed a temporary recessed embrasure; though such a device was obviously only practicable where upper-deck weapons were concerned, and proved questionably useful even for them. As protection the *Favorite* carried a complete covering of armour as high as the upper deck from bow to stern, like the early capital ships which were converted at the same time; with a maximum thickness of 4½ inch and in her case the special peculiarity that it extended still higher amidships to screen the 7-inch battery though not the chase guns.

The machinery of the *Favorite* was an ordinary direct-acting set made by Humphreys and Tennant, but it was a peculiarity that among the whole of the single screw-vessels in the British armoured sea-going fleet of 1870 the cylinders in her case and the *Lord Warden*'s alone lay on the starboard side of the shaft instead of the port, with the usual position of the condensers reversed to correspond. They drove a Mangin propeller that could neither hoist nor disconnect, wherein this ship was unique in the battlefleet. Four boilers supplied steam at 20 lb. pressure. Her bunkers had a capacity of 350 tons; a rather unusually large stowage in proportion to displacement, being more than a ninth of it.

Though unlike each other in nearly all other points of design, the three armoured corvettes were uniform in spar and sail dimensions except as regards bowsprits and head sails. All three were ship-rigged, but the masts were differently spaced in each and the lower rigging set up under a different method, the *Pallas* alone retaining the old combination of external chains with wooden deadeyes. As the *Favorite* had no chains her deadeyes were carried on the bulwarks. None of the three were fast sailers as compared with unarmoured corvettes, but whereas the *Favorite* and *Pallas* were very weatherly and handy under canvas the *Penelope* was leewardly and almost unmanageable when close hauled. Being the smallest, though with an equal canvas area to the others, the *Favorite* moved the fastest, and had she carried a hoisting propeller would have cut the water very well. But handicapped by screw drag her best record was only 10½ knots.

The corvette sail area was of course considerably less than that of the battleships, but the difference lay much more in the courses and topsails than in the topgallants and royals, which had the same hoist though rather less spread than the corresponding sails in the less heavily sparred armoured ships of capital rank such as the *Caledonia*, *Bellerophon* and *Defence*. The bowsprit ran in on the topgallant forecastle. No flying jib was carried.

The boat establishments of the British battleships of 1870 varied scarcely at all whatever the complement of men might be, so that vessels with 700 ranks and ratings or with only 450 both carried three boomboats, three double-banked davit boats and three single-banked, and a dinghy. But no such uniformity existed among the corvettes, each of which had a boat establishment peculiarly her own. The *Favorite*'s corresponded to that of the contemporary unarmoured corvettes, and comprised one small boomboat — which was a 30-foot twelve-oared pinnace — two 25-foot ten-oared cutters, a six-oared jolly boat, two gigs and a dinghy. The pinnace was stowed on the spar deck covering the battery, and the davits for the single-banked boats stood at the waist instead of abaft the main rigging on account of the bulwark gates mentioned above.

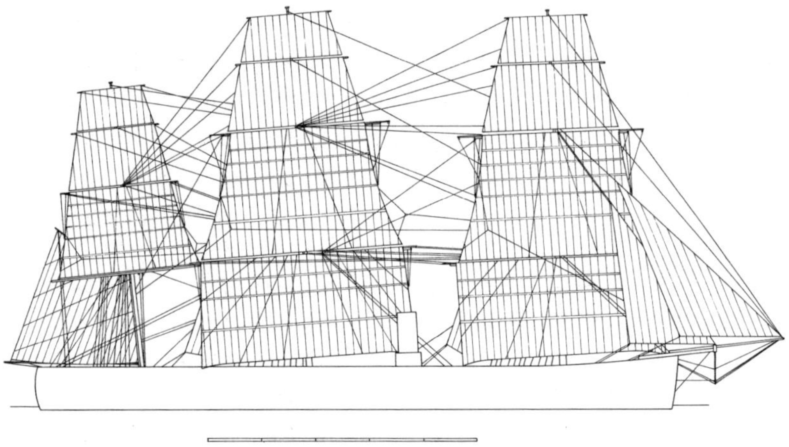

Sail plan of H.M.S. *Favorite* (from the original plans).

The ground tackle of the *Favorite* comprised the usual seven anchors in sizes proportionate to the ship by British naval standards of the period; that is to say two bowers and two sheets of 68 cwt. (iron stocked), one stream of 24 cwt., one kedge of 10 cwt. and one of 6 cwt. Her chain cables were 2-inch, with a 1-inch stream; her hemp cables one 17-inch and one 10-inch. She carried a patent capstan and two wooden catheads each side, the second being added when it was found that the bower anchors partially masked the arcs of training of the bow gun as stowed by original plan.

The *Favorite* was laid down at Deptford in 1860 as one of the *North Star* Class[2] of large wooden unarmoured corvettes, but ordered to be converted to an armoured design in May 1862, when the French decision to build a very numerous armoured fleet forthwith created an urgent situation among maritime Powers. Like all the converted ships, however, she was only regarded as a temporary unit of force to keep abreast of French expansion till enough iron-hulled armoured tonnage could be put afloat; and having answered that purpose was shelved with all the rest when only some ten years old. Launched in 1864 and ready for sea by the beginning of 1866, she commissioned at Sheerness under Captain Shortt in February of that year for the North America and West Indies station, where she formed a useful addition to the British squadron during the rather critical diplomatic period following on the United States Civil War. Three years later she returned home to pay off, and after a lengthy refit relieved the *Repulse* as First Reserve Guardship on the east coast of Scotland in 1872. On that service she remained employed for four years more, being paid off at Portsmouth in 1876 and laid up till sold to the breakers in 1885. Her name descended from a French prize and has never been revived.

The *Pallas* was the next to be launched. She was one of the only three British vessels purposely built to carry armour on a wooden hull (the remaining eight wooden ships with armour being all transformations) and less than half the size of the other two, the *Lord Clyde* and *Lord Warden*. Of that sister pair the *Pallas* was in many respects a dwarf or 'jackass' reproduction, showing the same extreme length of solid oak ram, the same form of stern, and the same pronounced slope of keel line. Ramming tactics having come into favour her design aimed more at quick turning powers under steam than in any previous model, and she mounted the smallest gun armament among British armoured ocean-going types. Her dimensions were 225 feet between perpendiculars with the comparatively great beam of 50 feet giving a larger ratio of breadth to length than any vessel in the armoured classes except the *Zealous* and *Repulse*. And in comparison to tonnage her draught was the heaviest in the whole Navy — greater in actual figures than in many capital ships — for though she only drew 18½ feet of water forward, that deepened to as much as 25 feet aft. She had three decks (like the battleships) but neither poop nor topgallant forecastle, and her bulwarks stood only 4 feet high, instead of 6 or 7 feet as in all the rest of the Navy large or small at the date of her launch. As originally completed her armament was all on the main deck, but two upper-deck chase guns were added at a later period. Officers and men berthed on the lower deck as in other wooden-hulled ships, except that the captain's quarters lay on the main deck right aft. For steering, a treble hand wheel stood on the quarter deck and another below. She was extremely handy under steam or sail, besides being an excellent heavy-weather seaboat and a steadier gun platform than many larger ships, being one of the few wooden hulls with bilge keels.

In her ordnance the *Pallas* underwent several changes. On first going to sea she only carried six guns, all of 7-inch calibre, and all on the main deck; two being Armstrong breech-loaders at bow and stern, and the other four rifled muzzle-loaders mounted in a central battery with four deeply recessed embrasures to increase the arcs of all round fire, each gun having an embrasure port to which it could be traversed as well as a port on the broadside. This armament, however, was soon recognized as inadequate for a vessel of her tonnage, and a few months later four 8-inch muzzle-loaders were substituted for the 7-inch in the battery and two 5-inch Armstrongs added as chase guns on the upper deck. Later still all four Armstrongs were replaced by 6-inch muzzle-loaders, classed officially as 64-pounders. In the end therefore she was armed with four 8-inch and four 6-inch. The latter calibre were all chase guns, but being mounted on two decks could all fire on either beam.

A saluting battery of small brass howitzers was carried on the quarter deck. Her complement was 250 ranks and ratings. Her armour only reached as high as the main deck, except amidships where it was carried up to protect the central battery and here it had its maximum thickness of 4½ inches. It was a peculiarity of its attachment to the body of the vessel that as the belt was not overlapped above the hull structure its upper edge formed a ledge or shelf running round the ship.

The machinery of the *Pallas* has a special historical interest as being the first in the British Navy to be designed to use steam on the compound principle, though the method of applying the varying pressures differed from that adopted in the engines of all new ships in the service built after the middle 'seventies. The 'reeving' of the steam was the same; that is to say from boilers to high-pressure cylinders, thence to low pressure and thence to condensers; but whereas in the later machinery designs the high-pressure cylinder operated one crank and the low another, in the *Pallas* each crank had both. By an arrangement known as 'Woolf's tandem', the two high-pressure cylinders were rearward extensions in reduced diameter of the two low, so that a high and a low piston worked on the same rod, as can be seen in the sketch attached. On the whole this type of machinery answered satisfactorily and economically in fuel; but it suffered from the drawback of requiring much space athwartships — which was so constant a problem for the designers of horizontal marine engines — and though in making those of the *Pallas,* the contractors, Humphreys and Tennant, overcame this by trunking the front side of the low-pressure pistons, such a remedy only sufficed in a hull of great beam compared to horse-power and was not applicable in the finer lined models of a dozen years later.

The screw was a four-bladed non-hoisting Griffiths, but it was found that she steamed just as well, and sailed better, with two blades permanently unshipped. Four boilers supplied steam at 30 lb.; and though her bunkers only stowed 260 tons of coal (90 tons less than the *Favorite)* her economy in consumption gave her nearly as great a radius of steaming action.

Except that she carried a flying jib the spar dimensions and sail area of the *Pallas* were identical with those of the *Favorite* already described; but the main mast was stepped on the forward side of the engines instead of abaft, which gave a different spacing between masts. Though her bluff lines and heavy screw drag were against fast movement — her best record under canvas being 9½ knots — she was very stiff, very weatherly when close hauled, and very handy in tacking and wearing. It was reported of her by one Admiral that in beating to windward she was the equal of a 36-gun frigate; that is to say on a level with the best.

Her ground tackle was on approximately the same scale as that of the *Favorite*, though with 74 cwt. bower and sheet anchors — and like that ship she carried two catheads a side — but the sheet anchors were stowed close abaft the bowers instead of at the waist. Her boat allowance, however, was much more liberal, comprising two 32-foot twelve-oared pinnaces, two 25-foot ten-oared cutters, two gigs and a dinghy. After her first commission a 25-foot steam cutter was added.

The *Pallas* was built at Woolwich, where her keel was laid in October 1863. She was launched in 1865 and commissioned next year at Portsmouth by Captain Connolly for the Channel Squadron, to which she belonged for nearly four years, being then transferred for a few months' temporary duty as First Reserve Guardship in the Kingston district of Ireland. After that she was paid off for a long overhaul and in 1872 commissioned for the Mediterranean, where she served continuously for seven years to follow. While on that station she was one of the four British armoured vessels employed to overawe the mutinous Spanish battleships participating in the Communist rebellion at

H.M.S. *Pallas*, as Guardship at Kingston district of Ireland about 1872. (NMM. HBA/1)

H.M.S. *Penelope* as she appeared when completed for sea in 1868. (NMM. HA2/1)

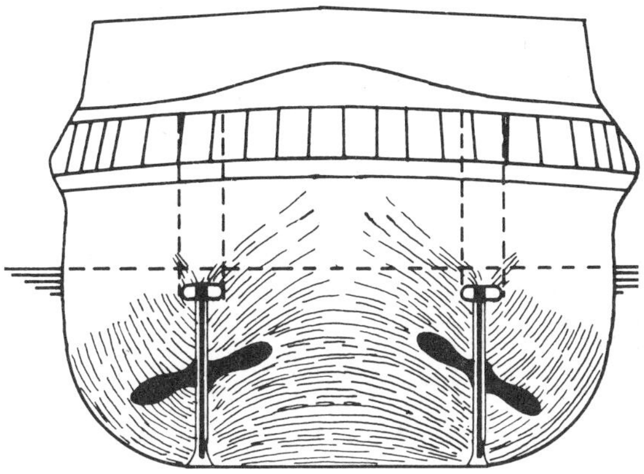

Diagram of *Penelope's* stern, showing central water arch. The vertical dotted lines indicate the screw wells up which the propellers were hoisted when sailing.

H.M.S. *Penelope*, in dry dock about 1912 showing the distinctive design of her hull below the waterline with her double stern posts, rudder posts and rudders. Although she was not the only vessel to be designed with twin screws which could be hoisted, she was certainly unique as far as the Royal Navy was concerned. (Photograph from the Cape Archives.)

Cartagena in 1873, which acted in semi-piratical fashion to the common danger of traffic at sea till driven back and abandoned to British custody. In 1879 she paid off finally; her last captain being the late Lord Fisher, a junior in that rank in those days quite unknown to the general public, but recognized in the service as an officer with a future. Having by that time completed twelve years' sea service and outlasted most of her wooden-built armoured contemporaries the *Pallas* was then relegated to the Fourth Class Reserve till sold for breaking up in 1885, the same year as the *Favorite*.

The *Penelope* was the last of the armoured corvettes to be built, and presented if possible even more individuality of model than the others. Of the three she had the heaviest tonnage and the only iron hull. She was the first armoured vessel with twin screws, and the only twin-screw vessel of any kind ever to carry hoisting propellers. In her proportion of beam to length and her single-decked embrasured eight-gun central battery she was a small-scale forerunner of the *Hercules*, but with the bows of the *Bellerophon*; and in her very square cross-section, shallow immersion and twin screws a small-scale forerunner of the *Audacious* class. When contrasted with the *Pallas* — launched in the year when the *Penelope* was laid down — the two differed so radically as to illustrate the uncertainty prevalent in responsible quarters with regard to the nature of future naval war. No clear instructions in writing for the Controller's department as to her design seem to exist, but the extremely light draught which was its dominant feature suggests that they were under the influence of a commonly prevalent theory of the day resulting from the events of the recent war with Russia, to the effect that the prospective functions of armoured fleets would lie mainly near the land. And it is rather curious to note that although the general unsoundness of such a theory as a guide to the true sea policy of a great maritime Power was afterwards recognized, it did seem to be vindicated by the career of this particular vessel at a subsequent period.

The *Penelope* carried her maximum beam over nearly three-quarters of her length — on a draught of only 16 feet forward and 17½ feet aft at full load. At her bow she carried a pointed ram of medium length and a short knee under the bowsprit for appearances; while her stern was unlike any other ever seen in the British Navy or probably any other in the world. To get screw drive adequate for a hull of more than 4000 tons with the slow-turning engines of the day a single propeller would have needed to be of larger diameter than the shallow draught of the ship permitted and hence the *Penelope* became the first armoured vessel to carry a pair.

And as reports had been received from the single-screwed *Favorite* and *Pallas* that their non-raising propellers spoilt good sailing, a decision was taken to fit the double-screwed *Penelope* with hoisters. It was not a success and never repeated with twin screws. The tail shafts were placed inside the hull instead of outside in the run, each terminating at a separate screw aperture under a hoisting well, with separate rudders on the posts supporting the after sides of their respective banjo frames. As all this, however, not only widened the quarters of the ship's underbody and spoilt her lines, but tended to produce a large dead water space between the rudders, the stern below the counter amidships was hollowed into an arch sloping downwards and forwards to meet the line of keel. A free upward passage of water from under the ship was thereby secured between the screws and rudders. Above the waterline the stern was simply that of an ordinary large frigate. Thus the *Penelope* was double-sterned beneath the water and single-sterned over it; a costly experiment in construction which failed to give her even moderately good powers of sailing. But with a flat-bottomed hull, shallow draught and high freeboard nothing would have done that.

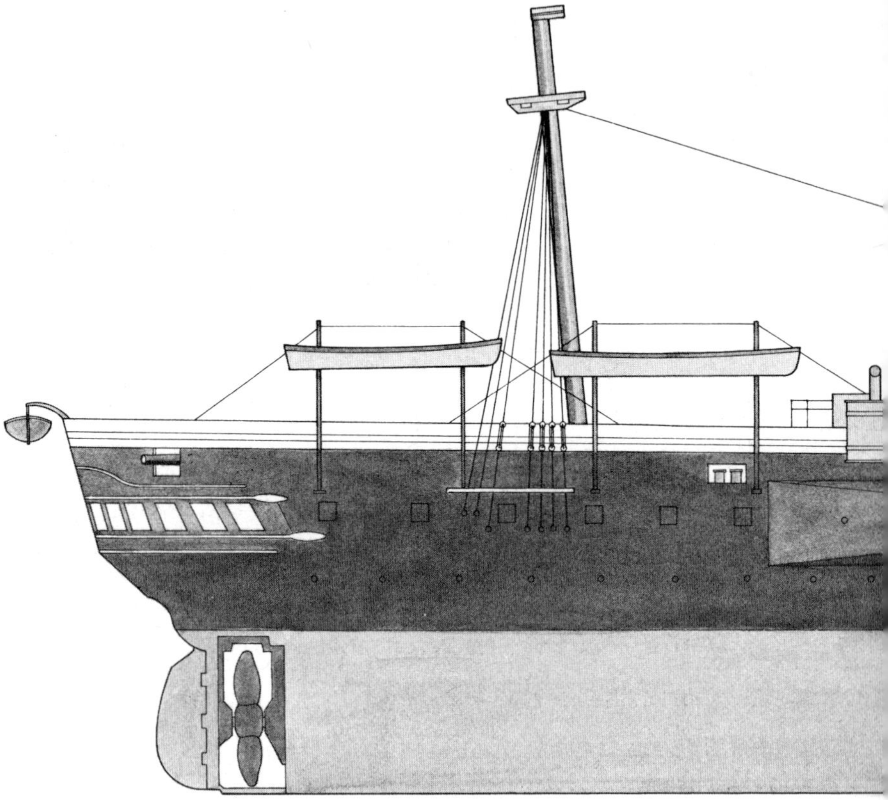

Hull profile of H.M.S. *Penelope*.

On the other hand those very features of design which were detrimental to sailing improved her as a fighting ship. A 'skimming dish' cross-section meant a high centre of gravity and therefore steady gun sights. The *Penelope* could ride with easy motion in heavy seas when many far larger capital ships in the same line were rolling their bulwarks under with closed ports; and the marked advance in that vitally important point which her model initiated came to stay through its adoption on a larger scale, first for the *Audacious* class and thereafter for all battleships British built. Internally her general structure was on the same plan as in all the iron hulled central battery ships; that is to say three decks and a topgallant forecastle, with the men living forward of the battery. On the whole the accommodation was good for the numbers berthed. The officers were on the lower deck except the captain. She had no poop.

For a vessel of her size the *Penelope* was well armed, being the smallest with 8-inch rifled muzzle-loaders when new except the *Pallas*; and whereas the latter only carried four of these the *Penelope* carried eight, all behind armour in a central battery with embrasure ports increasing the arcs of fire at each corner. As chase guns she carried two 5-inch Armstrongs under the topgallant forecastle and a third on the upper deck at the stern, all on truck mountings. Apart from coast-defence craft she was the first British armoured ship to carry the ordnance for which she was originally designed without alteration from beginning to end. Her Quarter Bill stationed a complement of 350 officers and men. As regards defensive provision her protective plating was one of Reed's early examples of thickening the armour of his designs by decreasing its spread. The waterline belt was very narrow. It reached up externally to the main-deck level as in every central battery ship; but the main deck of the *Penelope* lay exceptionally low, with the result that more of the side was left uncovered than in other vessels until the building of the *Audacious* class. The battery was enclosed by 6-inch side armour and 4½-inch transverse bulkheads; the lower strake on the sides being extended forward a short distance outside the battery to prevent a shell exploding near the crowns of the forward boilers which lay very little below the water-level as compared with other vessels on account of the *Penelope*'s light draught. Being iron built she had the great advantage of watertight partitioning.

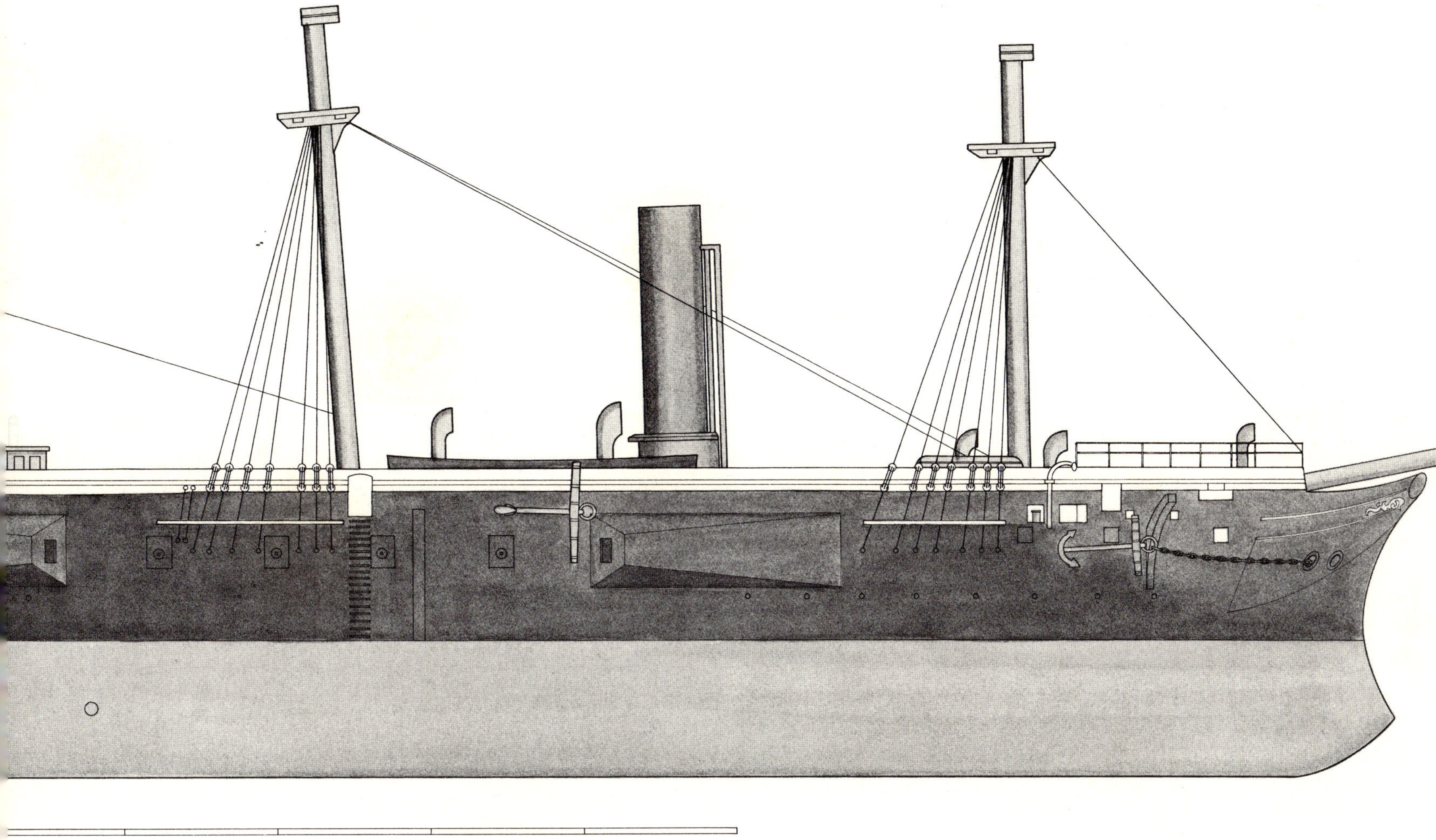

As this ship was the first of any considerable size to receive twin screws, for which Mandslay's firm obtained the contract, those makers were the first to be faced with the problem of laying two sets of large horizontal reciprocating engines (600 horse-power nominal) on a flooring less than 50 feet across. They put the cylinders amidships, with the condensers between the shafts and the angles of the bilges, and economized space athwartships by reducing the stroke to 2½ feet and distributing the steam between three small cylinders for each set instead of two large. That answered well enough for a good drive, but experience proved that to develop a given power, triple-cylindered machinery used more steam — and therefore more coal — than double-cylindered, and twin screws more than single. On both these counts therefore the *Penelope* made heavy demands on her fuel supply, proving in practice to be the most extravagant consumer for her size in the whole armoured fleet; and though her bunker capacity of 500 tons was also the greatest in comparison with any — being actually in excess of that in nine of the capital ships — her radius of action under steam was one of the lowest. Her four boilers worked at 30 lb. pressure (and stood in a separate compartment to the engines as in other iron-hulled vessels).

The uneconomical steaming of the *Penelope* might perhaps have been less disadvantageous for distant service had she been even a tolerable performer under sail; but she proved the worst in the whole Navy except the five-masted battleships. With her flat bottom and high seat on the water she drifted sideways like a crab when braced for close hauled sailing, and though answering helm well enough for tacking or wearing made so much leeway that it was useless to try working her to windward. Nor was she much better running free, in spite of hoisting propellers. Her after lines were not good for speed, and her sail area only equal to the *Favorite*'s though in hull she was half as heavy again. Her best logged hour under canvas only covered 8½ knots.

Though classed as a corvette her tonnage was equal to that of the larger wooden frigates and the more important articles of her equipment proportionate; the bower and sheet anchors being 76 cwt. and the chain cables 2½-inch link. She was the only vessel in the service whose boats (in smooth water) could hold her entire crew; for she carried two launches and a pinnace, three cutters and three gigs like a capital ship, with only the officers and men of a corvette; whereas the boats of the *Favorite* would barely contain one watch.

The *Penelope* was the first iron vessel to be built in the old national dockyard at Pembroke, where she was laid down in 1865 and launched in 1867. Her sea career commenced in the Channel Squadron in 1868 under Captain Marten, but she was transferred in 1869 to the First Reserve Squadron as Guardship of the Harwich district, on which duty she remained without a break for the next eighteen years except for a few months spent on the coast of Egypt during the war of 1882, where she did some very useful service. It chanced that the First Reserve Squadron called at Gibraltar on its summer cruise of that year just when the situation in Egypt was becoming highly dangerous; and as no armoured ship then in the Mediterranean station drew little enough water for possible operations in the Suez Canal the *Penelope* was detached by Admiralty wire to report to Admiral Sir Beauchamp Seymour at Alexandria.

On arrival there three days before the bombardment she was ordered by him to join the *Invincible* and *Monarch* in the inner harbour, whence European refugees were crowding in thousands on board any ships, naval or mercantile, that would take them. For forty-eight hours the *Penelope*'s boats assisted in this wholesale embarkation, until the last steamer had departed on the afternoon of the day before the expiration of the British ultimatum to the rebel army; and that evening at ten o'clock she followed the *Invincible* and *Monarch* out to the position allotted to them opposite the batteries covering the western approaches to the seaport, where on the morrow these three were to form the inshore attacking group under Seymour's personal supervision. As being the lightest in draught of the three the *Penelope* was thrust the closest in when the bombardment opened next morning, and in the course of the day fired 231 rounds against the western defences, receiving her share of attention in return. Among other items of subsequently necessary repair at the end of the exchanges she needed a new mainyard and a new 8-inch gun in place of one damaged by the explosion of an enemy shell in an embrasure port, which wounded the officer of the quarters and seven of the gun's crew, though these were her only casualties.

Soon afterwards, when the British plan of campaign required the temporary seizure of the Suez Canal for the free passage of the transports carrying the army to be landed at Ismailia for the march on Cairo, the *Penelope*'s light draught caused her selection as flagship of Admiral Hoskins, who was in charge of the operation. At the conclusion of hostilities she returned to her duty as a First Reserve Guardship for another five years, being paid off in 1887. After a long refit she was sent out in 1888 to the

H.M.S. *Penelope*, in dry dock about 1912, being stripped ready for disposal. (Photograph by courtesy of the Cape Archives)

H.M.S. *Penelope* as a receiving ship at Simonstown about 1897, with the cruiser *Eclipse* anchored just ahead of her. (Photograph by courtesy of the Cape Archives)

H.M.S. *Penelope* leaving Capetown in 1914 for her final destination, a breaker's yard at Genoa. (NMM. PX.21218)

Cape of Good Hope for service as harbour receiving ship at the Simons Bay station headquarters, and a few years afterwards converted for use as a local naval prison hulk till eventually sold for breaking up at Cape Town in 1912. Hers was a very exceptional case therefore of an armoured ship passing almost directly from seagoing commission on the effective list to permanent moorings as a floating depot, without the almost invariable long years in dismantled reserve awaiting appropriation or sale.

With this account of the only three British ships ever officially classed as 'armoured corvettes' is brought to an end the descriptions of the thirty vessels enumerated as constituting the 'Black Battlefleet of 1870'; the armoured seagoing force launched to take the place of the long dominant chequer-sided ships-of-the-line as prime guarantee for British security at home and influence abroad. Some were still building at that date; others in their first commission; none yet old. But so rapid were the changes in mid-Victorian naval architecture that within less than ten years to follow many were already awaiting sale; others in their last commission; all becoming obsolescent; and although most of those with iron hulls had very long lives still before them — one being afloat to the present day — by far the greater part of such existences were destined to be passed not as guardians of Empire but as floating storehouses, factories or barracks; or even as empty shells in charge of pensioned caretakers. All their successors under construction ten years later were steel built instead of oak or wrought iron, and not one of them carried broadside guns, single screws or square sails.

CHAPTER SEVENTEEN

H.M.S. 'TEMERAIRE', THE GREAT BRIG

H.M.S. *Temeraire* at Malta during her first commission in the Mediterranean when she was painted white. Sharing the picture is the *Minotaur* in dry dock. (NMM. 58/609)

N.B. This and the succeeding chapters did not form part of the original Black Battlefleet Series, but were published later.

The largest vessel ever built with a two-masted sail plan of sufficient scale to enable her to be handled without steam was the British iron-hulled, armoured and brig-rigged battleship *Temeraire*, of 8540 tons displacement and 7500 horse-power, launched at Chatham in 1875. She had the longest lower masts, lower yards and gaffs ever made for the British fleet in all its history, and though she bent no square mainsail her foresail was that fleet's largest square sail, and her main trysail its largest sail ever set fore-and-aft. Moreover, she possessed some quite unique features of a noteworthy character, for, in addition to certain exceptional details in general design, she was the only ship ever built in any country to carry heavy ordnance mounted on the 'Moncrieff' principle though even in her case it was confined to two of the eight guns forming her main armament. The other six were in a main-deck central battery.

As originally sketched her sheer draught proposed a length of about 320 feet, with a midship cross-section of 60 feet, those figures being a close approximation to the dimensions of the *Hercules* and *Sultan,* which in the early 'seventies were held to be the ideal conception of a capital ship with main-deck armament. But eventually the Admiralty selected the general proportions of the new mastless turret ship *Devastation* in preference (which had only 280 feet length to a 60 feet beam) on account of the increasing importance attached at that time to quick turning powers, when ramming tactics were supposed to be the key to victory. As actually built therefore the *Temeraire* had a waterline length at load draught of only 285 feet to a beam amidships of 62 feet; and this ratio of little more than 4½ to 1 was the lowest ever used in a British armoured and masted vessel. It did, as a matter of course, make her very handy, though no heavy-draught ship of that model could ever be really fast. And as the shortening of the original design was nearly all taken off the bow it brought the centre of gravity of armament and armour forward of its first

H.M.S. *Temeraire* at anchor in the eastern Mediterranean during her second commission in those waters. (NMM. 4675(A))

intended position, with a consequent tendency to force her down by the head and make her pitch very heavily in a head sea. At load draught she drew 27 ½ feet forward but half a foot less aft. On the other hand she rolled very little. Below water she was zinc-sheathed. She carried rather a short-pointed ram stoutly armoured and a very wide rounded stern with an outward slope from counter to taffrail like that of a wooden frigate, being the last stern of that type in any large British vessel.

Nor was that her only structural detail of an exceptional sort. In every other ship in the fleet large or small with a poop or topgallant forecastle or both these stood at full height for entrance at upper-deck level but in the *Temeraire* they only rose about 4 feet and to enter either of them it was necessary to use short steps leading downwards to the doors. Again, whereas the bulwarks and hammock nettings of all classes except turret ships were 6 or 7 feet high those of the *Temeraire* were only 4 feet, making it quite easy to see outboard all round, though affording less shelter against wind and sea. This gave her a low appearance in the water, which in combination with her shortness presented a smaller target than that of other vessels of comparable tonnage, turret ships excepted. Her bridge was before the main mast instead of abaft.

Her main deck carried a six-gun central battery and provided the living quarters for all ranks and ratings except the captain, commander and chaplain, whose cabins were under the poop. Forward of the battery and extending to the bows lay the fore mess-deck, where the majority of the ship's company were berthed. This was better than in the majority of vessels of the period both as to space, light and ventilation; and above all in head room. To pass aft from there it was necessary to go through the battery, entered by one of the heavy armoured doors in its forward enclosing bulkhead. This meant a step upward, since the main-deck level in most central battery ships was rather higher inside the battery than elsewhere, to put the guns as far above the water as possible. Moreover, in the *Temeraire* the rise was not uniform, because in her case the battery was divided internally into two parts, of which the forward part mounted heavier guns than the after, and needed more clearance overhead in consequence. No hands messed in the forward division, but some messes stood between the guns in the other. Armoured doors in the dividing bulkhead gave passage between the two, with a second step up in going aft. Exit from the battery towards the stern was through a third pair of armoured doors in its after enclosing bulkhead, leading to the after mess-deck where the marines were quartered. This again was separated by a light bulkhead from the officers' part of the ship, reaching right thence to the stern.

The officers' accommodation was much above the average of that in most of the older armoured battleships. In the *Temeraire,* all officers' messes and cabins were on main-deck level, and the wardroom stood round the stern above the counter, as in a wooden ship-of-the-line, with a few cabins opening from it on one side only, and a full row of square ports on the other side and aft. The remaining cabins with the gunroom, and the engineers' and warrant officers' messes, stood round the spacious half-deck, each with its square iron port opening outwards and backed by a glass sash sliding vertically. Of course sixty years ago no such luxuries existed in the Navy as bathrooms and smoking places. Officers tubbed in their cabins, and midshipmen in the open steerage. Wardroom rank smoked on the starboard side of the half-deck; all others on the port.

Below her main deck the *Temeraire* was divided into watertight compartments like all vessels of iron or steel structure; but as she was devoid of any unusual features in that respect it scarcely merits further description here. Whether she would have survived the flooding of the machinery spaces any better than the *Vanguard* may be doubted, as she had the same defect of a low-lying main deck without either watertight hatches or cofferdams. On the other hand, her machinery spaces had a far better protection against flooding than the *Vanguard*'s by her possession of wing compartments, which did in fact make all the difference on the occasion of her collision with the *Orion.*

It is in coming to the *Temeraire*'s offensive power that the most notable point of interest in her whole design arises, for she was a compromise between two opposing ideals as to the best system of arming a battleship. The *Thunderer* carried all her guns in upper deck centre-line turrets; the *Alexandra* carried all hers at the sides in a double-decked tier; and the *Temeraire* carried some one way and some the other. Both the last two were heavily masted vessels; but to prevent that from masking the fire of the *Temeraire*'s upper-deck centre-line guns they were placed far apart near the bow and stern with the mast between instead of being bunched together on the centre body as in the case of the *Thunderer.*

And when it had been decided to place two of the *Temeraire*'s heaviest weapons in those positions, the opportunity was taken of testing the suitability for naval service of a style of mounting much in favour for fortress use and known as the Moncrieff disappearing carriage. Its particular feature was a complete and continuous protection not only of the gun's crew but of the gun itself while being loaded — or when not engaged — by a mechanism whereby it was kept down in a shelter pit from which it was only raised to fire. As the *Temeraire*'s guns weighed 25 tons this was not a job possible for man handling, but it chanced that just at that time the application of hydraulic pressure to power use was beginning to attract scientific attention, and a device known as the Rendel water ram proved quite suitable here. Two sunken armoured pits, officially designated barbettes, were accordingly built on the midship line into the structure of the *Temeraire*'s topgallant forecastle and poop, with parapets rising about 3 feet above their surrounding decks. These were ovoid in contour, with one end much more widely rounded than the other, and lay lengthwise fore-and-aft, having an internal measurement of 30 feet and an all-round armour protection 10 inches thick for the fore barbette and 8 inches for the after. Such a shape was necessary to allow room not only for the gun when down inside for loading, but for a hydraulic rammer to lie opposite its muzzle almost as long as its bore. The wider end was open topped for the gun to rise and fire over, but the narrower — or rammer — end was roofed with iron sheeting. On the floor of the wider end stood a massive hollow turntable to carry the gun and its mounting, revolving on rollers by power drive in the same way as a turret, and about 2 ½ feet thick, so as to allow enough internal space for the gun-raising hydraulic ram. The mounting placed on it consisted of a pair of very stout cast-iron arms to bear the gun, movable in the vertical plane about their lower or elbow ends by junction to a heavy round cross-shaft or axle lying on trunnion bearings athwart the front of the turntable; in which bearings this shaft could be turned by a massive spur contact with the ram beneath. The turning movement so imported raised or lowered the upper ends of the arms in unison, and those held the gun trunnions that lifted or lowered the gun. In effect therefore it was a combined upper and lower trunnion movement keeping the gun horizontal throughout the course of its rise.

The loading and raising of the gun and the rotation of the turntable were all operated from a cased stand of four control levers, something like those in a railway signal box but on a smaller scale. These were ranged on the rammer side of the barbette immediately above the hydraulic valve box. One worked the ammunition lift, another the rammer, another raised or lowered the gun, and another turned the table. When the gun was up and trained on the target

Right: Longitudinal section of one of the *Temeraire's* barbettes with the gun raised for firing in the fore-and-aft line, or (in darker shade) lowered for loading. The scale will not allow of reproducing full details of the hydraulic mechanism, only the valve-box and main pipes are shown; some other minor details are also omitted. The ammunition hoist is shown as up for loading with the upper chamber of the carrier holding the powder charge. Dotted line in the covered part of the barbette shows the position of the hand turning gear which could be rigged in the event of a breakdown in the engine. Hand loading was also possible, but slow. The figures are shown at the stations in which the gun could be handled with a crew of three. One is at the gunlayer's position, one at the right-hand elevation wheel and one at the control levers whereby the gun was loaded, raised or lowered, and turned, and the ammunition hoist moved up or down; a full crew numbered six, but merely duplicated these duties. As the mounting was unique in the navy, the drill was never included in any book issued to the fleet, but was left to the Gunnery Lieutenant and Commanding Officer.

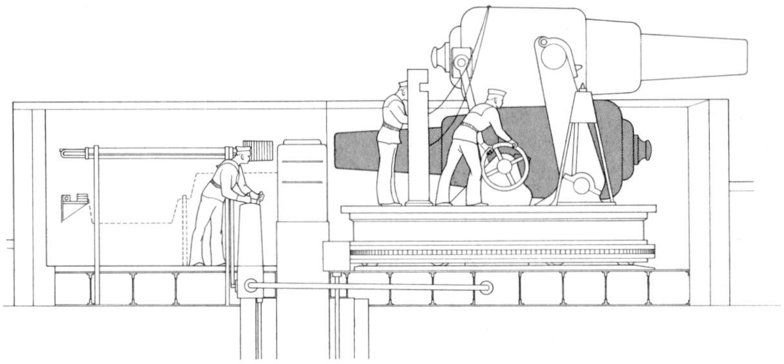

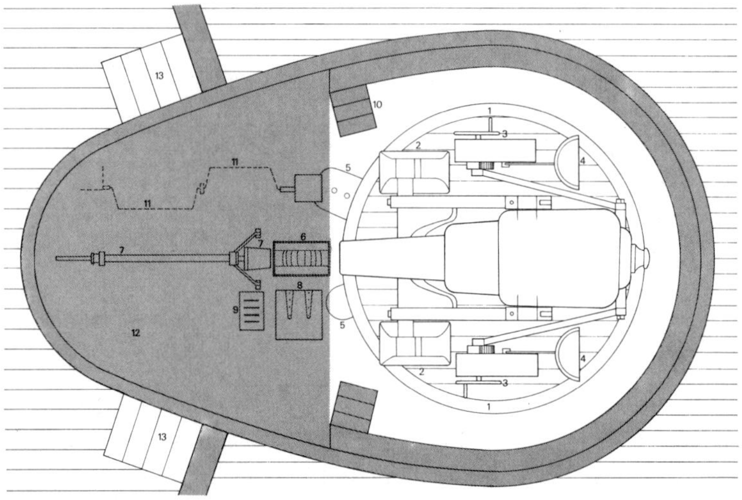

Left: Plan showing the main features of one of the *Temeraire's* barbettes with the portion covered by sheet-iron roof indicated by a darker shade. Minor details and the positions of the fore-sight tripods are omitted. The hydraulic rammer and all the other details under cover are indicated in outline only. Dotted lines show the position of the handles of the hand turning gear when in use. 1. Turntable with gun down in loading position. 2. Bearings for the cross shaft or axle of the gun-raising mechanism. 3. Elevation gear cases containing hand-worked multiple gearing and arcs of elevation graduated in degrees. 4. Gunlayer's sights and hindsight shields. 5. Turntable gear cases with single action power drive on one side and double geared hand drive on the other. 6. Ammunition lift up for loading with powder charge in upper chamber of carrier. 7. Hydraulic combination sponge and rammer. 8. Floor hatch to ammunition well. 9. Stand for levers controlling rammer, ammunition lift, gun-raising mechanism and turntable rotation engine. 10. Steps down into barbette from topgallant forecastle or poop. 11. Winch handles of hand rotation gear (unshipped when not required). 12. Space used for stowage of spare parts. 13. Steps up from deck to topgallant forecastle or poop.

H.M.S. *Temeraire*, the after 11-inch gun in the 'down' position for loading. (NMM. 59/2129)

H.M.S. *Temeraire*, the after 11-inch gun in the 'up' position for firing. (NMM. A.2254)

it was laid for elevation by a bar connexion on each side of the breech to elevation arcs, graduated in degrees and enclosed in gear cases on the turntable, with a double gearing operated by an external hand wheel. No other service of the gun was entirely dependent on manual power like this, though a hand winch could be rigged for turning the table if necessary.

Sighting was on the compound principle, in duplicate for use on either side of the gun. A pair of hind sight stands, each in a 6 feet rifle-proof frontal shield, were placed on the turntable a little in rear of the gun breech, in one of which the gunlayer took his position, while the other was usually occupied by the officer of the barbette. The fore-sights were on long-legged tripods near the front of the turntable. When the gunlayer had set his sight for range and alignment on the target it automatically indicated the proper degrees of elevation for the gun, which he read off and passed to the two numbers at the elevating wheels. And it is worthy of notice that in the very exhaustive report on the bombardment of the Alexandria batteries drawn up for the U.S.A. government by the officers of the American cruiser present to observe, the accurate placing of the *Temeraire*'s shells received special mention. As a general rule the gun was fired by a portable hand battery, but the ordinary lanyard and friction tube of the period could be used as an alternative. A full barbette's crew consisted of six numbers, but it was quite possible to keep the gun in action with three. Each barbette had an arc of fire of 240°.

From under each barbette a cylindrical armoured trunk went right down to below the waterline as a protection well for the ammunition lift, the hydraulic pressure main and the uptake shaft of the turntable rotation drive. Its top lay opposite and nearly under the gun muzzle when the gun was down for loading, so as to bring the ammunition carrier into position for ramming the charge and shell home; but as the diameter of the trunk was considerably greater than the casing of the lift a trap hatch covering the opening was necessary.

All the above mechanism worked without a hitch and almost without a sound. It was interesting to watch from an outside observer's standpoint. At first nothing was to be seen but its surrounding parapet about 3 feet hight. Then quite silently, except for a faint hissing, the gun would suddenly appear coming up from inside and swinging round simultaneously towards the target, in a fashion rather suggestive of an elephant getting on its legs and turning to make a charge. It checked and steadied, while the gun-layer passed the degrees of elevation, and got the reply 'Elevation correct': A momentary pause followed and then came the full-toned bellow of an old-fashioned

Training the after 11-inch gun of the *Temeraire* on to its target. The gunlayer behind the shield directs the two training hands, while the man under the covered portion of the barbette is at the loading position. (NMM. 58/4839)

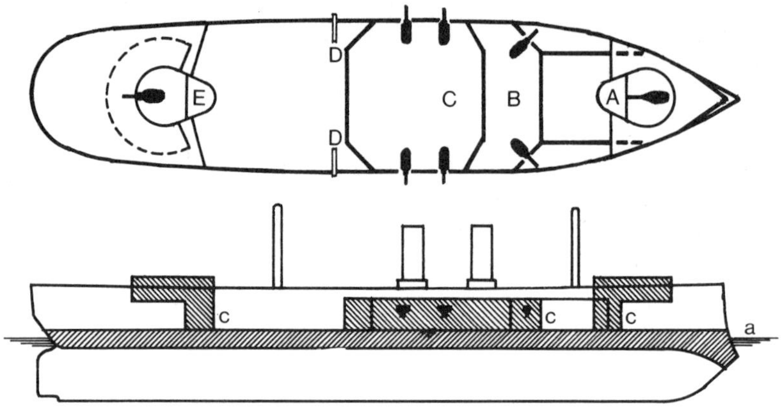

Main gun positions and armouring of the *Temeraire*. Internal armour indicated by broken shading. A. Fore barbette with 11-in. gun down for loading. B. Fore part of central battery with 11-in. guns. C. After part of central battery with 10-in. guns. D. Torpedo carriages. E. After barbette with 11-in. gun up for firing. a. 11-in. plating. b. 10-in. plating. c. 8-in. plating.

muzzle-loader using pebble powder, as it sent the shell away — easily seen in flight when clear of an immense rolling cloud of white smoke — while the hydraulic buffers in the raising mechanism checked the recoil. And almost immediately the gun was turning back and sinking down all in a combined movement to the loading position. In a few seconds it had totally disappeared.

In short the whole design answered completely for everything that the inventor claimed for it, even on a scale such as this where it was adopted for an 11-inch gun weighing 25 tons with a 540 lb. projectile. Yet in spite of its many good points it suffered from two major defects for sea service. One of these was weight, and though that mattered little on land, it was regarded as a sufficient deterrent against further use afloat. The armour of a barbette of this kind with one gun would have been more than sufficient for a circular turret with two of the same calibre. On shore the great majority of coastal batteries only needed protection against frontal attack. The second defect was the necessarily complete absence of any shelter against plunging fire which was inherent in the design. Here again shore batteries often had the advantage of height. But it must be said that in mid-Victorian times, when fighting ranges were measured in hundreds of yards instead of thousands, plunging fire received little attention; and even after breechloaders came into use the Moncrieff system remained in extensive adoption for coast defence. It was never again seen afloat, and thus all officers and men who ever belonged to the *Temeraire* had the experience of serving with a method of handling a gun absolutely unique in maritime history.

The greater part of her armament, however, was carried at her sides in a main-deck central battery of octagonal shape, armoured all round and divided by an internal armoured bulkhead into a forward and an after part — the first of which carried a pair of 11-inch 25-ton guns of identical pattern to those in the barbettes, but mounted on iron broadside carriages and slides and served entirely by hand, with a gun's crew of seventeen and two powdermen. Their ports were deeply embrasured to give an arc of fire from right ahead to the beam. For these the shell weighed 540 lb., and the full charge — in two halves — was 85 lb. of black pebble powder which gave a muzzle velocity of 1320 feet. Only one other vessel in the British service ever carried guns of this particular calibre, though several had guns of the same weight with a lower range and penetration. The after division of the battery contained four 10-inch guns of 18 tons in pairs on the broadsides, with a 400 lb. shell and a 70 lb. charge. These were on the same general type of mounting as the 11-inch on a proportionately lighter scale. The entire main armament consisted of course of rifled muzzle-loaders, since the heavy ordnance in the British services of the 'seventies and early 'eighties belonged entirely to that description. Her complement was 580 ranks and ratings.

At the time when the *Temeraire* was designed and under construction no British battleship was possessed of a secondary armament as now understood; and the very few small guns they had on board were only supplied for boat or field service, not being included in the plans of the vessels themselves. But it chanced that before the *Temeraire* was ready for sea great advances were taking place in various forms of torpedo attack. One of these was a development of the spar torpedo as carried in a steam pinnace; while another was a newly devised steel fish affair, running by itself under compressed air even up to several hundred yards, and developed by a British maker of marine engines named Whitehead. This seemed as if it might be dangerous if used from a boat; and so all British battleships received as an anti-torpedo defence a small number of 20-pounder Armstrong guns mounted on wooden chock carriages. Four went to the *Temeraire* and were put at ports in the waist and quarter deck. At the ends of her first and second commissions several additions and changes were made in this, whereby the Armstrongs were replaced by six 4-inch 25-pounders of the new breechloading kind which was then under adoption by degrees as a substitute for older guns of every calibre in the fleet. These were on Vavasseur mountings. To supplement them a mixture of 6-pounder quick-firing guns and Nordenfelt machine guns to the number of a dozen in all were ranged along the bulwarks. At that her anti-torpedo armament remained for the last six years of her service.

Similarly, it was only as an unforeseen addition to her powers as originally conceived that the *Temeraire* was ever provided with torpedoes to use herself. As first devised they were only meant as a weapon for a steam launch or other specially suitable boat, to be dropped from the gunwale. But the belief in ramming tactics so prevalent at that stage in naval changes led to the anticipation that torpedoes might prove a formidable deterrent to ramming attack as a supplement to battleship defensive equipment. No notion that they might be fired from an underwater discharge had been entertained anywhere, but a form of carriage in skeleton steel on low rollers was submitted by the Whitehead private factory for placing at an ordinary ship's broadside port; from which a torpedo could be ejected by air impulse from as much as 20 feet or more above the water. Two such were accordingly installed at main-deck ports of the *Temeraire* abaft the central battery, as there was no room for them anywhere behind armour; and by this she became the third battleship in British history to include torpedoes in her armament.

As a further defence against torpedo-boat attack by night she was also equipped with a pair of searchlights, another very recent invention of the period; so much so that when she arrived in the Mediterranean as a new ship in 1877 the only other vessel on the station carrying them was the recently arrived flagship. By modern standards these were toy affairs with only 30-inch projectors. To the end of her days the *Temeraire* was only candlelit internally.

As armour protection she possessed a complete waterline belt 11 inches thick amidships tapering to 10 inches at extremities; while above that her fore barbette had 10 inch plating and her after barbette, central battery, and both barbette ammunition trunks 8 inches. Armoured conning towers, however, were not considered essential when she was a new ship; nor was she ever given one at the succeeding period when their adoption became a feature in all armoured classes, very substantial though her protection was in all other respects by the standards of her early days. Needless perhaps to say it would be quite useless now; since the rolled iron plating of mid-Victorian times (however effective against the penetration of contemporary projectiles it proved at Alexandria) had only about one-third of the toughness of modern chemically hardened armour. It was when about half-way through her sea service that the *Temeraire* came under the order for all battleships on the British effective list to be fitted with boom-net defence, which was remaining in force till the end of her time, though nets were abolished throughout the fleet some thirty years after their general adoption.

In her motive power, as in a good many other things, she was among the very earliest battleships to illustrate a marked advance in practical science as utilized for maritime purposes, being third in order of receiving the compound engines which all battleships carried after her. The *Temeraire*'s engines by Humphrys and Tennant were also among the first few with the other novelty of inverted instead of horizontal cylinders, and drove a pair of outward-turning two-bladed propellers. Her trial speed on the mile was 14½ knots, which was about the average standard of her period. Her ten cylindrical boilers worked at what was then considered the high pressure of 65 lb. and her bunker stowed 620 tons of coal. For auxiliary steam purposes she carried the modern improvements of a steam steering engine and steam capstan but never a boat hoist.

The invention of compound engines added so greatly to the distance which could be covered by every ton of coal that it really made the complete disappearance of sails sooner or later inevitable in fighting vessels, if not indeed in all classes of ocean-going craft. At the time when the *Temeraire*'s keel was laid these conditions were not yet existent, for with their older type of machinery none of the battleships then afloat under any flag could cross the Atlantic under steam unaided by sail; and although by the time the *Temeraire* was ready for launching (about three years later) the first two British mastless battleships had in fact appeared — the *Devastation* and *Thunderer* — their total lack of canvas was only considered justifiable by nearly doubling their load of coal as compared with that of their masted squadron mates, and giving them twin screws as an insurance against complete inability to move.

Those two still had the old single-pressure engines, but

in giving the *Temeraire* compound pressure such an added bunker load was avoided, for she could steam nearly twice as far as the older battleships of her own size without it. Moreover, she also had twin propellers. In truth, therefore, she had no practical need for a sail plan of any kind. Yet so strong was the influence of Service custom in Victorian days that though the Admiralty had begun the process of dismasting the battle fleet under pressure they confined it to turret ships where masts masked the arcs of fire. And so the *Temeraire* was provided with a square-cut suiting. But it had to be restricted to two masts to avoid interference with the after barbette, and to make up for the absence of a mizen it was planned on an outsize in measurements for the fore and main. Being a very short ship in proportion to beam she did not need a mizen.

In her height of 169 feet from main truck to deck she was equalled by four other British ships of the time and even slightly exceeded by at least one French. But in spread she outreached them all; and the equality in hoist of those others was due to longer topmasts; for the *Temeraire*'s maintop was 6 ½ feet higher than any other in the fleet. Her fore and main yards were longer than the main yard of the *Victory,* and 6 feet more than those of the mid-Victorian *Warrior* and *Achilles*. Her topsail yards were by 4 feet the longest in the British service, except those of the *Alexandra*, and it was this last vessel also that alone carried upper yards equalling the *Temeraire*'s; and indeed in her case a little exceeding them, for though both ships were originally intended to have the same cut in upper canvas the topgallantmasts of the *Alexandra* and the upper yards of the *Temeraire* were all slightly shortened before actual fitting, leaving both ships with the same upper sail area but on different lines. The *Temeraire*'s had no flying jib, but her jib-boom measured 51 ½ feet, and her trysail gaffs were the longest in the fleet. Her foresail was the largest unit of canvas ever stitched in a British Admiralty sail loft, and her main trysail the largest on a gaff head containing 5100 and 3200 square feet of surface respectively before stretching. The foresail weighed over 2 tons. She carried no stunsails.

All the *Temeraire*'s spars were wooden except her iron lowermasts. All her standing rigging was wire except her chain bobstay, and all her running rigged tarred hemp with the Chatham thread except the wire pendants of the lower and topsail lifts, and of the jib and trysail sheets, and the Naval pattern chain lower trusses.

As boat complement she carried a pair of service pattern 18-oared 42 foot launches on the booms and a 37 foot steam pinnace, and at the davits three cutters, three gigs and a dinghy. One launch was armed with a 9-pounder muzzleloader slide mounted, and the other and the steam pinnace with a 3-pounder quick-firing gun. The cutters carried Gardner machine guns. To avoid interference with the arc of fire of the after barbette all boat davits were forward of the poop. The boomboat hoisting-out purchases were peculiar in being rigged with their masthead pendants and yardarm outhauls on the foremast instead of as in all other ships on the main; the reason being that the fore was exceptionally placed in being the nearer of the two to the boomboat crutches in the waist.

The *Temeraire*'s six anchors were all of Victorian 'Admiralty' pattern, her pair of 6-ton iron-stocked bowers being stowed on the topgallant forecastle abreast the fore barbette, and the 5 ½-ton wooden-stocked sheets in the then usual position abaft the fore rigging. She carried 450 fathoms of 2⅝-inch chain cable, and was the last British-built battleship in which the hawsepipes led in below. The cable holders lay horizontally on the fore mess-deck immediately over the navel pipes and were driven by an engine right down in the flats. For use with a 35 cwt. stream anchor the chain lockers also contained 100 fathoms of 1⅝-inch stream cable. Her hawsers were one 6-inch wire, one 5-inch wire and a pair of heavy hemps.

The Great Brig was built at Chatham, and there hoisted her first pennant under Captain Culme Seymour in 1877 to join the fleet in the Mediterranean; on which station she remained continuously for the succeeding 14 years of her sea-going service, except during the winter of 1887-8 when for a few months she belonged to the Channel Squadron. So much public interest did she attract as a new ship, that in addition to long columns of special notice in the general Press, an entire double page of the *Illustrated London News* — in those days one of the only two picture papers — was devoted to engravings of the vessel and of her equipment.

On arriving out on her station she joined the flag of Admiral Hornby in Besika Bay, about the middle of the 21 months' period during which that open and windy roadstead formed the headquarter anchorage of the Mediterranean Command, in order to watch the Dardanelles throughout the progress of the Russo-Turkish war then being waged. Thence the *Temeraire* and five others went up the straits to Constantinople. That was in 1878. For about a year to follow, the squadron, including the *Temeraire*, was retained in the vicinity of Constantinople to hold a strong position on the political chessboard during the lengthy international proceedings leading up to and including the Berlin Congress. After that they went back to the Mediterranean to assist in taking over Cyprus from the Turks, and the first commission of the *Temeraire* came to an end at Malta in 1881, where she was joined by a new crew from home under Captain Nicholson.

Next year the Egyptian war broke out and she was one

H.M.S. *Temeraire* putting out to sea about 1887 during her second commission in the Mediterranean. (NMM. B.8809)

of the eight battleships that attacked the sea defences of Alexandria. On account of their wide arcs of fire she was placed together with the *Inflexible* as a connecting link between the eastern and the western separated but simultaneous bombardments of the flank defences. A United States corvette happened to be present lying in a central observation position in the offing, the report of whose officers gives a far better description of the action than any published in Britain or elsewhere. Two stated points impressed these strictly neutral spectators, of a sort that somehow sounds familiar. One was the poor quality of the British shells, which often failed to burst. The other was the excellent quality of the British shooting, which never failed to score. And though all were included in this appreciation only the practice of the *Temeraire* came in for individual mention as being that of what the report referred to as evidently a 'very efficient ship'.

She was paid off a second time at Malta in 1884, where again she received a new ship's company sent out from home for a third commission, this time under Captain Domvile. In the course of this commission the *Temeraire* acted for several months as station flagship for the Commander-in-Chief while his own was under a long refit. He happened to be H.R.H. the Duke of Edinburgh, and thus although it was the only occasion on which she flew an Admiral's flag it put her among the few vessels that have done so for the son of a reigning British sovereign.

In 1887 she was ordered home to pay off at Portsmouth, which brought her back to England after an absence of ten years, though not yet to lie idle. She was now to join the Channel Squadron for a change, and the day after her third ship's company went ashore at Portsmouth north jetty her fourth came on board. But even so she spent most of her time in Mediterranean waters, for the greater part of that particular winter cruise of the Channel Squadron went as far 'up the Straits' as Genoa, after the usual routine visits to Vigo and Lisbon. Nor was that destined to end it. She did return to England with the others; but the continual strengthening of the French fleet based on Toulon required a proportionate increase in our own, and the *Temeraire* was despatched out accordingly once more to pass the three concluding years of her service in the same surroundings as the first. She was then under Captain Drummond.

At an early stage in this fourth and concluding commission it came very near to being final before being properly half-expired. The squadron were at sea in close order with the *Temeraire* last ship in the starboard column and the *Orion* last but one in the port, which at two cables between columns put her four points on the former's port bow at short manoeuvring distance. The two were signalled to change stations, and under the instructions of the signal book then in use such movements had to comply with the rule of the road by passing port to port; but placed as these two were that necessitated stopping the *Orion*'s engines on turning to starboard. Both captains were late on the bridge, and when the flagship's signal was hauled down the officer of the watch in the *Orion* left his engines running on putting the helm over, and with full way on had too little room to clear. Fortunately the officer on watch in the *Temeraire* had a quick eye, and seeing collision inevitable instantly gave helm orders for turning towards the *Orion* instead of away from her. Thus when the latter's ram struck the *Temeraire*, abreast her engine room and below her armour belt, her stern was already swinging away from the blow, which glanced off after ripping the outer skin and flooding a wing compartment, but that was all. It was a very close thing to a disaster that might have equalled the famous *Victoria* episode.

Before that happened Gerard Noel had joined the *Temeraire* as her fifth and last commanding officer. By that time she had become the very last of all the innumerable British capital ships which had hoisted topsails in the Middle Sea since Blake smoked out the Tunisian pirates in Commonwealth days. The other seven now making up the station battle squadron were all turret ships without a single square yard of canvas among them. And Captain Noel was very well known throughout the Navy of his day as a cast-iron disciplinarian, of a firm opinion that the handling of a square-rigged ship under sail was the highest of human occupations, and obviously the reason why the world had been created with an abundance of sea room. A long prospect of employment still lay ahead of him, but he was well aware that it contained no promise of a chance for him to cast an eye aloft after his time in the *Temeraire*, and made the most of it accordingly as all hands had reason to know.

Opportunities for manoeuvring under canvas in a squadron where no other ship had any were necessarily confined to occasions when the *Temeraire* was by herself. But towards the end of her service, when the squadron was cruising in the Levant and about to visit Suda Bay in Crete, Noel signalled for and obtained permission from the Admiral to enter and anchor under sail after the others had gone ahead under steam. Suda Bay is a long, straight and not very wide stretch of water, completely landlocked by hills except for a narrow entrance at one end about 4 or 5 miles distant from the anchorage at the top. The wind was steady, but dead foul and too squally for carrying royals. Engines were stopped at the entrance and sail made on the starboard tack shortly after one o'clock to begin the long beat up that lasted till six; during which those enormous yards were boxed fourteen times through the full round of their swing, with all the attendant business on tacks, sheets, bowlines, weather braces, lifts and trusses.

The *Temeraire*'s log of the passage up the Dardanelles. The letters in the 'weather' column during noon and early afternoon indicate snow and poor visibility, and it was 'Too thick to observe bearing'. The narrative of the passage is in the 'remarks' column; note especially '1.30 HMS *Alexandra* (Flag) grounded south of Chanak'. Ballard and his fellow midshipmen had to keep up the deck log of which this is a fair copy. (Public Record Office, ADM. 53/11343)

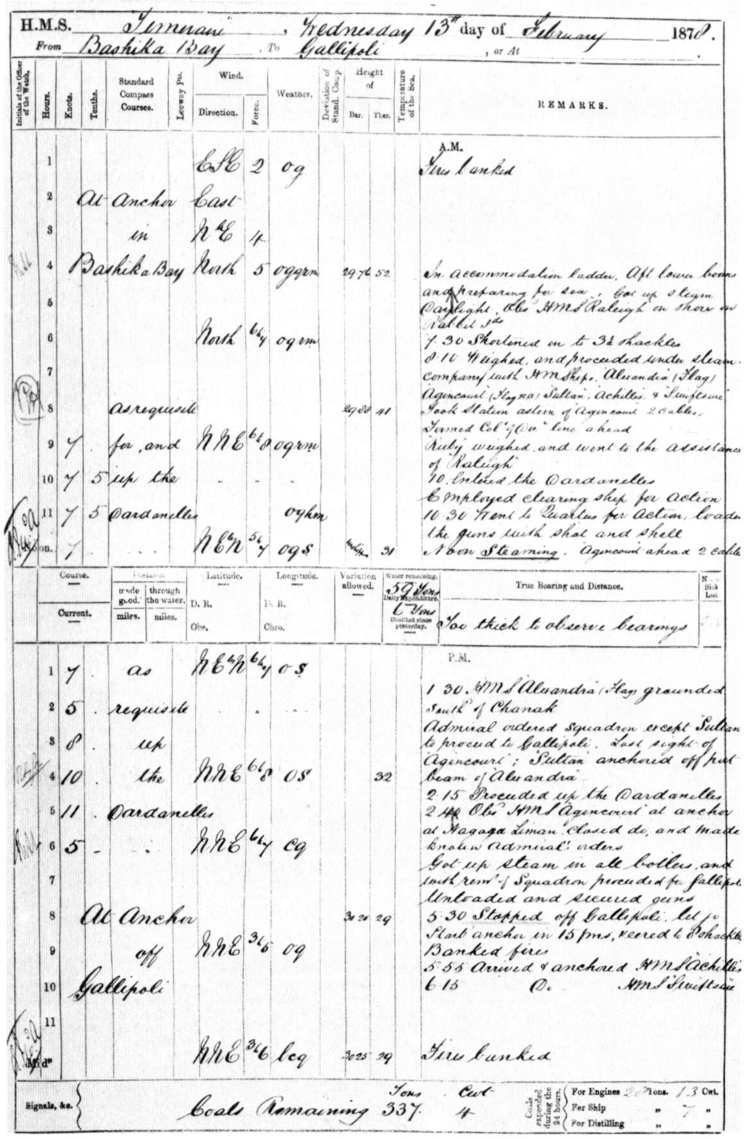
H.M.S. Temeraire, Wednesday 13th day of February 1878.
From Bashika Bay, To Gallipoli
REMARKS.
A.M.
Fires banked
In accommodation ladder, Aft lower boom and preparing for sea. Got up steam
Daylight, Obs HMS Raleigh on shore on Rabbit Isd
7.30 Shortened in to 3½ shackles
8.10 Weighed, and proceeded under steam. Company with HM Ships, Alexandra (Flag) Agincourt (Flag RA) Sultan, Achilles & Swiftsure
Took station astern of Agincourt 2 cables.
Formed Col of Div line ahead
Ruby weighed, and went to the assistance of Raleigh
10. Entered the Dardanelles
Employed clearing ship for action
10.30 Went to Quarters for Action, loaded the guns with shot and shell
Noon Steaming. Agincourt ahead 2 cables
Too thick to observe bearings
P.M.
1.30. HMS Alexandra (Flag) grounded South of Chanak
Admiral ordered Squadron except Sultan to proceed to Gallipoli. Last sight of Agincourt; Sultan anchored off port beam of Alexandra
2.15 Proceeded up the Dardanelles
2.40 Obs HMS Agincourt at anchor at Nagara Liman closed do, and made known Admiral's orders
Got up steam in all boilers, and with rem of Squadron proceeded for Gallipoli
Unloaded and secured guns
5.30 Stopped off Gallipoli. Let go Starb anchor in 15 fms, veered to 8 shackles
Banked fires
5.55 Arrived & anchored HMS Achilles
6.15 Do. HMS Swiftsure
Fires banked
Coals Remaining 337 Tons 4 Cwt

Admiral Sir Gerard Noel, who as captain of the *Temeraire* on her final commission had the distinction of commanding the last armoured ship to enter harbour under sail in 1891. (NMM. 6138)

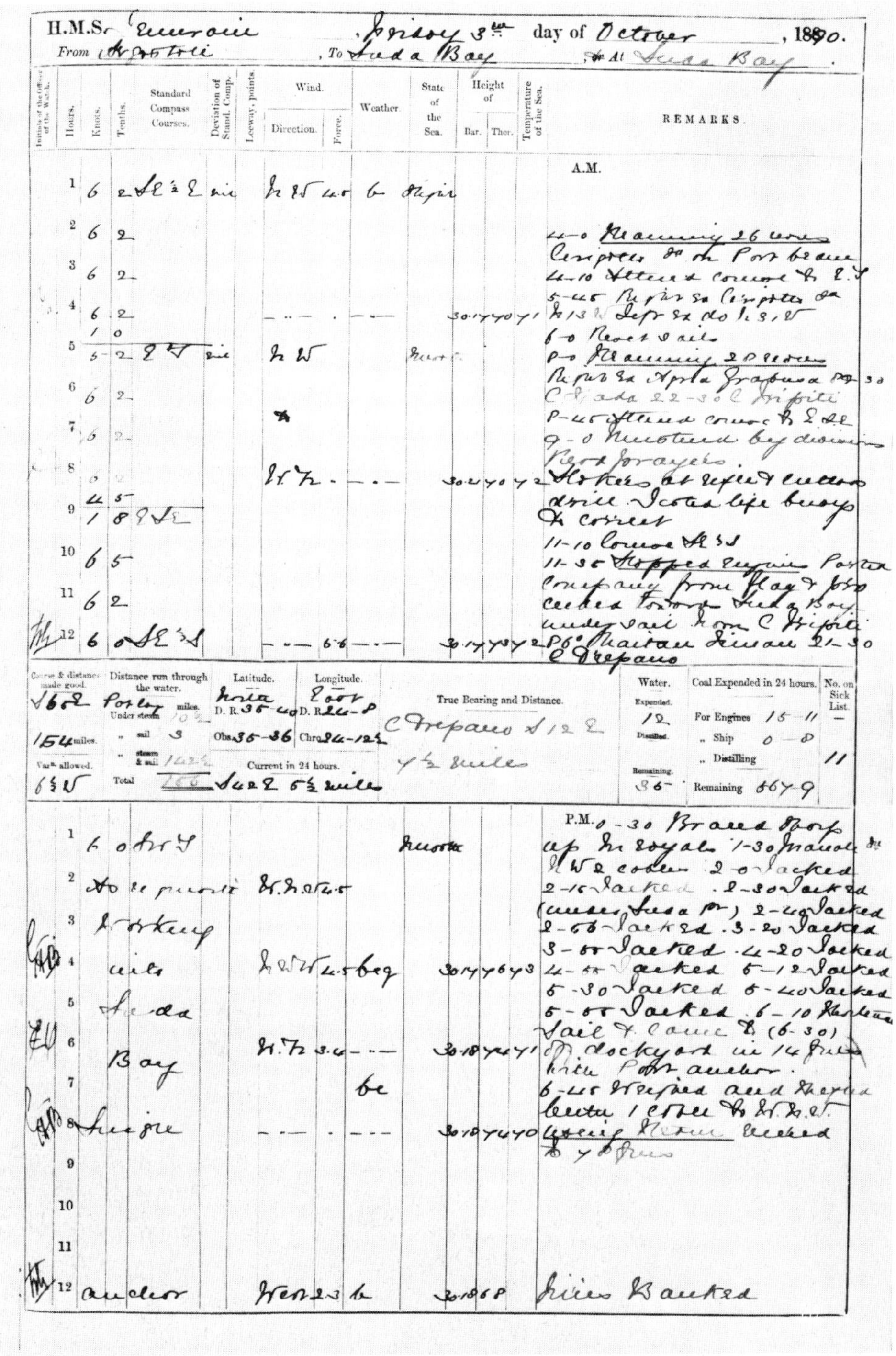

H.M.S. Temeraire — day of October, 1890.
From Argostoli, To Suda Bay, At Suda Bay

Hours	Knots	Tenths	Standard Compass Courses	Deviation of Stand. Comp.	Leeway, points	Wind: Direction	Wind: Force	Weather	State of the Sea	Height of Bar.	Height of Ther.	Temperature of the Sea	REMARKS

A.M.

Course & distance made good. Distance run through the water. Latitude. Longitude. True Bearing and Distance. Water. Expended. Distilled. Remaining. Coal Expended in 24 hours. For Engines. Ship. Distilling. Remaining. No. on Sick List. Current in 24 hours. Total.

P.M.

The log of H.M.S. *Temeraire* for October 3rd, 1890, the last day on which a British battleship ever proceeded under sail alone. In the 'remarks' column is the entry '11.35 *Stopped Engines* Parted Company from flag and proceeded towards Suda Bay under sail', and later the repeated entry 'tacked', concluding '6.10 shortened sail and came to (6.30) off dockyard in 14 fathoms with Port anchor'. In the lefthand margin are the initials of the officers of the watch. 'G.A.B.' is Admiral Ballard himself, who was officer of the watch for part of this historic afternoon. (Public Record Office, ADM. 53/15997)

The rest of the fleet had arrived in the morning and everyone who had not already gone ashore was watching the *Temeraire*, including the Admiral. She presented a spectacular finish to the centuries of such movement under the pendants of the British Mediterranean Command, which was talked of for many years after on that station.

Not many months later came her orders for home to pay off. She had wintered at Salonica with the fleet and was cheered away by all present. Her destination was Plymouth, though hers was a Portsmouth ship's company, and there in the Keyham basin the Great Brig unbent her sails for good and was stripped to a gantline. And there the writer of this book, after four and a half years' experience of her bridge, quarter deck and central battery, passed over her gangway to serve elsewhere in May 1891.

Some plans for modernizing her had been prepared, though they were eventually abandoned on account of expense. But though relegated to the Fourth Reserve she was retained on the effective list in that category for twelve years more before finally going the way of all ships large or small. She became H.M.S. *Indus II* in 1904, *Akbar* in 1915, and was sold to Dutch shipbreakers in May 1921.

CHAPTER EIGHTEEN

THE COAST DEFENCE SHIPS

H.M.S. *Royal Sovereign* as she appeared after conversion into a turret ship, with her bulwarks lowered and her guns cleared for action. (NMM. 58/2263)

H.M.S. *Hecate* at sea during manoeuvres in the mid 1880s. These vessels were claimed to be safe enough in a seaway, but from this picture it cannot be denied that they must have been extremely uncomfortable. (NMM. 58/6094)

Through Coles's campaigns and Ericsson's work in the American Civil War, the first British turret ships had come into existence as the *Royal Sovereign* and *Prince Albert*. The former, although little advanced in construction from the time of her launch some five years earlier as a three-decker of 130 guns, was selected to be cut down for conversion into a mastless armour-clad hull carrying 5 guns on one deck as her future offensive potential. Tactically this made her more formidable, but strategically it reduced her value as a fleet unit of strength very materially; for whereas in her originally intended form she could have lent powerful support to British interests in any sea in the world, in her completed form she could exercise no influence on events at more than two days' steaming from a home base. It did, however, invest her with immense importance inside that limited distance; and though her bunkers only carried coal for three days' steaming, and the longest sea passage she ever made was on a trip to attend a French national fête at Cherbourg, she created a great impression there, not altogether free from a secret discomfort at the existence of such a floating fortress just across the ditch.

Though adopting the general system of armament advocated by Coles, the Admiralty introduced several considerable changes in the details of its installation. Thus, instead of having the cupola shape, the gun shields of the *Royal Sovereign* were circular open-topped 'turrets' in form, partly to increase the likelihood of deflecting enemy shot, partly to allow more space for shell racks inside, and partly to reduce weight by omitting overhead cover as unnecessary. These were trained for direction manually by long-handled turning winches below deck, as ordered through a voice-pipe by the captain of the turret, standing inside it just high enough to look over the top from the side opposite to the gun ports. The sights were on the turret edge and the guns laid for elevation by a degree pointer according to range.

H.M.S. *Royal Sovereign* as she appeared following her launch as a screw vessel at Portsmouth in 1857. Originally laid down as a sailing three-decker, she was lengthened by the stern and converted into a screw vessel in 1855. After her launch, and running steaming trials in 1858, she remained incomplete until selected for conversion into a cupola, or turret ship. (NMM. 58/1036)

The *Royal Sovereign* with the *Scorpion* 'attacking' shore installations during manoeuvres off Dover Castle in 1868. (NMM. 58/3103)

H.M.S. *Royal Sovereign* at anchor in Plymouth Sound as originally converted with a tall funnel. The tall box-like structure right forward houses the ventilator for the heads, and her lion figurehead can be seen just forward of that. Today that same figurehead can be seen gracing the entrance to H.M.S. *Vernon*. (NMM. HBN/1)

H.M.S. *Prince Albert* at Plymouth about 1875 after she was fitted with a flying bridge and charthouse. (NMM. HBO/2)

Ready racks in the turret provided shot for first supply, and powder charges were passed up through an opening in the floor. Four such turrets were placed along the midship line fore-and-aft the deck, of which that farthest forward mounted a pair of guns and was slightly larger than the others mounting one gun each. All five guns were 9-inch 12½-ton rifled muzzle loaders. With this armament and no rig her originally intended complement of 1260 ranks and ratings was reduced to 296; yet even so, if she could have got at an enemy of her own original 130-gun design, she would easily have made a bonfire of the latter at no harm to herself. She carried a lion medallion, as figure head. But it was a beast on a short length of chain.

As protection the *Royal Sovereign* was completely armoured with 4 ½ inch plating. Her displacement was about 5700 tons and her speed 11 knots. She was completed for sea at Portsmouth in 1864 and though remaining afloat for more than fifteen years had only one commission as an independent command, even that merely lasting three months when a new ship. For the remainder of her existence she either acted as a training tender to the *Excellent* or was laid up in reserve before being sold in May 1885. She was the only wooden-hulled turret ship in the British fleet. The second British turret ship was the *Prince Albert*, though she was the first to be designed in that form instead of converted to it. And, moreover, she was iron in hull; built under contract by the Samuda firm at Poplar. In general features she resembled the *Royal Sovereign* on a two-thirds scale of displacement at 3880 tons, with the same style of single-screw, square-stemmed, medium freeboard sheer draught, the same four turrets similarly constructed and armed — except that the bow turret had only a single gun, and that all her guns were on iron carriages instead of wood — and the same complete 4½ -inch armour on the sides doubled on the turrets. And lastly she shared the same extremely limited scope of action as a mastless vessel with only sufficient coal about 100 hours' steaming in calm weather; being launched as specifically stated for 'coast defence' though more formidably armed than any ship afloat under a foreign flag.

The *Prince Albert* was ready for service by 1866, and though purposely kept afloat for thirty-five years — that is to say till the death of Queen Victoria, who had originally selected the name — she was only in fully manned commission for a few weeks at a time on three separate occasions, including one for trials as a new ship. For all the rest of her existence she either did duty as a gunnery tender at Portsmouth, or Plymouth, or was laid up in reserve for war emergency until, in March 1899, she was sold to Ward at Preston for breaking up. It is worth a passing mention that both she and the *Royal Sovereign* were subjected to an unusually severe test to ascertain whether their entirely novel method of mounting a heavy gun would suffer in power of rotation by an extremely violent blow. For this purpose one turret in each ship was actually fired on by 100 lb. projectiles at a hundred yards' range. The plates that were struck were damaged beyond repair but otherwise the tur-

H.M.S. *Cyclops* at anchor on the Mersey about 1890. As compared with when she was first completed, her breastwork has been extended to the full width of the hull and light guns added to her flying deck, while the light pole signal mast at the forward end of her breastwork has been removed. (NMM. F. 26658)

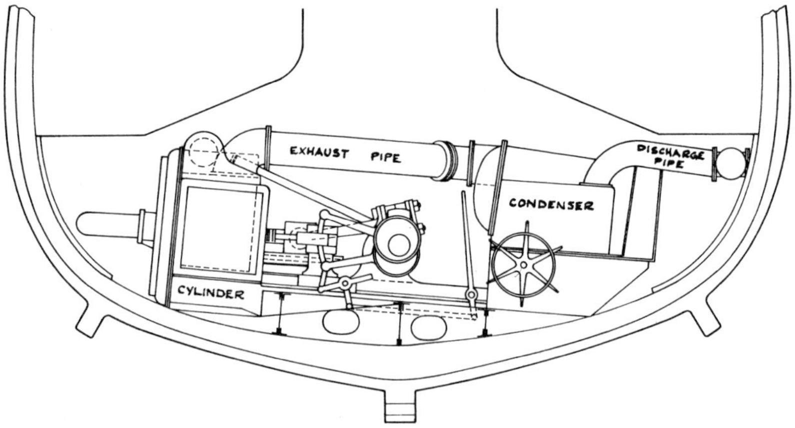

Direct acting steam engine, as fitted in the coast defence ships.

rets suffered no appreciable harm. In spite of great cost neither of these two powerful craft could be classed as capital ships within the long-established meaning of the term; but the strategic conception nevertheless which left the initiative in movement to an enemy — and which they were the first of British design to embody — retained at least sufficient official acceptance to absorb an annual and appreciable proportion of the national expenditure on new fleet construction for several more years.

The naval estimates during the five years following the completion of the *Prince Albert* included provision for laying down the keels of as many as ten armoured vessels, having a tonnage approximating to hers, and all classed as suitable only for coast defence; in the general design of which Coles had a large voice.

All were ready between 1870 and 1874, and all had, what was then the novelty for armoured ships, twin screws. Seven of the ten were identically armed and closely similar in their main features. The remaining three were each of solitary design. Before the plans of any had been prepared the rapid increase in the size of guns had resulted in a 10-inch of 18 tons by 1865 and a 12-inch of 25 tons by 1866, while armour kept up more than a parallel rise in thickness.

The seven 'near to sisters' were the *Cyclops, Gorgon, Hecate, Hydra, Magdala, Abyssinia* and *Cerberus*, all of 3500-4000 tons displacement with very low freeboard to meet a strong preference of Coles for such, sharp underwater rams, two circular and covered turrets, separated by a high superstructure, each mounting a pair of 10-inch 18-ton guns and turned by steam instead of hand, over an arc of fire each of 270°; side armour up to 8 inches with 10 inches on the turrets, horizontal simple-pressure engines giving a speed of 11 knots, and coal for four days' steaming in smooth water. Though not actually unfit to face heavy weather, their decks were well awash in even a moderate sea. Ranking as Commander's Commands they were manned by 190 officers and men, for whom the accommodation was the worst in the fleet, being usually referred to in lower-deck language as 'ratholes with tinned air'.

Though these seven lasted on an average for over thirty years, none of them were ever commissioned with full complements for more than a few weeks at a time, to participate in uncommon special occasions such as State reviews or large-scale manoeuvres. All the rest of their careers were passed in reserve, and the first four in the list above never left the Home station. But the other three departed for distant shores as soon as their trials were complete, and never returned, the *Magdala* and *Abyssinia* going to Bombay in permanence for the Indian Government and the *Cerberus* as far as Melbourne for the State Goverment of Victoria. On the passage out they followed coastal routes as much as practicable, accompanied all the way by colliers to enable them to recoal at any convenient anchorage, and to be towed when necessary. On the ocean stages they proved seaworthy though uncomfortable; and though it never fell to the lot of any of the seven to meet an enemy, there can be little doubt that no contemporary opponent of their own tonnage during at least the first fifteen years of their period afloat would have stood much chance against them. They might be said to resemble full-armed knights riding on donkeys, easy to avoid but bad to close with. Minor additions were made to their armaments at later stages by mounting a few quick-firing guns on their superstructure, as defence against torpedo craft, but otherwise they underwent no weapon changes. It was a novelty in their design that the turrets and superstructure stood on an armoured redoubt rising centrally like an island about 6 feet above the deck of the low main hull, so that their armour waist protection was not on a single vertical plane as in broadside high freeboard ships, but in two halves divided horizontally, with the upper withdrawn about 6 feet inboard of the lower. All seven were contract built in iron.

* * *

The *Cyclops, Gorgon, Hecate* and *Hydra* were all sold, followed in 1903 by the *Magdala* and *Abyssinia*. Only the *Cerberus* continued to serve, being renamed *Platypus II* in 1918. In 1924 she was sold to the Melbourne Salvage Company, but instead of being broken up she was in 1926 beached as a breakwater at Williamstown, a suburb of Melbourne. There she still survives. She is unkempt and dirty, but much of her superstructure remains, and her 10-inch guns are still mounted, although are swung inboard.

H.M.S. *Glatton* proceeding to sea for gunnery exercise. The photograph taken about 1896 clearly shows her low freeboard, and apart from the addition of some light armament, she has not been substantially altered in appearance. (NMM. AFM/1)

* * *

The three coastal service vessels not included in the above class were the *Glatton, Hotspur* and *Rupert*, which had no sisters, nor shared any details of design except that each had only one turret, and the two latter a high freeboard as distinct from the whole of the rest. The *Glatton* was Admiralty built and considerably larger than the *Cyclops*, having a displacement of nearly 5000 tons and twice the bunker space of the smaller vessel, with an appropriately extended area of action. But her deck was less than 4 feet above the water at load draught, and although her single turret carried 12-inch 25-ton guns she had no stern fire whatever. As a combatant conception, therefore, she somewhat resembled the *Monitor* on a four times greater

H.M.S. *Hydra* as originally completed during the mid 1870s. Her breastwork amidships supporting her turrets and superstructure does not extend over her full beam, and the triple hand steering wheel can be seen just forward of her charthouse. (NMM. HAT/1)

H.M.S. *Hotspur*, after her reconstruction in 1883, with her original fixed turret replaced by one which revolved. Additional superstructure, and a flying deck forward carrying light quick firing guns. (NMM. AFK/2)

scale, and when new carried the most massive protection of any craft afloat with side armour up to 12 inches and 18 on the turret. Like the *Cyclops* class she passed almost her whole service until she was sold in 1903 in reserve or as a drill tender for gunnery training, with occasional very brief spells in commission. Her lack of properly adequate berthing for 220 ranks and ratings hardly permitted otherwise except in an emergency.

This was different in the *Hotspur* and *Rupert*; for, though designed only for coastal service, with the very limited steaming radius which that implied, they were built with the normal height of freeboard of an ocean ship, giving them a deck extra to the *Cyclops* class or *Glatton*, with correspondingly increased living spaces for all hands. As a result the conditions of their employment were the reverse of those in the others, commissioned time being the rule and reserve time the exception. Both passed some years on the Mediterranean station as well as on the coastguard districts in the United Kingdom. The *Hotspur* was launched under contract on the Clyde in 1870. Her displacement was 4000 tons and, as first completed, she carried one very heavy and two medium calibre guns, instead of the two or four heavy only of all the low freeboard craft. Her one so-called 'turret' had no counterpart structurally throughout the fleet, being merely a fixed circular armoured enclosure built amidships on deck just forward of the funnel casing, with four alternative ports giving a total arc of fire of about 220° to a 25-ton gun mounted inside on a hand-worked

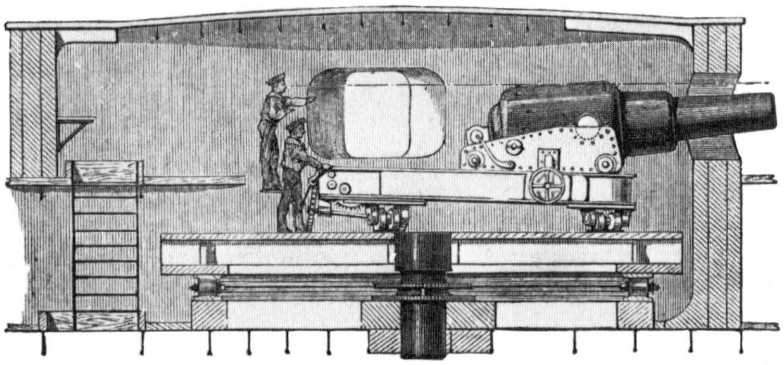

A simplified diagram depicting both the laying and the training of the 25-ton gun of the *Hotspur*, and the circular platform supporting the gun allowing it to be brought to bear at any of the four ports. (NMM. 58/3683)

The 12-inch, 25-ton gun in the fixed turret of the *Hotspur* on a Scott Slide mounting. Apart from being of heavier construction, when compared with the Elswick Compressor for controlling recoil, which consisted of iron bars running lengthwise within the slide, the Scott Compressor applied the friction externally, and was controlled by the notched wheel on the side of the gun carriage, instead of a lever. The implements in the foreground are the 'sponge' for damping down any unburnt powder after firing, the 'ram' for forcing home the new charge and shell, and the 'worm' for gouging out any unburnt remains of the previous charge. (NMM. 7115)

A plan view of the original 'fixed turret' of the *Hotspur* with the 12-inch 25-ton gun run out ready for firing. The simplified diagram of the gun mounting and its slide, with the compressor plates controlling the recoil omitted, shows clearly how the gun was trained with the winches at the rear of the slide, and the controls for running the gun outboard. (NMM. 58/3682)

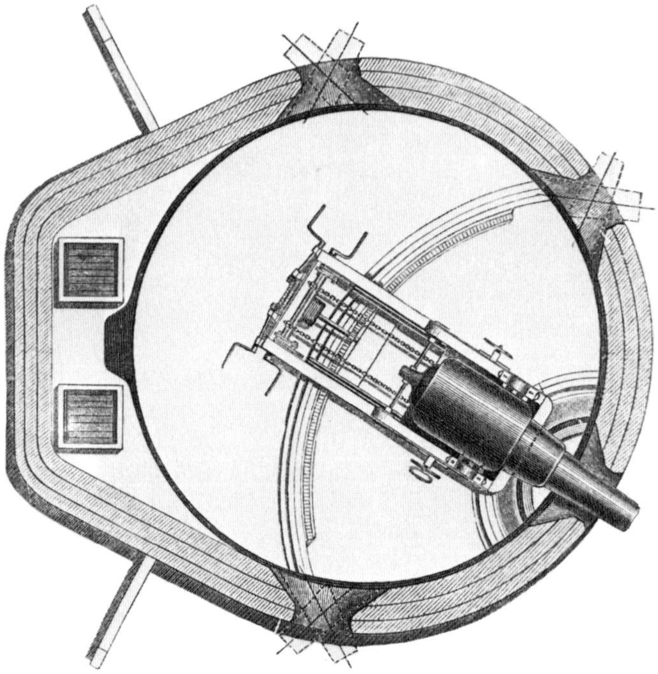

H.M.S. *Hotspur* as guard ship at Berehaven during manoeuvres in 1888, showing the early form of net defence against torpedo attack completely surrounding the ship making any movement on her part out of the question, while the picket boat patrols the anchorage. (NMM. 58/6536) (Below.)

H.M.S. *Rupert* shortly before her major reconstruction in 1891. Here she is shown substantially as she first appeared except for the light armament on her superstructure and the fighting top added to her mainmast. Torpedo net booms, added during the 1880s can be seen stowed along her sides. (NMM. AFL/1) (Top right.)

H.M.S. *Rupert* leaving Portsmouth after her reconstruction of 1893, now armed with breech loading guns, and the array of cowls around her funnel indicate much better conditions for those serving on board. (NMM. L4/75)

turntable. Unlike the *Glatton* she was not entirely devoid of stern fire, though it only took the moderate form of what was really a secondary battery of a pair of truck mounted 64-pounders. When about twelve years old, her single heavy weapon was replaced by two of the same calibre in a revolving turret, and two of the then newly adopted 6-inch breechloaders were substituted for the truck guns aft. Light quickfirers were added. Her side armour was in one vertical plane, varying from 7 to 11 inches in thickness, and 10 on the turret. She could steam 12 ½ knots, but only for about 70 hours in fine weather, which was the minimum for any vessel of her tonnage under the White Ensign. With a complement of only 230 ranks and ratings her berthing was ample. She was sold in 1904.

The *Rupert* was launched about three years later, and in main features of design resembled the *Hotspur* on a larger scale, having a displacement of 5400 tons, two knots more speed and coal for about 100 hours' fine-weather steaming. She was the last British vessel specially built for coast defence, nor was it until the war against Germany broke out forty years later that any British craft were designed for purely coastal operations, although by that time the object was attack. The *Rupert*'s single turret carried a pair of 18-ton muzzle-loaders for the first fifteen years, which were then changed for the new 9.2-inch breechloaders of 22 tons. Aft she had originally a pair of truck 64-pounders, but these also were later changed for breechloaders. She was thus the only one of the twelve Victorian coast-defence ships ever to be fully armed on the breechloading principle. Apart from larger size her chief structural difference from the *Hotspur* lay in the possession of the same type of armoured redoubt rising amidships as in the low freeboard vessels, carrying the turret and a superstructure. The plating on this was 8 inches, on the turret 12, and on the sides of the main hull 8 to 11. Her complement of 295 all told was much the largest among the coast defence craft except the *Royal Sovereign*; but whereas in the case of the latter the extra numbers were required for the guns, in the *Rupert* they were required in the stokehold, as she had double the horse-power in proportion to armament.

As the last of these the *Rupert* was sold in 1907 at Bermuda, having been struck off the effective list the previous year, thirty-two years after completion for sea and forty-two after the completion of the *Royal Sovereign*. That passage of time, therefore, represented the period during which a vessel suitable only for passive war effort was still regarded as possessing enough value to Britain to be worth the cost of maintenance in efficient condition.

CHAPTER NINETEEN

THE 'DEVASTATION', THE FIRST MASTLESS CAPITAL SHIP

H.M.S. *Devastation* about 1885 after she was fitted with torpedo net booms passing the boys' training ship *St. Vincent* in Portsmouth Harbour. (NMM. HCB/3)

The *Devastation* was the first true capital ship under any flag without a single sail, and she represented the first complete application to a sea-going battleship of the principle of mounting the main armament on top of the hull instead of inside it.

After the loss of the *Captain*, the *Monarch* had remained for some years the ideal of a masted turret-armed capital ship, being in fact taken as the general model for certain others laid down abroad. But even in her case the combination of turrets with masts suffered from the unavoidable drawback of preventing any part of the main armament being used for chase fire; which was only in part remediable by mounting medium calibre guns at the bow and stern. The stern fire of some much smaller turret ships such as the *Cyclops* class was heavier and better protected than the bow fire of the *Monarch*; and the relative size of the targets was in an inverse proportion to that of the guns.

It was this realization that at last decided the final abolition of masts in all future British-built turret ships in favour of a trebled coal stowage, even though that would mean larger and more expensive hulls in proportion to armament, and in the matter of general expenditure on naval administration the maintenance of increased fuel reserves at home and abroad as an addition to the cost of national defence. And thus it was that the first mastless vessel to combine all the qualities of a true capital ship came to be built. Having cut the masts away to give more scope to the guns, the Admiralty had to decide how to make the most of the gained freedom for arranging the disposition of armament, but that question had already been considered and answered in designing mastless coast-defence craft. In the *Cyclops* class at least a full half of the armament could be brought to bear on any point in the whole arc of the horizon, and the entire armament on about two-thirds of that circuit; which was roughly the maximum practicable for any vessel large or small. It was now to be adopted again. Moreover the armoured central redoubt on which her turrets stood offered a structural advantage in raising the height of the guns and distributing the weight of body armour more evenly athwartships on the frame of the hull. This was also adopted again. But the very low flotation of the *Cyclops* and her extremely limited coal bunkerage were quite unsuitable for reproduction in an ocean-going vessel. The sketch design for a mastless capital ship, therefore, as approved by the Board, took the form of an enlarged *Cyclops*, with about twice and a half her displacement tonnage, twice her weight of armament, seven times her coal stowage, four times her engine power, and a normally high freeboard amidships instead of an abnormally low. On these general lines the *Devastation* was iron built at Portsmouth. She had a flat bottom, square bilges and full lines at both ends, a pointed ram, and a submerged counter. Her sides had an absolutely straight rise from bilge to deck. The armoured redoubt standing amidships to carry the two turrets was 156 feet long and 50 broad, but in the *Devastation* not visible externally, being enclosed to its full 7 feet height by an unarmoured extension reaching to the ship's sides transversely to give a good freeboard, and about 30 feet towards the stern to improve the accommodation for officers. The transverse extension provided extra and well-lit berthing for the men.

Even before the keel of the *Devastation* was laid the turret guns of all her predecessors including the *Monarch* had been surpassed in weight by nearly half as much again in the design of a new 12-inch-muzzle-loader of 35 tons, firing a 700 lb. shell. It was to carry an armament of four of these on board an ocean-going hull that the construction of the ship herself was mainly planned. These 35-ton guns were the heaviest in British use afloat or ashore ever worked by hand alone except for the steam turning of the turrets, and comprised the entire armament of the *Devastation* apart from a 9-pounder boat gun when she was first completed for the pennant. The circular turrets had a diameter of 31 feet, and an arc of fire of 280° each. At later stages, during the ship's periodical large-scale refits, a light battery of 6-pounder quick-firers was added on the superstructure generally known as her 'hurricane deck', and a pair of torpedo carriages for 14-inch Whiteheads placed below the forecastle. And finally, in 1891, when she was twenty years old, her main armament itself was changed by the substitution of four 10-inch 29-ton breechloaders for her original turret M.L. guns. The new weapons only fired a

From the forecastle of the *Devastation* as originally completed showing her forward turret with its two 12-inch 35-ton guns. (NMM. A. 2319)

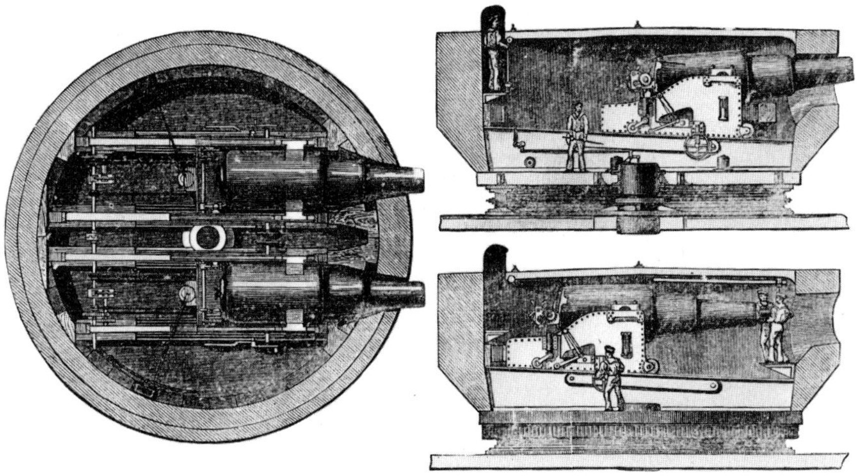

Plan and sectional illustrations of the 12-inch guns in a turret of the *Thunderer*, sister to the *Devastation*. In the upper sectional drawing the gun is run out ready for firing, in the lower, it has been run in after firing, it is being reloaded while a third hand is standing by to release the compressor, and allow the gun and its carriage to run down the slide by its own weight to be ready for firing. (NMM. 58/3700)

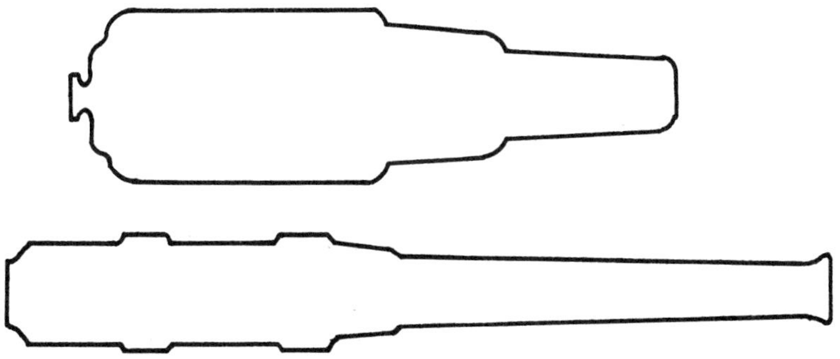

Turret guns of H.M.S. *Devastation*. First armament: 12-in., 35-ton muzzle loading gun with 714 lb. projectile, 110 lb. charge and 1340 ft muzzle velocity. Second armament: 10-in. 29-ton breech loading gun with 500 lb. projectile, 252 lb. charge and 2040 ft muzzle velocity. The noteworthy difference in the design of these two guns was due to the different chemical nature of their respective charges. With the old black or brown powders used in muzzle loading guns full combustion was almost instantaneous, causing a pressure that might rise to as much as 30 tons to the square inch in the powder chamber, but fell to less than 2 tons at the muzzle. For that a short bore sufficed. With the introduction of slow-burning propellents requiring a long bore for full use the slow combustion never caused pressures of more than 18 tons to the square inch in the chamber, but they remained high all along the bore and were still as high as 4 tons at the muzzle. This was suitable for breech loading guns.

500 lb. shell, but had twice the range, nearly twice the penetration, and at least three times the rate of fire.

The *Devastation* was very extensively armoured. Her turrets had plating of 12-14 inches throughout, and though the protection of other parts might seem from the diagrams attached to this account to have been on a complicated system, it was not in reality very much so. It consisted of two courses, that is to say, a lower one outside the main shell of the ship, and an upper one standing internally to the shell. The former, with a thickness of 8-12 inches, ran the entire length of the waterline, down to 5 feet below the surface, and upwards over the after parts and waist. The latter, with a thickness of 10-12 inches, protected the redoubt enclosed within the centre body as a stand for the turrets. Apart from plunging fire — little apprehended in those short-range days — the *Devastation* presented an almost completely protected target, on which, with the minor exceptions of the hurricane deck and a short section above water at the bows, no enemy shot could strike except on armour, whether externally on first impact, or so closely in rear of the external surface that its penetration was merely a skin puncture of no consequence.

She was not the first British capital ship to have twin screws, but the four that came a little before her were masted and side-armed. Twin screws gave her an extra insurance against the total inability to move that inevitably follows on an engine breakdown in a vessel without canvas. Her original machinery comprised two sets of Penn's trunk engines; which, as in all battleships of the period of her launch, functioned on the old extravagant single-pressure steam system, with rectangular boilers at 30 lb. load. But because, in order to make up for her lack of any sails, she was built with bunkerage for 1800 tons of coal, she could travel far nevertheless; as that was more than double the capacity of any contemporary battleship except her own younger sister vessel *Thunderer*, and indeed more than that of any British battleship coming less than fifteen years later. And when the *Devastation*'s original engines were replaced (at the same time as she was re-armed) with triple expansion machinery and cylindrical boilers supplying steam at 60 lb., her steaming reach was doubled yet again. She could then cross the Atlantic and return without coaling on the far side. For auxiliary machinery outside the engine room she had the novelty of steam steering gear, as well as a steam capstan and steam derrick hoist on her solitary mast.

The *Devastation* was launched in July 1871 at Portsmouth, twenty-one months after the laying of her keel, and completed for sea by the end of the year following. She hoisted her first pennant there for a period of independent

Armour diagrams of H.M.S. *Devastation*.

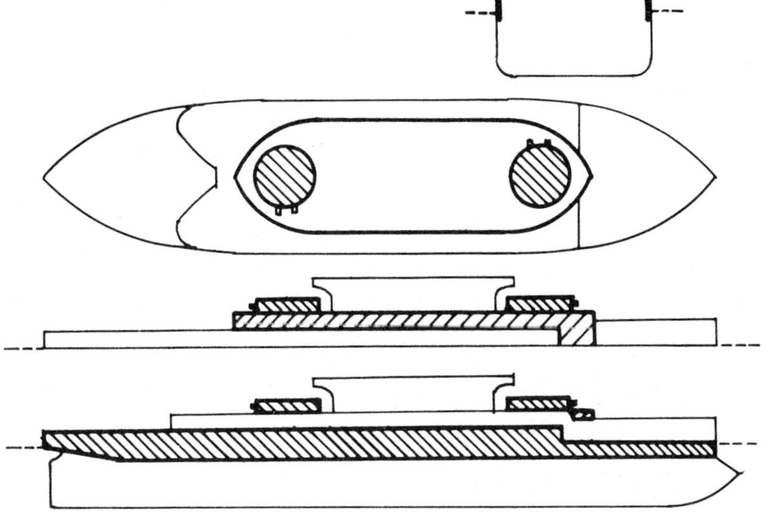

cruising, as a general test of her qualities, lasting ten months, under Captain W.N.W. Hewett, an officer with the extremely rare distinction of a double award of the Victoria Cross, for special gallantry on two separate occasions. Her complement of only 400 officers and men was the smallest in any British capital ship from at least two centuries before right down to the present day, with the solitary exception of her sister vessel; and it was peculiar also in its uniquely large proportion of engine-room ranks and ratings, for whereas both before and after the *Devastation* the battleship ratio between combatant and engine-room staffs has never been less than four to one, and often very much more, with her it was only about two. And, again, the total number presented an anomaly only removable by Order in Council, for though she was the most powerful fighting vessel afloat, it placed her below the number entitling her captain to the higher of the two scales of pay of his rank.

In her testing cruise she proved a good seaboat and steady gun platform with an easy roll, but bad to drive against a head gale on account of comparative lack of bow freeboard. At the end of it Hewett handed over the command to Captain Frederick Richards — in after years First Sea Lord — under whom the ship was somewhat later sent to join the fleet in the Mediterranean, remaining on that station for a second commission till the end of 1878 and so completing five years' continuity of sea service. As her quadrennial survey had thus become considerably overdue, through the exigencies of the political situation in the Middle East during the Russo-Turkish war, she was ordered home to pay off for a two years' overhaul and refit, at the conclusion of which she was left unemployed in reserve for the next four years; that happening to be one of the recurrent phases in domestic State affars when it is politically profitable to denounce expenditure on national defence as an inexcusable misuse of national money. For several years in succession the naval estimates had been kept rigidly below eleven millions, and Lord Northbrook, as First Lord of the Admiralty — and himself very wealthy — made the remarkable statement that if more money had been voted for the fleet he would not have known what to do with it.

The interior of the conning tower of the *Thunderer*, sister to the *Devastation*. While it might appear almost empty, and the instruments crude as compared with a warship of today, none-the-less, bringing all the essential controls together under the protection of armour plating for when the ship went into action, was an improvement over the design of the first armoured vessels which at that time were scarcely more than ten years old. (NMM. 58/4943)

But, as sometimes happens, this attitude was suddenly startled, by unexpectedly vigorous Russian action in the approaches to the Indian frontier in what was known as the Penjdeh incident; and a number of ships of all sorts, including the *Devastation*, were hurriedly commissioned from reserve. This action had a reassuring effect and the public excitement faded; but having pulled the *Devastation* out of the corner of the dockyard basin, where she had reposed during six years, the Admiralty decided to keep her out, and detailed her to relieve one of the older vessels in the force of seagoing commissioned ships known in those days as the First Reserve Fleet, an organization that came to an end when the civil duties in peace or war of the land coastguard stations ceased to be an Admiralty responsibility. Under that system the entire shores of the British Isles were divided into eight districts, within each of which the coastguard stations were manned by naval ratings permanently transferred after six years' service in the active fleet, but remaining its prime man-power reserve for war. Each district was under the administrative orders of the captain of a battleship with her headquarter anchorage in its principal harbour, only partly manned for most of her time, but made up to full complement very quickly for war or practice by embarking the necessary number of men from the shore stations of her district. This was done as an exercise every year, when all eight ships assembled under the Admiral Superintendent of Coastguard and Reserves, either to participate in the annual naval manoeuvres or to proceed for a month's separate practice cruise, as far as the Mediterranean or Baltic; for it must be borne in mind in this connection that 'coastguard' had nothing to do with 'coast defence' as regards the operations of the First Reserve Fleet. The *Devastation*'s district embraced the east coast of Scotland, with headquarters at Queensferry — long before Rosyth had even been suggested as a naval base — and there or thereabouts she was to be seen for the next five years. By the end of that time her engines were getting worn out and her armament obsolete, but she was otherwise still in such good condition as to be considered worth modernizing in both respects as described above. She was

H.M.S. *Devastation* in her final appearance, painted grey, and a wireless gaff added to her mast head. (NMM. AAX/2)

relieved and paid off accordingly in 1890, to receive new types of guns and engines.

As the expense of this process, combining refit with modernization, was spread over the naval estimates of three successive years, it was not completed till the end of 1893; but though by then the *Devastation* was already twenty-one years old, and far surpassed in power by many battleships launched since her early days, she was sufficiently reconditioned for eight more years of commissioned time, and fifteen more before going to the scrap heap. She was now selected as one of several for what had become a revived form of service — suitable for middle-aged tonnage — arising as a consequence of the building of shore barracks to take the place of the former port guardships at Portsmouth, Plymouth and Chatham as depots for ratings awaiting draft for sea. The old and now abolished hulks so used had long since ceased to be 'guardships' in anything but name, though in the far-distant past the duty had fallen on war-effective capital ships, and the Admiralty decided to re-establish the custom of stationing effective commissioned vessels as genuine guardships at the three principal ports, using such as were among those fully fit for war, though not in first-line squadrons on account of age. These were always ready, for immediate full duty, but, when not actually ordered out, only retained officers, principal ratings, and enough other hands on board for peace requirements at sea or in harbour; the war complement being made up from the local barracks at a few hours' notice if necessary. They carried out all the routine firings and other practices, and took part in all major scale manoeuvres. The *Devastation* was commissioned as port guardship at Plymouth near the end of 1893, and continued as such during the next five years. At the end of 1898 she was recommissioned for a similar duty at Gibraltar under an extension of the port guardship system to the principal Crown Colonies. That was her final period of flying the pennant, which came to an end when she was ordered home a little more than three years later to be paid off for the last time in 1902, and relegated to the third class reserve. In 1908 she was sold to Ward for breaking up.

From the above account it will be seen that, like so many other British capital ships during the long Victorian peace, the *Devastation* never participated in any of the more spectacular events of history. But that does not prove that she or they had been uselessly built, by any means. Being always available to prevent trouble, they never had to deal with it. All the maritime Powers at the time of the *Devastation*'s launch were well aware that she could bring the then most formidable weapons in the world to exercise their persuasive influence at any spot in the navigable waters of the globe, and knowledge of that kind has a decidedly sobering effect on hotheads in high places. Under the flag of a robber nation she would have been a perpetual threat to civilized peace, but under the White Ensign she was one of the preservers of peace in no small degree.

And, apart from all that, her place must be forever unique, in one respect at least in the story of the sea, as the first true capital ship by the standards of her day that never ever had any use for the wind as even an auxiliary source of movement.

CHAPTER TWENTY

THE 'ALEXANDRA', THE LAST BRITISH CENTRAL BATTERY SHIP

H.M.S. *Alexandra* at Malta during the late 1880s, dressed overall. In addition to her hull being painted white and light armament added to her topsides, she had been given torpedo nets, the booms for which can be seen stowed below her main deck level. (NMM. A.2235)

As a lone example of an obsolete, though once universally accepted principle in fighting-ship design, among the vast number of modern warships assembled at Spithead for the Diamond Jubilee of Queen Victoria in 1897, lay the last British capital ship in seagoing commission of all the hundreds in the past that had carried the majority of their heavy guns under the three centuries old system, whereby they were placed internally between decks instead of in turrets or barbettes above the main hull. Moreover, she was the only one present on that occasion among those officially termed 'battleships' to be armed mainly with muzzle-loading weapons. Yet she still flew the flag of an admiral commanding his own force — as she had indeed done throughout her whole long service — though much inferior in power now to several of the other ships included in its composition.

The *Alexandra* was built at Chatham of malleable iron. Like so many of her contemporaries entering the water during the mid-Victorian period of wholesale transformation in naval architecture she was a vessel of complete individuality in design with nothing even approaching to a sister ship. She came as the largest and fastest of the eighteen central battery ships forming the backbone of the British armoured fleet during the intermediate stage between the full broadside vessels with numerous light calibre guns and the turret ships, the other seventeen being the *Bellerophon, Hercules, Superb, Temeraire, Belleisle, Orion, Repulse, Penelope, Pallas, Favorite, Audacious, Swiftsure, Vanguard, Triumph, Iron Duke, Sultan* and *Invincible*, the last-named seven of which, as well as the *Alexandra* herself having the battery double-decked, though only in her case were both the decks embrasured to give right ahead fire. With what was then a medium battleship length of 325 feet, combined with

H.M.S. *Alexandra*. Top: Sheer draught outline showing external armoured surfaces. Centre: Main deck showing lower battery of eight 10-in. guns and berthing decks forward and aft. Bottom: Upper deck showing topgallant forecastle, poop, and upper battery of two 11-in. and two 10-in. guns.

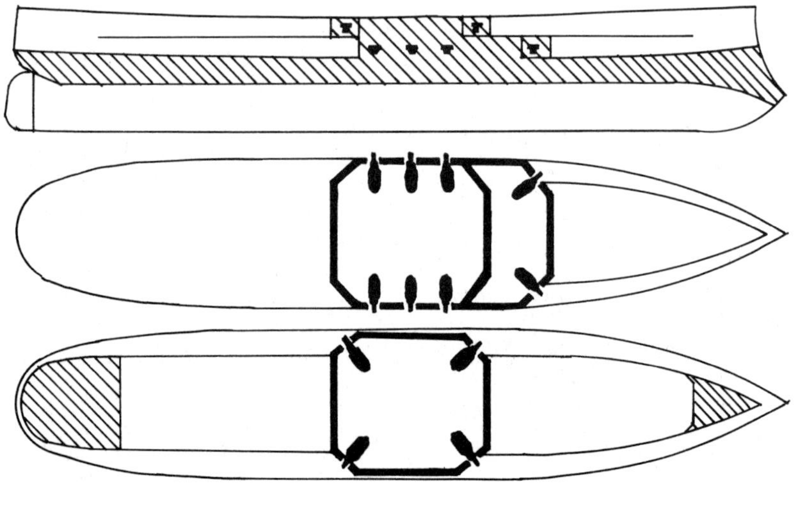

Small arms racks and one of the four 14-inch torpedo carriages of the *Alexandra*. (NMM. 59/366)

medium lines under water, a rather full midship cross-section and a nearly vertical side, her calculated displacement was 9490 tons. She had a pointed 6-foot ram, and a 9-foot overhang of counter topped by a narrow rounded taffrail. She steered easily and as a seaboat was remarkably steady. Her deck-space, however, was very small for her size, partly on account of the narrowing restriction caused by the recessing of its breadth to allow arcs of fire ahead and astern for the upper central battery, and partly because its area was very greatly reduced amidships by the battery bulkheads, and at its ends by a poop and topgallant forecastle. And with 7 foot bulwarks surrounding what was left outside these enclosures the cramping effect of the aspect meeting the eye on stepping through the gangway gave the impression of a much smaller vessel than she actually was; in which matter the contrary held good in the broad, clear, and spacious decks of such central battery ships as the *Hercules, Superb* or *Temeraire*, of less than her own size, but with no upper battery.

Externally it must be admitted, that very unlike her royal godmother, the *Alexandra* was no beauty. Her figure as she sat the water was high and square, and the very extensive embrasuring of her sides to increase end-on fire completely broke up the surface curves so necessary to a handsome hull in the eye of a seaman. It was the same on looking aloft, where the spar plan showed squat and square as compared with that of many other battleships, owing to the combination of a short hoist with a great spread, and the big funnels did not rake symmetrically at the same angle as the masts. But good looks do not always accompany good qualities in naval architecture any more than in other things.

The central battery principle disrupted the traditional berthing arrangements of British battleships. It absorbed nearly a third of the main deck, dividing its berthing space into two widely separated portions, and in many cases breaking the continuity of its level. This suggested the long-overdue improvement of moving the officers' quarters up from the lower deck.

But when the obvious desirability of an improvement came to be recognized the change went almost to the opposite extreme for a time, as is sometimes likely to occur; of which the *Alexandra* was a conspicuous example. In her case the officers' messes and cabins were far better than any ever seen in the Service before, whereas the accommodation for the men was inferior to that in not a few older ships. Longitudinally the available deck-space was about equally divided between ranks at one end and ratings at the other — the marines lived in the battery — but owing to the long forward embrasuring of her sides the bow half of the *Alexandra*'s main deck was narrowed to only 40 feet (a corvette's beam) whereas abaft the battery it had the full width of more than 60 feet. Through this contraction the stokers' messes had to be put on the deck below which had no light except candle lanterns.

In ships of long broadside armament the main deck ports were of equal size throughout, whether used for guns or only for ventilation, and any could be made available for a gun as far as size went. But in central-battery ships the ventilation ports running the whole length of the main deck outside the battery limits were much smaller than the gun ports inside; although rectangular and iron-lidded, to open outwards like gun ports, instead of inwards like scut-

H.M.S. *Alexandra*, a corner of her main deck battery towards the end of her active career. Although this picture is specially posed for the photographer, it does give some indication of conditions aboard the earlier armoured ships of the Victorian navy. (NMM. 59/313)

British 10-inch rifled muzzle loader of 1870 on Naval broadside mounting with gun's crew of seventeen numbers and two powdermen. A: Firing position. B: Sponging and loading scuttle (disc of scuttle not shown).

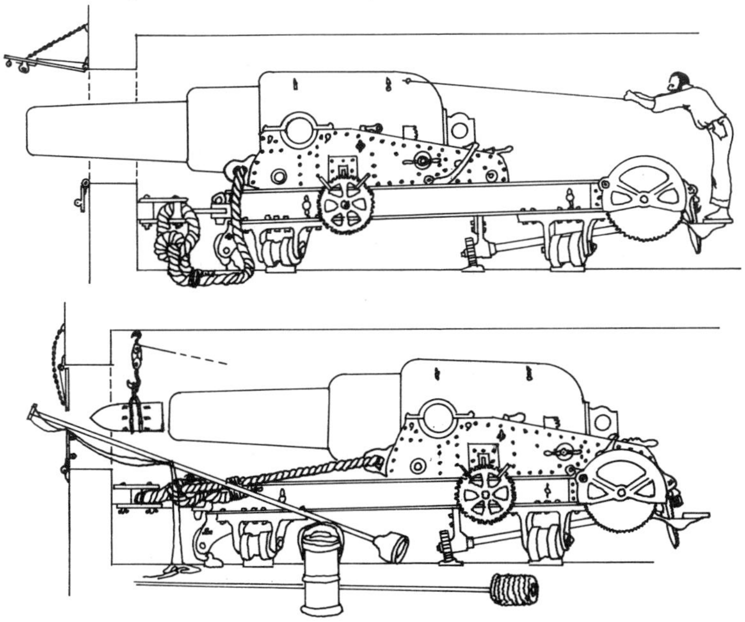

The main deck battery of the *Alexandra* in action. The gun, loaded and run out ready for firing, while the captain of the gun signals to alter the elevation. It is worth noting the rope mantlets suspended from the deckhead beams in order to restrict the damage from flying splinters etc. Such a mantlet used about 300 fathoms of rope, and one about 16 feet by 7 feet weighed about 820 pounds. (NMM. 58/5970)

tles, and in officers' cabins backed by a glass sash sliding up or down window fashion.

Below the main deck the *Alexandra*'s system of watertight compartments was on the same general plan as in the other central-battery ships of iron hull. The lower deck stood almost exactly abreast the load waterline; over which the bulkheading reached up to the main deck, and as in her case that was 8 feet higher, the margin of safety against overall flooding from one compartment to another, of the kind that proved fatal to the *Vanguard*, was much greater than in earlier battleships. In the main compartments the double bottom was carried up to meet the wing passages.

As first completed for sea the *Alexandra* was armed with a pair of 11-inch 25-ton rifled muzzle-loaders and eight 10-inch of 18 tons, all mounted on Scott carriages and slides. Hers was the last central battery designed for a British ship, and in plan remained unique, though it might perhaps have provided a standard model for further use, had her keel not been laid just before the final Admiralty decision in favour of turrets for all battleships to be built thereafter.

As will be seen from the diagram that part of it at upper-deck level took the form of a rectangular oblong, measuring 69 feet internally fore-and-aft and 55 feet athwartships, with a slice cut off at each corner for a gun port, providing an arc of fire from the line of keel to a little beyond the beam. The two ports at the forward end were occupied by the two 11-inch guns, and the two aft by 10-inch.

The remaining eight 10-inch guns formed the main deck (or lower) tier of armament. For them the battery at that level was divided athwartships into two compartments of different size, by a transverse armoured bulkhead with armoured communication doors between them; presumably as an extra protection against raking fire from ahead, since it added a lot of weight having no other discoverable use. Of these compartments the aftermost — and larger — was an underlying repetition in shape and dimensions of the upper battery that stood immediately over it, like a two-storied building, but instead of the four corner ports of the latter it had three on each side, at which stood six 10-inch guns with broadside arcs of fire. The smaller compartment was a forward extension or addition to the larger, beyond the dividing bulkhead, having a deeply embrasured corner port on each side to give a 10-inch gun an arc of training from right ahead to 7° abaft the beam, and increase the volume of chase or bow fire, but only under normally favourable conditions, as the embrasures shipped such green water when steaming against a heavy head sea that the ports could not be opened.

As a central battery much reduced the extent of side armour required for gun protection it offered a great structural advantage to a designer; but from a gunnery standpoint it was not without drawbacks, for although it did perhaps to some extent facilitate control and concentration of fire, it complicated and congested the arrangements for ammunition supply very considerably. Being unavoidably situated above the boilers, the position of the battery prevented the magazines and shell rooms from being placed directly below the guns, or even conveniently near to them; with the result that all but a small proportion of the ammunition had to travel through long passages to reach the battery hoists. This increased the risk of a stoppage by the possible flooding of the line of conveyance through under-water injury to the ship, besides adding to the heavy labour of supply, which was man-handled throughout. What this latter meant may perhaps receive enough evidence from the *Alexandra*'s contribution to the smashing of the powerfully armed Egyptian coastal forts, when she fired (in addition to smaller stuff) 269 heavy calibre shells weighing in the aggregate more than 50 tons. A minor proportion of this expenditure came from the ready racks in the battery, but these were soon empty, and all the rest had first to be trolleyed by hand in the passages, and then hoisted by hand at least 30 feet to reach the battery floors. Moreover, about 9 tons of powder charges had to be passed as well. In some at least of the central-battery ships present on that occasion — and quite likely in all — it was found necessary to use the guns' crews of the disengaged side as a reinforcement to the ammunition party feeding the guns in action.

It will be seen from the above account that the *Alexandra* had arcs of fire for four guns ahead, six on each beam — even with both sides engaged — and two astern; but never more than half of her total armament could be concentrated on one target, as she had no central-line gun. She was the first British battleship to be fitted with swinging covers to the gun ports instead of hanging, and the full weight of her broadside as she was originally armed came to 2540 lb. But when she came home after twelve years of continuous service abroad she arrived during the period when the fleet was gradually changing over to breech-loaders, and she was selected as one of the more effective older vessels fit for a partial conversion. The difficulty of mounting the long new guns satisfactorily between decks decided against any change for her at that level; but in her upper battery, with its corner ports, they could be easily secured in the fore-and-aft line with their chase lying along the embrasures instead of projecting externally to the ship's side. The four muzzle-loader guns in the upper battery were accordingly changed for the same number of 9·2-inch breech-loaders of 22 tons on Vavasseur carriages; and although these only fired a shell of 380 lb. they had double the range and rate of fire of the weapons they replaced, and

H.M.S. *Alexandra*, the damage sustained in the Commander's cabin during the bombardment of Alexandria. (NMM. 58/5969)

H.M.S. *Alexandra* in dry dock after the bombardment of Alexandria. The damage shown in this picture was probably from the same shot which entered the Commander's cabin. (NMM. 58/2445)

at least half as much more force of penetration. None was ever power worked.

Secondary batteries, in their present meaning, were unknown in the early days of the British armoured fleet; but most of its vessels carried a few small guns as a defence against boats armed with spar torpedoes, which in the *Alexandra* took the form of six 20-pounder Armstrongs on wooden chock carriages placed on the top of the upper battery enclosure. They had a much more accurate trajectory than any contemporary ordnance of comparable calibre, and at the attack on Alexandria proved very useful for range-finding: as at that time no such thing as a regular range-finder had been devised. After the ship had been about seven years in commission these were landed to give place to an equal number of the new 4-inch breech-loaders with 25 lb. shells, of the type which initiated the interrupted screw breech-closing principle now in general use; and at the same time ten Nordenfelt machine-guns firing solid steel bullets of 1 lb. weight were installed along her bulwarks, as a defence against the appearance of the seagoing torpedo-boats carrying Whiteheads; which eventually became the real forerunners of destroyers twenty times their size. At later periods still the 4-inch were replaced by 4·7-inch and the Nordenfelt by 6-pounder and 3-pounder shell guns.

The *Alexandra* was the first battleship to carry Whitehead torpedoes as part of her own armament; of which she stowed a dozen, of the original inventor's own design and production, having 16-inch diameter, a speed of 8 knots and a range of 700 yards, fired by air impulse through a port from an open frame iron carriage and charged with 60 lb. of gun-cotton. But after her first commission these were changed for what was then a new mark, having only a 14-inch diameter and 30-lb. charge, but a far superior speed and range. And like all British battleships of that period she was also equipped with a dozen 500-lb. mines, fired electrically by observation through an armoured cable circuit; as a means of protection if lying at an otherwise undefended anchorage in war-time, or conversely for clearing a passage through a minefield by countermining.

As armament for her boats she first had three 9-pounder and one 7-pounder rifled muzzle-loader guns and two rifle calibre Gatlings. Later the four muzzle-loader guns were changed for 3-pounder quick-fire breech-loaders, and the Gatlings for Gardners. And this completes the total weapon list of the *Alexandra* apart from small arms. Her full complement was 665 ranks and ratings (not including a Flag Officer and his staff) which was raised to 685 on the final change in light guns, to increase the ammunition supply party.

In the form of passive defence which is represented by armour, she was the most massively protected British vessel ever built among those which carried the majority of their guns between decks; with the sole exception of the *Superb*. In point of fact, the *Alexandra*'s sides were better shielded against any form of attack known when she was new, than those of some later and much larger successors. In partaking for many hours — with the *Sultan* and *Superb* — in the engagement with the most formidably armed section of the Alexandria coastal defences, she was hulled thirty-four times by enemy fire without any loss in her fighting efficiency; and only suffered four casualties to her hands, though the average range was no more than 1500 yards. Her water-line belt varied from 12 inches thickness amidships to 6 inches at bow and stern, with a width of 10 ½ feet, increased below water forward sufficiently to encase her pointed ram. Abreast the battery it rose right to the top, 25 feet above water, with 8-inch plating as its highest layer to cover the upper tier. Transversely the upper guns were protected by 6-inch armoured bulkheads, and the lower by an 8-inch bulkhead forward and a 6-inch aft, with another 6-inch dividing the battery internally. A unique feature in this ship was the position of an after 5-inch armoured bulkhead lying athwartships at waterline level 65 feet from the stern (like an internal connexion of the two sides of the outer belt), with the object of affording extra protection to the engines against a raking from aft. But the *Alexandra* had no conning tower till she underwent a partial conversion when fifteen years old. During her third commission, when ten years old, she was fitted with the new swinging-boom net defence against torpedoes, of the pattern at that time in process of issue to all British battleships.

During her acceptance trials on the Maplin mile she established a new record for battleship speeds, by exceeding 15 knots as the mean of four runs; and for ten years thereafter no new battleship in the world did better. Moreover, during that same period only three of them (all British) had a superior steaming radius of action. This last was due to the fact that the *Alexandra* was one of the world's first two battleships to receive compound engines and cylindrical boilers, which doubled all former standards of steam pressure in large vessels. The other was the turret ship *Dreadnought*, having nothing whatever in common with the *Alexandra* in general design, but identical machinery by the same makers from the same place. Those days were a full generation before the advent of turbines and multiple propellers, and twenty years before water-tube boilers; but the drive of the above two battleships did indicate progress in more ways than one, as apart from the adoption of compound steam pressure, they introduced to the navy a system of placing the machinery (already in use in some new merchant steamers) whereby the cylinders stood above the crank shaft instead of lying beside it. Hitherto the objection to this arrangement had arisen from the belief in the earlier days of steam that the engines of a fighting ship should have the protection of lying entirely below the waterline. But with the increasing thickness of high-side armour, it was considered a sufficient shield to justify raising the machinery in this way to lengthen its stroke, and obviate the troublesome wearing down of the sliding parts inherent in a horizontal lay-out. In later years this became universal in the Navy.

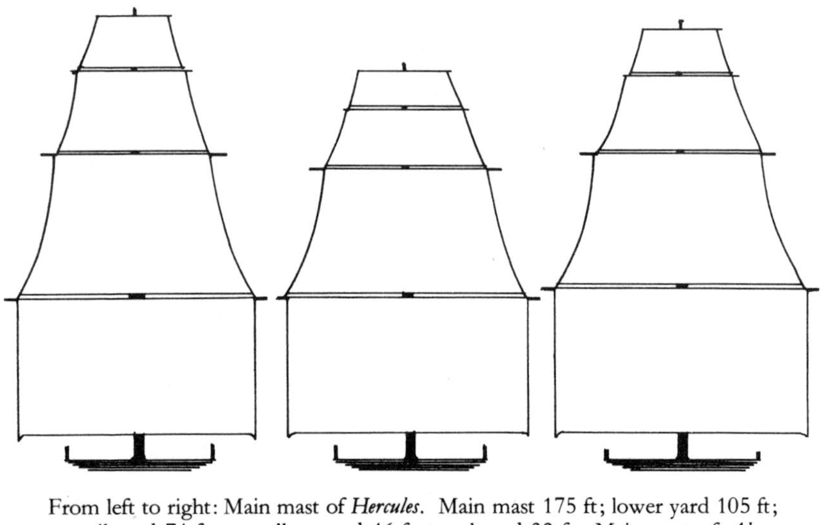

From left to right: Main mast of *Hercules*. Main mast 175 ft; lower yard 105 ft; topsail yard 74 ft; topgallant yard 46 ft; royal yard 32 ft. Main mast of *Alexandra*. Main mast 153 ft; lower yard 105 ft; topsail yard 78 ft; topgallant yard 52 ft; royal yard 38 ft. Main mast of *Temeraire*. Main mast 170 ft; lower yard 111 ft; topsail yard 78 ft; topgallant yard 49 ft; royal yard, 34 ft.

The *Alexandra* had twin screws, and engines with three cylinders each, a high-pressure one put between two low. The contract for her machinery went to the once great works of Humphrys and Tennant, and it proved of such excellent quality, that when her boilers were reviewed sixteen years after her first trials, her speed showed no falling off. Her twelve cylindrical boilers were ranged fore-and-aft back to back in rows of six each, with the furnaces facing outwards. A centre-line bulkhead ran between the rows, and another bulkhead athwartships divided them into two groups transversely, so they were in four compartments. The safety valves were loaded to 60 lb. pressure which exactly doubled former naval practice. Two telescopic funnels carried the draught up 76 feet from the furnaces. During her fourth successive commission of three years her boiler pressure had to be reduced by wear, but after that she received a new boiler set. As auxiliary machinery she carried a steam capstan and steam steering engine, supplemented by hand gear for both. A steam derrick was added as a boom-boat hoist when she was deprived of her square rig at fifteen years old.

Reference has been made several times before to the necessity for masting the earlier armoured vessels if they were to be available for oceanic service, on account of the extravagant coal burning of the engines of their day, and the limited space for bunkerage unavoidable in their class. Yet even when a much more economical type of machinery had been devised and installed on board the *Alexandra*, which sufficed for reaching distant destinations, the conservatism so often dominant at Whitehall led to her being masted with a heavily sparred barque rig, partly on the

The launch of the *Alexandra* by Princess Alexandra, Princess of Wales at Chatham on April 7th, 1875. (NMM. 58/4217)

ground that it might save fuel, and partly as a help in case of a machinery breakdown. The first of these considerations came to nothing as she never made a voyage at any time except under steam, and the second was discounted by her possession of double screws. Eventually therefore, after her twelve years abroad — when she had become the last British masted battleship except the *Temeraire* — the whole square-rig was scrapped and pole masts substituted.

The cut of her sail plan differed considerably from that of long-established British Naval pattern, in its great proportion of breadth or spread to hoist, as may be seen in the accompanying diagram. In their length of 105 feet her lower yards were in some cases equalled by others in the fleet, but surpassed by none except those of the *Temeraire*. Her 78 feet topsail yards were unequalled by those of any other vessel except again those of that same ship.

Finally, her 52 foot topgallant yards and royal yards of 38 were the longest ever made for the Navy. On the other hand her main truck only stood 153 feet above the deck, whereas that of the *Hercules* — which she relieved as Mediterranean flagship — stood at 175 feet; a figure equalled or very closely approached by the main truck heights of all the larger masted vessels on the station. It was remarked by an oldtime observer, on first seeing the *Alexandra* leading a line, that the flag of the Commander-in-Chief flew 20 feet below the pendants of the whole column in his wake. He thought the flagship looked jury rigged. Except for iron lowermasts all her spars were wood. The fore-and-aft canvas was of normal scale and cut for a ship of her size, apart from an immense jib on a jibboom of 64 feet, the longest in British naval records; with the unique peculiarity of measuring 10 feet more than the main topmast. But she carried no flying jib. The head gear was of course of the running-in description borne by all masted vessels with a ram bow. After her second commission her jibboom was discarded.

The total area of the *Alexandra*'s square sail was about 27,000 square feet, but like other twin-screw ships she was a very dull sailer, and though answering to handling under canvas never logged more than 6 knots. When stripped of the rig her lowermasts were cut down about 15 feet, and the fore and mizen fitted with fighting tops for 3-pounder quick-firers.

At that period the old allowance for sailing ships-of-the-line of four heavy anchors and four light was still in force, although to their steam and armoured successors such matters as kedging out of harbour, or dependence on anchors for safety if embayed on a dangerous coast, no longer applied. But traditional standards of sea gear changed very slowly in the mid-Victorian Navy, and the *Alexandra* was supplied with eight anchors of various weights accordingly; though the experience of her long service more than suggests that half the number would have fully sufficed.

Her cables, however, followed the then new fashion of leading in at upper-deck level instead of below, and were worked by a steam capstan. The bower and sheet anchors were catted and fished by hand and stowed on beds on the topgallant forecastle.

Up to the time of her launch the British naval standard of boat equipment for a vessel with a complement of 450

H.M.S. *Alexandra* hit by an Egyptian shell during the attack on Alexandria. It is perhaps worth noting the sextant carried by the officer. At this time it was frequently used for estimating the range of the target, the method introduced by Sir Howard Douglas about 100 years earlier. (NMM. 58/5972)

Shipping beef and supplies on board the *Alexandra* during operations in the Sea of Marmora in 1878. (NMM. 58/4853)

or more ranks and ratings had consisted of three boomboats — one being a steamer — and seven at harbour davits. But in her case this was raised to four boomboats and eight at davits, including a second and smaller steamboat. And as an addition to the ship's allowance proper she was provided with a 14-oared barge, and two gigs for an admiral and his staff. Of the boomboats a steam pinnace and an 18-oared launch were carried close to the ship's sides on the top of the upper battery, being hoisted and turned in by special-sized davits. The other two were 14-oared pinnaces carried amidships on high crutches over the deck forward of the battery and hoisted in by yard tackles. At a later stage one of the new type picket boats took the stowage of the steam pinnace which was shifted to the position of one of the forward pinnaces and the latter was returned to yard. Of the eight harbour davit boats three were stowed inboard for sea.

The *Alexandra* was built at Chatham from the designs of Barnaby and launched on April 7, 1875 by the then Princess of Wales who was later Queen of that name, which made it a highly ceremonial occasion. All Chatham was decorated for the event. Queen Victoria was not there, as in the early years of her widowhood she withdrew almost entirely from public appearance; but nearly all the rest of the Royal family were present, as well as the whole Board of Admiralty, the Diplomatic Corps, most of the Cabinet, the Commander-in-Chief of the Army, and about a hundred members of Parliament. The religious service was conducted by the Archbishop of Canterbury in person, and began the custom that has been observed at the more important naval launchings ever since. By the beginning of 1877 she was ready for sea and commissioned under Vice-Admiral Hornby to relieve the *Hercules* as flagship in the Mediterranean, being seen no more in home waters for the next twelve years. She arrived out to find the fleet lying at Besika Bay, and with them she went up the Dardanelles, and remained in the Sea of Marmora while the Congress of Berlin was convened. Of the original six all were individually relieved one by one by fresh arrivals except the *Alexandra*, which led them out again when the White Ensign was withdrawn to be seen no more in the Dardanelles for nearly forty years.

That was early in 1879, when she had completed two years of her first commission. The third and last year was passed in political cruising about the Levantine end of the station, where the changes arising from the Treaty of Berlin were gradually taking effect including the lease of Cyprus to Britain by Turkey. It was while thus employed that she narrowly escaped disaster by a collision with the *Achilles*. Both were in a measure to blame, but the handiness of the *Alexandra* under helm enabled her to swerve from impact just in time to take the bump broadside to broadside instead of being rammed amidships. And as far as damage to the main hull went, it was the *Achilles* that suffered most, for as the *Alexandra* was swinging clear the tandem blades of her starboard propeller cut two deep parallel gashes in that ship's bilge, while she herself escaped with some crumpled port lids and a clean sweep of her boats at the starboard davits.

She was paid off at Malta early in 1880 and recommissioned with a new ship's company sent out from home; Hornby being succeeded by Sir Beauchamp Seymour. The first two years of this second commission passed uneventfully; but in the third the *Alexandra* was destined to bear a prominent part in the first occasion on which British armoured vessels were ever in contact with an enemy; and the last which any British battle fleet of whatever description fought with and against muzzle-loading guns. But it may be added that it was also the last in which the ships had nothing to fear of attack either under water or from the air. The enemy had no torpedoes, and though well supplied with mines never got a chance to lay them. Aircraft did not exist till a full generation later. Nor was there any of the musketry so common in the fleet battles of the Napoleonic period. Attack and defence — except for some large calibre machine-gun fire — was alike confined to the heavy gun which is the battleship's distinguishing weapon.

This second commission was finished in the January following, when the ship paid off at Malta on the arrival of a new ship's company from home, and Lord John Hay succeeded Seymour, who had been created Baron Alcester, as Commander-in-Chief. Some of her officers and men were involved in the Sudan campaign, but for the ship this third commission was otherwise uneventful, and she was paid off and recommissioned at Malta for a third time with a new crew sent out from home in March 1886, having by that time completed nine years on the station.

The *Alexandra* now renewed the association with Royalty that had attended her launch by the appointment of the Duke of Edinburgh to succeed Lord John Hay as station Commander-in-Chief, and of Prince George of Wales — afterwards King George V — as one of her lieutenants. It was during this period that the Navy was passing from wooden to iron and steel-sided ships, which were much hotter internally under the glare of a tropical sun. But the discomfort which this caused in their crowded berthing spaces could be appreciably reduced by painting them white outside instead of black, and all vessels on tropical stations were changed to white accordingly by order. This order, however, did not include the battleships in the Mediterranean, as only the smaller craft on that station had to take turns of duty in the Red Sea. Nevertheless, as the *Alexandra* had a large ship's company living on a none too spacious mess deck, the Admiralty acceded to a request from the Duke that she should be painted white. So she changed her complexion accordingly to an albino when ten years of age, but reverted to her maiden hue on returning home. Her fourth and final commission of her long foreign service passed in the normal peace duties of a station flagship, except during one period. The ground swell of racial antagonism in the Levantine area never subsided altogether, and international action was again deemed expedient in 1887 to stop the Greeks from starting a war against the Turks, for complicated reasons not necessary to describe here. Intervention took the form of a blockade of the Greek ports by a combined squadron from four Powers, placed under the temporary command of the Duke of Edinburgh; which display of force answered its purpose without firing a shot. Before the end of the commission

Alfred Ernest Albert, Duke of Edinburgh, the second son of Queen Victoria, and Admiral of the Fleet, who saw service with many of the vessels of the Black Battlefleet, and was responsible for his Flagship, the *Alexandra*, being painted white. (NMM. 59/11)

H.M.S. *Alexandra* as she appeared in 1901 right at the end of her active career. Stripped of her canvas, military fighting tops added to her fore and mizen masts, and as an addition to signalling, a mast head semaphore at the truck of her main mast. Above all, as a foretaste of things to come, she was one of the first ships in the Royal Navy to be fitted out for wireless transmission, and is shown with a wireless gaff on her mizen. (NMM. AAN/3)

the *Alexandra* had become the last square-rigged British battleship in the Mediterranean, except her squadron mate of very long standing the *Temeraire*, and much inferior in power to many vessels in those waters of more modern type, both British and foreign. In May 1889 she sailed for England, under orders to pay off for the first time in a home instead of an oversea port; after twelve years and four months unbroken service abroad, during the whole of which she was never west of Gibraltar nor east of the Syrian shores.

For twenty-one months thereafter she was in dockyard hands at Chatham, for extensive reconditioning and the various alterations in armament, rig, etc. already detailed. Then in 1891 she began another long spell of continuous service — this time all in home waters and lasting ten years — by hoisting the flag of Rear-Admiral Fitzroy in command of the First Reserve Fleet. Her headquarters were at Portland, whence she only went out for the usual summer cruise of that force, or for any general fleet manoeuvres, or routine gunnery practice. Fitzroy was succeeded in turn by Vice-Admirals E. Seymour, Domvile and Noel. All the contemporary battleships of her early days had either been broken up or converted to hulks long before the last of these Flag Officers lived under her poop, and as mentioned already she was a lone figure in class at the Diamond Jubilee Review.

In 1901 she paid off at Chatham for the last time, and was disarmed and placed in the lowest category of reserve as a mechanical training ship lying in a dockyard basin. As that she remained till sold to Garnham for breaking up in October 1906. She had been the last to be launched of the eighteen vessels in the British Navy of the central battery or 'box and belt armour' type, and ended by being many years the last to fly her colours in seagoing commission.

STATISTICAL APPENDICES

INTRODUCTION

1870 was the date selected by Admiral Ballard for his review of the Black Battlefleet: therefore so far as is practicable we have kept as close as possible to that date in preparing the following tables. Most of the material has been taken from official sources of which the principal are: —

Iron Cased Ships of England — a half yearly return of ships in the Royal Navy giving dimensions, armour, armament, cost etc. Later the title was changed to *Steam Ships of England.*

Results of Trials made in Her Majesty's Screw Ships and Vessels — published over a number of years and dating back to the 1850s, this gives not only the results of steaming trials but also much information and details on the machinery etc. Where several trials took place we have selected that trial where the results came closest to the official record.

Masting and Rigging Dimensions Books — several manuscript volumes of reports on the majority of vessels in service with the Royal Navy from *c.* 1850 until the end of the century. Where reports are missing details can frequently be found in the official Admiralty draughts now held by the National Maritime Museum.

Figures in these contemporary records vary frequently over the ensuing years. This is particularly the case with 'displacement' when it was introduced in place of 'tons burden' in the early months of 1873. Although the official *Navy List* displacement is quoted as at the time it was introduced, a more accurate figure can be obtained by adding together the weights of the hull and the equipment etc.

Again completion dates can be disputed: those given are the official ones even if in some instances we find vessels commissioned before completion: these dates must therefore be regarded as approximate.

Although the official reports quote speeds to three decimal points, these must be taken with reserve. In some instances they had been worked out mathematically without allowance for propeller slip. Moreover trials run off Maplin, for example, can be open to correction since the relatively shallow depth of water leads to considerable inaccuracy.

In the case of trunk the effective diameter of the cylinder is given instead of the actual diameter; i.e. the diameter of the remaining area after the end area of the trunk has been subtracted.

It is nearly impossible to accurately state a ship's endurance. At full speed and with full bunkers most ships had an endurance of about 3 ½ to 4 ½ days. Whereas a merchant ship, and especially a North Atlantic liner, may be expected to make the whole crossing at approaching her top speed, in the case of a warship such conditions would only obtain for a limited period and her actual endurance would be far greater at a more modest speed.

For example the *Minotaur*, with a clean bottom on good quality coal and in fair weather, would burn some 143 tons of coal per 24 hours to develop 2,400 IHP to produce her full speed of 10 ½ knots giving a paper endurance of about 5 ¼ days: at 1,600 IHP producing 9 ¼ knots she would burn about 98 tons per 24 hours thus increasing her endurance to 7 ¾ days steaming.

Regarding the old coast defence ships covered in Chapter XVIII, it is not realistic to give comparable details of masts and spars: only a few were rigged in the true sense of the word. What we have done is to cite the rig of the *Cerberus* with which she was specially fitted out for her voyage to Australia. Of course the *Devastation* was never rigged for sail!

APPENDIX A

CONSTRUCTION DATES

	Where Built	Order to Build	Order to Convert	Laid down	Launched	Completed	Commissioned
MINOTAUR	Blackwall	2 Sept. 1861	—	12 Sept. 1861	12 Dec. 1863	1 June 1867	April 1867
AGINCOURT	Birkenhead	2 Sept. 1861	—	30 Oct. 1861	27 Mar. 1865	19 Dec. 1868	June 1868
NORTHUMBERLAND	Millwall	2 Sept. 1861	—	10 Oct. 1861	17 April 1866	8 Oct. 1868	Oct. 1868
ACHILLES	Chatham	10 April 1861	—	1 Aug. 1861	23 Dec. 1863	26 Nov. 1864	Sept. 1864
WARRIOR	Blackwall	11 May 1859	—	25 May 1859	29 Dec. 1860	24 Oct. 1861	Aug. 1861
BLACK PRINCE	Glasgow	6 Oct. 1859	—	12 Oct. 1859	27 Feb. 1861	12 Sept. 1862	June 1862
BELLEROPHON	Chatham	23 July 1863	—	28 Dec. 1863	26 April 1865	11 April 1866	Mar. 1866
HERCULES	Chatham	Feb. 1865	—	1 Feb. 1866	10 Feb. 1868	21 Nov. 1868	Oct. 1868
LORD CLYDE	Pembroke	3 July 1863	—	29 Sept. 1863	13 Oct. 1864	15 Sept. 1866	June 1866
LORD WARDEN	Chatham	25 May 1863	—	24 Dec. 1863	27 May 1865	30 Aug. 1867	July 1867
MONARCH	Chatham	15 Jan. 1866	—	1 June 1866	25 May 1868	5 Aug. 1869	May 1869
CAPTAIN	Birkenhead	30 Jan. 1867	—	30 Jan. 1867	27 Mar. 1869	Jan. 1870	April 1870
PRINCE CONSORT	Pembroke	—	31 May 1861	13 Aug. 1860	26 June 1862	6 Feb. 1864	Jan. 1864
CALEDONIA	Woolwich	—	31 May 1861	1 Oct. 1860	24 Oct. 1862	6 July 1865	April 1865
OCEAN	Devonport	—	5 June 1861	23 Aug. 1860	19 Mar. 1863	6 Sept. 1866	July 1866
ROYAL OAK	Chatham	—	14 May 1861	1 May 1860	10 Sept. 1862	28 May 1863	April 1863
ROYAL ALFRED	Portsmouth	—	5 June 1861	1 Dec. 1859	15 Oct. 1864	23 Mar. 1867	Jan. 1867
ZEALOUS	Pembroke	—	2 July 1862	26 Oct. 1859	7 Mar. 1864	4 Oct. 1866	Sept. 1866
REPULSE	Woolwich	—	9 Oct. 1866	29 April 1859	25 April 1868	31 Jan. 1870	Mar. 1870
HECTOR	Glasgow	25 Jan. 1861	—	8 Mar. 1861	26 Sept. 1862	22 Feb. 1864	Jan. 1864
VALIANT	Poplar	25 Jan. 1861	—	1 Feb. 1861	14 Oct. 1863	15 Sept. 1868	Sept. 1868
DEFENCE	Jarrow	14 Dec. 1859	—	14 Dec. 1859	24 April 1861	12 Feb. 1862	Dec. 1861
RESISTANCE	Poplar	14 Dec. 1859	—	21 Dec. 1859	11 April 1861	5 Oct. 1862	July 1862
AUDACIOUS	Glasgow	29 April 1867	—	26 June 1867	27 Feb. 1869	10 Sept. 1870	Oct. 1870
INVINCIBLE	Glasgow	29 April 1867	—	28 June 1867	29 May 1869	1 Oct. 1870	Oct. 1870
IRON DUKE	Pembroke	26 Sept. 1867	—	23 April 1868	1 Mar. 1870	21 Jan. 1871	Jan. 1871
VANGUARD	Birkenhead	21 Oct. 1867	—	21 Oct. 1867	3 Jan. 1870	28 Sept. 1870	Oct. 1870
FAVORITE	Deptford	—	29 May 1862	23 Aug. 1860	5 July 1864	17 Mar. 1866	Feb. 1866
PALLAS	Woolwich	23 July 1863	—	19 Oct. 1863	14 Mar. 1865	4 April 1866	Mar. 1866
PENELOPE	Pembroke	Feb. 1865	—	4 Sept. 1865	18 June 1867	27 June 1868	June 1868
TEMERAIRE	Chatham	—	—	18 Aug. 1873	9 May 1876	31 Aug. 1877	Aug. 1877
ROYAL SOVEREIGN	Portsmouth	—	3 April 1862	17 Dec. 1849	25 April 1857	20 Aug. 1864	July 1864
PRINCE ALBERT	Poplar	8 April 1862	—	29 April 1862	23 May 1864	23 Feb. 1866	June 1866
CYCLOPS	Blackwall	5 Aug. 1870	—	10 Sept. 1870	18 July 1871	4 May 1877	Jan. 1872
GORGON	Jarrow	5 Aug. 1870	—	5 Sept. 1870	14 Oct. 1871	19 Mar. 1874	April 1872
HECATE	Poplar	5 Aug. 1870	—	5 Sept. 1870	30 Sept. 1873	24 May 1877	April 1872
HYDRA	Glasgow	5 Aug. 1870	—	5 Sept. 1870	28 Dec. 1871	31 May 1876	Aug. 1872
MAGDALA	Blackwall	—	—	6 Oct. 1868	2 Mar. 1870	Nov. 1870	—
CERBERUS	Jarrow	—	—	18 Sept. 1867	2 Dec. 1868	Sept. 1870	—
ABYSSINIA	Poplar	—	—	23 July 1868	19 Feb. 1870	Oct. 1870	—
GLATTON	Chatham	5 June 1868	—	10 Aug. 1868	8 Mar. 1871	24 Feb. 1872	May 1872
HOTSPUR	Glasgow	21 Sept. 1868	—	2 Oct. 1868	19 Mar. 1870	17 Nov. 1871	Nov. 1871
RUPERT	Chatham	11 Feb. 1869	—	6 June 1870	12 Mar. 1872	1 July 1874	May 1876
DEVASTATION	Portsmouth	11 Feb. 1869	—	12 Nov. 1869	12 July 1871	19 April 1873	Jan. 1873
ALEXANDRA	Chatham	—	—	5 Mar. 1873	7 April 1875	31 Jan. 1877	Jan. 1877

Note. In some instances vessels were commissioned for trials before completion.

APPENDIX B

DIMENSIONS OF HULL

	Tonnage		Weights		Length	Beam	Draught	
	Burden	Displ.	Hull	Equipment	B.P.		Fwd	Aft
MINOTAUR	6621	10627	7170	3220	400'0"	59'4¾"	26'3"	27'0"
AGINCOURT	6621	10627	7559	3305	400'0"	59'5"	27'3"	27'9"
NORTHUMBERLAND	6621	10584	7245	3339	400'4"	59'5"	26'4"	27'9"
ACHILLES	6121	9694	6671	3023	380'0"	58'3½"	26'6"	27'2"
WARRIOR	6109	9137	6072	3034	380'2"	58'4"	26'2"	26'9"
BLACK PRINCE	6109	9137	6150	3019	380'2"	58'4"	26'4"	26'10"
BELLEROPHON	4270	7551	4946	2605	300'0"	56'1"	22'10"	26'7"
HERCULES	5234	8677	5500	3177	325'0"	59'0½"	22'11"	26'6"
LORD CLYDE	4067	7842	5026	2716	280'0"	58'11"	23'9"	27'2"
LORD WARDEN	4080	7842	5183	2759	280'0"	59'0½"	23'11"	27'11"
MONARCH	5102	8322	5423	2899	330'0"	57'6¼"	22'7"	26'0"
CAPTAIN	4272	6950	5293	2623	320'0"	53'3"	24'6"	25'6"
PRINCE CONSORT	4045	6832	4190	2525	273'1"	58'5"	24'0"	26'10"
CALEDONIA	4125	6832	4307	2548	273'0"	59'2"	24'5"	27'0"
OCEAN	4047	6832	4393	2534	273'1"	58'5"	24'5"	27'6"
ROYAL OAK	4056	6366	3997	2369	273'0"	58'6"	23'11"	25'2"
ROYAL ALFRED	4068	6707	4384	2323	273'0"	58'7"	23'8"	27'0"
ZEALOUS	3716	6096	3796	2300	252'0"	58'7"	25'0"	25'9"
REPULSE	3749	6190	4096	2094	252'0"	59'1½"	25'4"	26'2"
HECTOR	4089	6713	4653	2044	280'2"	56'5"	25'1"	26'0"
VALIANT	4063	6713	4602	2128	280'2"	56'4"	25'2"	26'2"
DEFENCE	3720	6070	4330	1744	280'0"	54'2"	24'4"	26'1"
RESISTANCE	3710	6070	4337	1728	280'0"	54'1"	23'7"	26'7"
AUDACIOUS	3774	6034	4035	2065	280'0"	54'0"	22'0"	23'0"
INVINCIBLE	3774	6034	3930	2109	280'0"	54'0"	21'11"	22'11"
IRON DUKE	3774	6034	3849	2075	280'0"	54'0"	21'7"	22'7"
VANGUARD	3774	6034	3921	2155	280'0"	54'0"	22'0"	23'0"
FAVORITE	2094	3232	2089	1143	225'0"	46'9½"	19'11"	22'9"
PALLAS	2372	3787	2437	1350	225'0"	50'0"	19'1"	24'3"
PENELOPE	3096	4394	2822	1572	260'0"	50'0"	15'9"	17'4"
TEMERAIRE	—	8415	5730	2810	285'0"	62'0"	26'8"	27'2"
ROYAL SOVEREIGN	3765	5080	3798	1282	240'7"	62'2"	21'6"	24'11"
PRINCE ALBERT	2529	3905	2981	924	240'0"	48'1"	18'8"	20'4"
CYCLOPS	2107	3430	2539	797	225'0"	45'0"	16'0"	16'0"

	Tonnage		Weights		Length B.P.	Beam	Draught	
	Burden	Displ.	Hull	Equipment			Fwd	Aft
GORGON	2107	3430	2539	797	225′0″	45′0″	16′0″	16′0″
HECATE	2107	3430	2539	797	225′0″	45′0″	16′0″	16′0″
HYDRA	2107	3430	2539	797	225′0″	45′0″	16′0″	16′0″
MAGDALA	2107	3344	2640	700	225′0″	45′0″	15′3″	15′3″
CERBERUS	2107	3344	2640	700	225′0″	45′0″	15′3″	15′3″
ABYSSINIA	1854	2908	2200	700	225′0″	42′0″	14′6″	14′6″
GLATTON	2709	4912	3658	1182	245′0″	54′0″	18′2″	19′4″
HOTSPUR	2637	4010	2803	1207	235′0″	50′0″	19′0″	20′8″
RUPERT	3159	5358	3712	1646	250′0″	53′0″	21′6″	23′6″
DEVASTATION	4407	9188	5827	3363	285′0″	62′3″	26′4″	26′6″
ALEXANDRA	—	9490	6160	3330	325′0″	63′8″	26′0″	26′6″

APPENDIX C

Thickness	Weight Tons	Depth below designed load water line	Timber backing	Armament: Main deck	Armament: Upper deck
MINOTAUR Sides 5½″	1,776	5′9″	Teak 10″	4-9″ 12½ ton MLR 18-7″ 6½ ton MLR	4-7″ 6½ ton MLR
AGINCOURT Sides 5½″	1,776	5′9″	Teak 10″	4-9″ 12½ ton MLR 20-7″ 6½ ton MLR	4-7″ 6½ ton MLR
NORTHUMBERLAND Sides 5½″	1,549	5′9″	Teak 10″	4-9″ 12½ ton MLR 18-8″ 9 ton MLR 2-7″ 6½ ton MLR	4-8″ 9 ton MLR
ACHILLES Sides 4½″	1,259	5′0″	Teak 18″	4-8″ 9 ton MLR 18-7″ 6½ ton MLR	4-7″ 6½ ton MLR
WARRIOR Sides 4½″	975	5′0″	Teak 18″	4-8″ 9 ton MLR 20-7″ 6½ ton MLR	8-7″ 6½ ton MLR
BLACK PRINCE Sides 4½″	975	5′0″	Teak 18″	4-8″ 9 ton MLR 18-7″ 6½ ton MLR	6-7″ 6½ ton MLR
BELLEROPHON Sides 6″. Skin 1½″	1,093	6′0″	Teak 10″	10-9″ 12½ ton MLR 3-7″ 6½ ton MLR	2-7″ 6½ ton MLR
HERCULES Sides 9″, 8″, 6″, 5″ Skin 1½″	1,339	6′0″	Teak 12″ and 10″	8-10″ 18 ton MLR 2-9″ 12½ ton MLR	4-7″ 6½ ton MLR
LORD CLYDE Sides 5½″, 4½″ Skin 1½″	1,379	6′0″	Wooden ship, side 31½″ thick	1-9″ 12½ ton MLR 12-8″ 9 ton MLR 2-7″ 6½ ton MLR	1-9″ 12½ ton MLR 2-8″ 9 ton MLR
LORD WARDEN Sides 5½″, 4½″ Skin 1½″	1,379	6′0″	Wooden ship, side 31½″ thick	1-9″ 12½ ton MLR 12-8″ 9 ton MLR 2-7″ 6½ ton MLR	1-9″ 12½ ton MLR 2-8″ 9 ton MLR
MONARCH Sides 7″, 6″, 5″, 4½″, 4″ Turrets 10″, 8″. Skin 1½″	1,364	5′0″	Teak 12″	1-7″ 6½ ton MLR	4-12″ 25 ton MLR 2-7″ 6½ ton MLR
CAPTAIN Sides 8″, 7″, 4″, 3″. Deck 1″. Turrets 10″, 9″. Skin 1½″	1,562	5′0″	Teak 12″	—	4-12″ 25 ton MLR 2-7″ 6½ ton MLR
PRINCE CONSORT Sides 4½″	941	5′6″	Wooden ship, side 29½″ thick	4-8″ 9 ton MLR 16-7″ 6½ ton MLR	4-7″ 6½ ton MLR
CALEDONIA Sides 4½″	930	5′6″	Wooden ship, side 29½ thick	4-8″ 9 ton MLR 16-7″ 6½ ton MLR	4-7″ 6½ ton MLR
OCEAN Sides 4½″	941	5′6″	Wooden ship, side 29½ thick	4-8″ 9 ton MLR 16-7″ 6½ ton MLR	4-7″ 6½ ton MLR
ROYAL OAK Sides 4½″	930	5′6″	Wooden ship, side 29½ thick	4-8″ 9 ton MLR 16-7″ 6½ ton MLR	4-7″ 6½ ton MLR
ROYAL ALFRED Sides 6″, 4½″	950	5′6″	Wooden ship, side 29½″ thick	10-9″ 12½ ton MLR 4-7″ 6½ ton MLR	4-7″ 6½ ton MLR

Thickness	Weight Tons	Depth below designed load water line	Timber backing	Armament: Main deck	Armament: Upper deck
ZEALOUS Sides 4½″	791	6′0″	Wooden ship, side 30½ thick	16-7″ 6½ ton MLR	4-7″ 6½ ton MLR
REPULSE Sides 6″, 4½″	1,043	6′0″	Wooden ship, side 31″ thick	8-8″ 9 ton MLR	4-8″ 9 ton MLR
HECTOR Sides 4½″	912	5′6″	Teak 18″	12-7″ 6½ ton MLR	2-8″ 9 ton MLR 4-7″ 6½ ton MLR
VALIANT Sides 4½″	912	5′6″	Teak 18″	12-7″ 6½ ton MLR	2-8″ 9 ton MLR 4-7″ 6½ ton MLR
DEFENCE Sides 4½″	607	4′0″	Teak 18″	2-8″ 9 ton MLR 10-7″ 6½ ton MLR	4-7″ 6½ ton MLR
RESISTANCE Sides 4½″	607	4′0″	Teak 18″	2-8″ 9 ton MLR 8-7″ 6½ ton MLR	6-7″ 6½ ton MLR
AUDACIOUS Sides 8″, 6″ Skin 1¼″	924	5′0″	Teak 10″	6-9″ 12 ton MLR	4-9″ 12 ton MLR 4-64 pdr. 64 cwt. MLR
INVINCIBLE Sides 8″, 6″ Skin 1¼″	924	5′0″	Teak 10″	6-9″ 12 ton MLR	4-9″ 12 ton MLR 4-64 pdr. 64 cwt. MLR
IRON DUKE Sides 8″, 6″ Skin 1¼″	924	5′0″	Teak 10″	6-9″ 12 ton MLR	4-9″ 12 ton MLR 4-64 pdr. 64 cwt. MLR
VANGUARD Sides 8″, 6″ Skin 1¼″	924	5′0″	Teak 10″	6-9″ 12 ton MLR	4-9″ 12 ton MLR 4-64 pdr. 64 cwt. MLR
FAVORITE Sides 4½″	560	3′0″	Wooden ship, side 26″ thick	—	8-7″ 6½ ton MLR 2-64 pdr. 64 cwt. MLR
PALLAS Sides 4½″	560	4′0″	Wooden ship, side 22″ thick	4-8″ 9 ton MLR 2-64 pdr. 64 cwt. MLR	2-40 pdr. 32 cwt. BL
PENELOPE Sides 6″, 5″	688	4′0″	Teak 10″ and 11″	8-8″ 9 ton MLR	3-40 pdr. 32 cwt. BL
TEMERAIRE Sides 11″, 10″, 9″, 8″ Bulkheads 8″, 5″. Barbettes 10″, 8″. Armour tubes 4″, 3½″, 1½″. Decks 1½″, 1″. Skin 1½″, 1″	1,990	5′0″	Teak 12″ and 10″	2-11″ 25 ton MLR 4-10″ 18 ton MLR	2-11″ 25 ton MLR 4-20 pdr. 13 cwt. BL
ROYAL SOVEREIGN Sides 5½″. Turrets 5½″. Deck 1″	960	3′0″	Wooden ship, side 36″ thick	—	5-9″ 12½ ton MLR
PRINCE ALBERT Sides 4½″. Turrets 5½″, 4½″. Deck 1⅛″ to ¾″	879	4′0″	Teak 18″	—	4-9″ 12½ ton MLR

Thickness	Weight Tons	Depth below designed load water line	Timber backing	Armament	
				Main deck	Upper deck
CYCLOPS Sides 8″, 6″. Turrets 10″, 9″. Breastwork 9″, 8″. Deck 1½″. Skin 1¼″, 1″	1,130	4′3″	Teak 11″ and 9″	—	4-10″ 18 ton MLR
GORGON Sides 8″, 6″. Turrets 10″, 9″. Breastwork 9″, 8″. Deck 1½″. Skin 1¼″, 1″	1,130	4′3″	Teak 11″ and 9″	—	4-10″ 18 ton MLR
HECATE Sides 8″, 6″. Turrets 10″, 9″. Breastwork 9″, 8″. Deck 1½″. Skin 1¼″, 1″	1,130	4′3″	Teak 11″ and 9″	—	4-10″ 18 ton MLR
HYDRA Sides 8″, 6″. Turrets 10″, 9″. Breastwork 9″, 8″. Deck 1½″. Skin 1¼″, 1″	1,130	4′3″	Teak 11″ and 9″	—	4-10″ 18 ton MLR
MAGDALA Sides 8″, 6″. Turrets 10″, 9″. Breastwork 9″, 8″. Deck 1½″, 1″. Skin 1¼″, 1″	—	4′0″	Teak 11″ and 9″	—	4-10″ 18 ton MLR
CERBERUS Sides 8″, 6″. Turrest 10″, 9″. Breastwork 9″, 8″. Deck 1½″, 1″. Skin 1¼″, 1″	—	4′0″	Teak 11″ and 9″	—	4-10″ 18 ton MLR
ABYSSINIA Sides 7″, 6″. Turrets 10″, 8″. Breastwork 8″, 7″. Deck 1½″, 1″. Skin 1¼″, 1″	556	4′0″	Teak 11″ and 9″	—	4-10″ 18 ton MLR
GLATTON Sides 12″ to 10″. Turret 14″, 12″. Deck 3″. Skin 2″	1,732	4′0″	Teak 20″ and 18″	—	2-12″ 25 ton MLR
HOTSPUR Sides 11″, 8″. Deck 3″, 2″. Skin 1¼″	1,242	5′0″	Teak 15″ and 12″	—	1-12″ 25 ton MLR 2-64 pdr. 64 cwt. MLR
RUPERT Sides 12″, 11″, 9″. Turret 14″, 12″. Deck 3″, 2″. Skin 1¼″	1,619	5′0″	Teak 14″, 12″ and 10″	—	2-10″ 18 ton MLR 2-64 pdr. 64 cwt. MLR
DEVASTATION Sides 12″, 10″. Turrcts 14″, 12″, (in two thicknesses). Decks 3″, 1¼″. Skin 1½″	2,411	5′0″	Teak 18″ and 16″	—	4-12″ 35 ton MLR
ALEXANDRA Sides 12″, 10″, 8″, 6″. Bulkheads 8″, 7″, 6″, 5″. Decks 1½″, 1″. Skin 1½″	2,060	5′6″	Teak 12″ and 10″	2-11″ 25 ton MLR 2-10″ 18 ton MLR	8-10″ 18 ton MLR 6-20 pdr. 13 cwt. BL

APPENDIX D

	Engine								Screw	
	Engine builder	Type	No. of Cyl.	Nominal H.P.	Diameter of Cylinder	Stroke	Pressure lbs	Coal tons	Dia.	Pitch
MINOTAUR	Penn	Trunk	2	1350	104.23″	52″	25	750	24′0″	22′2″
AGINCOURT	Maudslay	Double piston rod	2	1350	101″	54″	25	750	24′6″	23′4″
NORTHUMBERLAND	Penn	Trunk	2	1350	108.9″	52″	25	750	24′0″	23′4″
ACHILLES	Penn	Trunk	2	1250	104.23″	48″	25	750	24′0″	25′4″
WARRIOR	Penn	Trunk	2	1250	104.23″	48″	20	700	24′6″	30′0″
BLACK PRINCE	Penn	Trunk	2	1250	104.23″	48″	25	700	24′6″	28′6″
BELLEROPHON	Penn	Trunk	2	1000	104.23″	48″	30	620	23′6″	20′1″
HERCULES	Penn	Trunk	2	1200	118″	54″	30	600	23′6″	24′0″
LORD CLYDE	Ravenhill	Double piston rod	2	1000	116″	48″	28.5	600	23′0″	23′6″
LORD WARDEN	Maudslay	Horizontal	3	1000	91″	54″	31.25	600	22′9″	21′4½″
MONARCH	Humphrys	Horizontal	2	1100	120″	54″	31.6	600	23′4″	26′4″
CAPTAIN	Laird	Trunk	2×2	900	80.125″	39″	29.625	500	17′0″	21′8⅛″
PRINCE CONSORT	Maudslay	Horizontal	2	1000	92″	48″	22	560	21′0″	24′9″
CALEDONIA	Maudslay	Double piston rod	2	1000	92″	48″	21	600	21′0″	25′2½″
OCEAN	Maudslay	Double piston rod	2	1000	92″	48″	22	570	21′0″	23′0″
ROYAL OAK	Maudslay	Horizontal	2	800	82″	48″	22	500	19′0″	27′6″
ROYAL ALFRED	Maudslay	Double piston rod	2	800	82″	48″	22.25	560	19′2½″	27′10″
ZEALOUS	Maudslay	Horizontal	2	800	82″	48″	22	600	19′1″	27′6″
REPULSE	Penn	Trunk	2	800	82″	48″	20	460	19′0″	28′0″
HECTOR	Napier	Horizontal	2	800	82″	48″	22	450	20′0″	26′6″
VALIANT	Maudslay	Horizontal	2	800	82″	48″	25	450	20′2″	27′6″
DEFENCE	Penn	Trunk	2	600	70.75″	42″	20	450	18′0″	21′0″
RESISTANCE	Penn	Trunk	2	600	70.75″	42″	23	450	18′0″	21′0″
AUDACIOUS	Ravenhill	Horizontal	2×2	800	77″	36″	31	450	16′2″	18′6¾″
INVINCIBLE	Napier	Horizontal	2×2	800	72″	36″	34	450	16′2″	20′9⅞″
IRON DUKE	Ravenhill	Horizontal	2×2	800	77″	36″	32	450	16′6″	19′0″
VANGUARD	Laird	Horizontal	2×2	800	72″	36″	31.75	450	16′2″	20′9″
FAVORITE	Humphrys	Horizontal	2	400	64″	32″	22	350	16′0″	18′0¼″
PALLAS	Humphrys	Woolf type comp	4	600	(2)51″(2)99½″	39″	32	250	17′10″	19′6″
PENELOPE	Maudslay	Horizontal	2×3	600	55.25″	30″	30.5	330	14′0″	15′6″
TEMERAIRE	Humphrys	Inverted compound	2×2	—	(2)70″ (2)114″	46″	60	400	20′0½″	22′6″
ROYAL SOVEREIGN	Maudslay	Horizontal	2	800	82″	48″	21	280	19′1½″	26′8¾″
PRINCE ALBERT	Humphrys	Horizontal	2	500	72″	36″	24	300	17′0″	20′10″
CYCLOPS	Elder	Inverted compound	2×2	250	(2)31″(2)57″	27″	60	250	12′0″	8′3″
GORGON	Ravenhill	Inverted simple	2×2	250	45″	24″	34	270	12′0″	9′2¼″

Abstract of trial data

Date	Place	State of vessel at time of trial	Draught Fwd	Draught Aft	Displ. Tons	r.p.m. of engine	I.H.P.	Average speed knots
10 May 1867	Stokes Bay	Complete	25′8″	26′10″	10.185	61.473	6949	14.328
12 Dec. 1865	Plymouth	Lower masts only	23′1½″	25′2″	9.060	52.75	4426	13.547
15 Sept. 1868	Plymouth	Rigged and stored	26′2″	27′9″	10.548	58.385	6558	14.132
15 Mar. 1865	Stokes Bay	Rigged and stored	25′11	26′11″	9.487	54.583	5722	14.322
1 Apr. 1868	Stokes Bay	Complete	25′9″	27′8″	9.231	53.14	5267	14.079
10 Sept. 1862	Stokes Bay	Rigged and equipped	26′4″	27′3″	9.300	56.09	5772	13.604
17 Aug. 1864	Stokes Bay	Rigged	21′6″	26′5″	7.236	73.86	6521	14.171
5 Feb. 1869	At sea	Complete	23′0″	26′5″	8.685	67.189	7187	13.802
9 Feb. 1867	Stokes Bay	Rigged and stored	23′0″	27′0″	7.602	64.754	6064	13.433
13 Sept. 1867	Plymouth	Rigged and stored	23′7″	27′9″	7.839	63.327	6706	13.496
20 July 1869	At sea	Complete	21′8″	25′9″	8.070	62.667	7470	14.715
23 Mar. 1870	At sea	Complete	24′1½″	25′6″	7.664	74.15	5772	14.202
8 Mar. 1865	Plymouth	Rigged and stored	24′1″	26′7″	6.700	55.000	3953	12.77
16 Feb. 1867	Plymouth	Rigged and stored	23′8″	27′3″	6.753	57.583	4538	13.078
25 June 1864	Plymouth	Rigged and stored	23′9″	26′2″	6.535	58.000	4244	12.896
15 June 1863	Stokes Bay	Complete	23′8″	25′8″	6.416	60.833	3704	12.529
5 Apr. 1867	Stokes Bay	Complete	23′7″	27′2″	6.720	60.562	3434	12.359
24 Nov. 1866	Plymouth	Rigged and stored	24′10″	25′11″	6.102	59.5	3623	11.706
13 July 1870	Stokes Bay	Rigged and armed	24′4″	25′10″	6.010	58.74	3347	12.284
23 Feb. 1864	Stokes Bay	Complete	24′2″	25′8″	6.455	59.083	3256	12.36
18 Sept. 1865	Stokes Bay	Rigged and stored	23′6″	24′6″	6.229	61.167	3560	12.649
10 Feb. 1868	Plymouth	Rigged and equipped	24′3″	26′1″	6.077	65.917	2343	11.228
25 Aug. 1873	Plymouth	Rigged and armed	24′7″	26′11″	6.300	69.61	2329	11.409
21 Oct. 1870	Plymouth	Rigged and armed	22′0″	23′0″	6.170	72.220	4021	12.829
26 Sept. 1870	Plymouth	Rigged and armed	20′6″	21′6″	5.563	70.790	4832	14.093
2 Nov. 1870	Plymouth	Rigged and armed	20′10″	21′6″	5.620	80.860	4268	13.644
9 Apr. 1870	At sea	Rigged and armed	20′4″	21′5½″	5.528	72.220	4914	14.651
16 Mar. 1866	Nore	Complete	19′7″	22′7″	3.169	67.833	1773	11.825
4 July 1867	Plymouth	Complete	18′9″	24′6″	3.780	78.025	3210	12.452
1 July 1868	Plymouth	Rigged and stored	15′7″	17′4¾″	4.368	102.95	4703	12.764
17 Sept. 1877	Inside Wight	Rigged and armed	26′8″	27′4″	8.571	73.764	7516	14.651
23 June 1864	Stokes Bay	Rigged, partly stored	21′2″	24′8″	4.965	51.25	2463	11.003
7 Mar. 1866	Plymouth	Rigged and stored	17′9″	19′10″	3.687	61.167	2128	11.652
30 Dec. 1871	At sea	Not rigged or armed	14′5″	15′5″	3.100	124.62	1528	10.790
24 Aug. 1872	Plymouth	Not rigged or armed	14′5″	15′5″	3.100	109.59	1709	11.158

MACHINERY DATA

	Engine								Screw	
	Engine builder	Type	No. of Cyl.	Nominal H.P.	Diameter of Cylinder	Stroke	Pres-sure lbs	Coal tons	Dia.	Pitch
HECATE	Ravenhill	Inverted simple	2×2	250	45″	24″	34	270	12′0″	9′0″
HYDRA	Elder	Inverted compound	2×2	250	(2)31″(2)57″	27″	60	250	12′0″	8′6″
MAGDALA	Ravenhill	Horizontal	2×2	250	48″	21″	32	220	12′0″	9′0″
CERBERUS	Maudslay	Horizontal	2×2	250	43″	27″	30	220	12′0″	9′0″
ABYSSINIA	Dudgeon	Diagonal simple	2×2	200	34″	21″	30	90	9′6″	9′6″
GLATTON	Laird	Horizontal	2×2	500	60″	27″	31.4	240	14′0″	13′9½″
HOTSPUR	Napier	Horizontal	2×2	600	64″	33″	20	300	14′0¾″	14′9″
RUPERT	Napier	Horizontal	2×2	700	72″	36″	35	480	15′6″	17′0″
DEVASTATION	Penn	Trunk	2×2	800	80″	39″	30	1350	17′6″	19′6½″
ALEXANDRA	Humphrys	Inverted compound	2×3	—	(2)70″(4)90″	48″	60	400	21′0″	22′2¾″

(a) The quantity of coal quoted is authorised capacity of bunkers, not in all cases the maximum amount.

(b) The trial selected is the one on which official data appears to have been based.

Abstract of trial data								
Date	Place	State of vessel at time of trial	Draught		Displ. Tons	r.p.m. of engine	I.H.P.	Average speed knots
			Fwd	Aft				
10 Apr. 1872	At sea	Not rigged or armed	14′1″	15′6″	3.080	105.33	1579	10.55
4 July 1872	Greenock	Not rigged or armed	14′6″	15′8½″	3.144	120.41	1472	11.201
21 Oct. 1870	Maplin	Not rigged or armed	13′0″	14′8″	2.997	108.26	1436	10.666
14 Aug. 1869	Maplin	Not rigged or armed	13′0″	14′2½″	2.920	100.47	1369	9.75
1 Nov. 1870	Maplin	Not rigged or armed	13′6″	14′8″	2.816	117.46	949	9.595
21 Dec. 1871	Maplin	Armed, not rigged	18′5″	10′5½″	4.900	82.098	2868	12.109
14 June 1876	Plymouth	Rigged and armed	19′7″	21′4″	4.182	86.59	1981	11.291
8 Dec. 1873	Stokes Bay	Rigged and armed	21′6″	23′6″	5.356	75.735	3177	12.456
15 Apr. 1873	At sea	Armed	26′0″	26′6″	9.118	72.832	5652	13.309
28 Feb. 1877	Inside Wight	Rigged and armed	26′3″	26′0″	9.432	63.365	8312	14.805

DIMENSIONS OF MASTS AND YARDS — APPENDIX E

Taken from the bound volumes of the official Form D-301 now held by the National Maritime Museum, Greenwich

AGINCOURT, MINOTAUR, NORTHUMBERLAND

Five masts, rigged on second class scale

Lower masts, deck to cap		
Fore and mizen	77 ft. 6 ins.	diameter 36 ins.
Second and main	83 ft. 0 ins.	diameter 37 ins.
Jigger	67 ft. 8 ins.	diameter 24 ins.
Topmasts		
Fore, second and main	55 ft. 0 ins.	diameter 18 ins.
Mizen, including pole	69 ft. 0 ins.	diameter 13½ ins.
Jigger, including pole	57 ft. 0 ins.	diameter 11 ins.
Topgallant masts, including pole		
Fore, second and main	42 ft. 0 ins.	diameter 10 ins.
Dimensions of yards		
Lower yards	91 ft. 0 ins.	diameter 22 ins.
Topsail yards	68 ft. 0 ins.	diameter 15 ins.
Topgallant yards	44 ft. 0 ins.	diameter 10 ins.
Royal yards	30 ft. 6 ins.	diameter 6 ins.
Gaffs		
Fore and second	40 ft. 6 ins.	diameter 10½ ins.
Main	48 ft. 0 ins.	diameter 12 ins.
Mizen	44 ft. 0 ins.	diameter 11 ins.
Jigger	41 ft. 0 ins.	diameter 9 ins.
Gaff topsail yards		
Mizen	22 ft. 0 ins.	diameter 7 ins.
Jigger	20 ft. 0 ins.	diameter 6 ins.
Bowsprit		
Length outboard	20 ft. 0 ins.	diameter 28 ins.
Length, including housing	45 ft. 6 ins.	
Jibboom		
Length	45 ft. 6 ins.	diameter 14½ ins.

Reduced to three masted barques.
Northumberland 1875, *Agincourt* and *Minotaur* 1893/94.

Their existing second and mizen masts lifted out, those remaining taking on the conventional names of fore mast, main and mizen. With both the *Agincourt* and *Minotaur* the only other change of note was the increase in the size of the lower yard on the main to 105 ft. 0 ins., with a diameter of 25 ins.

The changes on the *Northumberland* were more extensive. In addition to giving her a new mizen, of which both the lower mast as well as the topmast had a length of 81 ft. 0 ins., the area of canvas on her other masts was increased by giving her topmasts and yards of the same dimensions as those of the *Achilles*, as a three-master.

Gaff		
Mizen	45 ft. 0 ins.	diameter 10½ ins.
Gaff topsail yard		
Mizen	24 ft. 0 ins.	diameter 7½ ins.

ACHILLES, BLACK PRINCE, HERCULES, WARRIOR

Ship rig, on first class scale

Lower masts, deck to cap		
Foremast		
Achilles	86 ft. 3 ins.	diameter 40 ins.
Black Prince, Warrior	79 ft. 3 ins.	diameter 38 ins.
Hercules	81 ft. 0 ins.	diameter 33 ins.
Mainmasts		
Achilles, Black Prince, Warrior	86 ft. 3 ins.	diameter 40 ins.
Hercules	86 ft. 0 ins.	diameter 35 ins.
Mizen		
Achilles, Black Prince, Warrior	67 ft. 6 ins.	diameter 26 ins.
Hercules	67 ft. 0 ins.	diameter 24 ins.
Topmasts		
Fore and main	65 ft. 0 ins.	diameter 22 ins.
Mizen	50 ft. 6 ins.	diameter 16 ins.
Topgallant masts		
Fore and main	52 ft. 6 ins.	diameter 12½ ins.
Mizen	39 ft. 6 ins.	diameter 9 ins.

In later years the fore and main topmasts of the *Hercules* was increased to 65 ft. 6 ins. in length, her mizen topmast to 51 ft. 0 ins., and the diameter of her mizen topgallant mast to 10½ ins.

Dimensions of yards		
Fore and main masts		
Lower yards	105 ft. 0 ins.	diameter 25 ins.
Topsail yards	74 ft. 0 ins.	diameter 16 ins.
Topgallant yards	46 ft. 0 ins.	diameter 11 ins.
Royal yards	32 ft. 6 ins.	diameter 6½ ins.
Mizen		
Crossjack	71 ft. 0 ins.	diameter 17 ins.
Topsail yards	51 ft. 6 ins.	diameter 11½ ins.
Topgallant yards	33 ft. 6 ins.	diameter 8 ins.
Royal Yards	24 ft. 6 ins.	diameter 5 ins.
Gaffs		
Foremast		
Achilles	50 ft. 2 ins.	diameter 11½ ins.
Black Prince, Warrior	41 ft. 0 ins.	diameter 11 ins.
Hercules	43 ft. 6 ins.	diameter 11 ins.
Mainmast		
Achilles	50 ft. 2 ins.	diameter 11½ ins.
Black Prince, Warrior	49 ft. 0 ins.	diameter 11 ins.
Hercules	47 ft. 6 ins.	diameter 11½ ins.
Spanker		
Achilles	49 ft. 0 ins.	diameter 11 ins.
Black Prince. Warrior	49 ft. 0 ins.	diameter 11 ins.
Hercules	40 ft. 0 ins.	diameter 8½ ins.

Both the *Achilles* and *Hercules* had spanker booms for a time.

Spanker booms		
Achilles	70 ft. 0 ins.	diameter 16 ins.
Hercules	55 ft. 6 ins.	diameter 13 ins.
Bowsprit		
Length outboard		
Achilles	19 ft. 6 ins.	diameter 27 ins.
Black Prince, Warrior	40 ft. 2 ins.	diameter 35 ins.
Hercules	20 ft. 0 ins.	diameter 28 ins.
Length including housing		
Achilles	46 ft. 0 ins.	
Black Prince, Warrior	63 ft. 0 ins.	
Hercules	40 ft. 7½ ins.	
Jibbooms		
Achilles	63 ft. 3 ins.	diameter 16½ ins.
Black Prince, Warrior	44 ft. 6 ins.	diameter 15½ ins.
Hercules	51 ft. 5 ins.	diameter 15½ ins.
Flying jibbooms		
Black Prince, Warrior	44 ft. 9 ins.	diameter 8½ ins.
Hercules	44 ft. 0 ins.	diameter 8 ins.

When the *Achilles* was rigged with four masts, the dimensions of her bow mast and its yards were the same as those of her mizen.

CAPTAIN

Ship rig, on first class scale

Lower masts, deck to cap		
Foremast	91 ft. 6 ins.	diameter 32 ins.
Mainmast	96 ft. 0 ins.	diameter 32 ins.
Mizen	73 ft. 0 ins.	diameter 24 ins.
Tripod legs, deck to centre point of mast		
Foremast	72 ft. 8 ins.	diameter 18 ins.
Mainmast	79 ft. 9 ins.	diameter 19 ins.
Mizen	61 ft. 0 ins.	diameter 15 ins.
Topmasts		
Fore and main	63 ft. 0 ins.	diameter 22 ins.
Mizen	50 ft. 0 ins.	diameter 15 ins.
Topgallant masts		
Fore and main	51 ft. 0 ins.	diameter 13½ ins.
Mizen	40 ft. 6 ins.	diameter 9½ ins.
Dimensions of yards		
Fore and main masts		
Lower yards	103 ft. 0 ins.	diameter 24 ins.
Topsail yards	76 ft. 0 ins.	diameter 17 ins.
Topgallant yards	48 ft. 0 ins.	diameter 11 ins.
Royal yards	34 ft. 0 ins.	diameter 7 ins.
Mizen		
Crossjack	80 ft. 0 ins.	diameter 18 ins.
Topsail yard	60 ft. 0 ins.	diameter 13 ins.
Topgallant yard	37 ft. 0 ins.	diameter 9 ins.
Royal yard	26 ft. 0 ins.	diameter 6 ins.
Gaffs		
Fore and main	44 ft. 0 ins.	diameter 11 ins.
Spanker	45 ft. 0 ins.	diameter 10½ ins.
Spanker boom	60 ft. 0 ins.	diameter 15 ins.
Bowsprit		
Length outboard	20 ft. 0 ins.	diameter 24 ins.
Length with housing		'Not recorded'
Jibboom	31 ft. 0 ins.	diameter 13½ ins.
Flying jibboom	17 ft. 0 ins.	diameter 7½ ins.

MONARCH

Barque rig, on first class scale

As originally completed her upper masts, yards etc. were identical in dimensions with those of the *Achilles*. Such differences that existed were limited to her head gear and gaffs. At this period she did not have a flying jibboom. When reduced to a barque, her lower masts remained unaltered, along with her bowsprit and jibboom.

Lower masts, deck to cap		
Foremast	79 ft. 0 ins.	diameter 40 ins.
Mainmast	81 ft. 0 ins.	diameter 40 ins.
Mizen	63 ft. 0 ins.	diameter 26 ins.
Topmasts		
Fore and main	59 ft. 6 ins.	diameter 19½ ins.
Mizen, including pole	59 ft. 0 ins.	diameter 11 ins.

By about the year 1878 her mizen topmast had been lengthened to 63 ft. 6 ins.

Topgallant masts		
Fore and main	49 ft. 0 ins.	diameter 12 ins.
Dimensions of yards		
Fore and main masts		
Lower yards	96 ft. 0 ins.	diameter 23 ins.
Topsail yards	68 ft. 0 ins.	diameter 15 ins.
Topgallant yards	43 ft. 0 ins.	diameter 10 ins.
Royal yards	30 ft. 6 ins.	diameter 6 ins.
Gaffs		
Fore and main	37 ft. 6 ins.	diameter 8½ ins.
Spanker gaff	49 ft. 0 ins.	diameter 10½ ins.
Gaff topsail yard on mizen	24 ft. 0 ins.	diameter 7½ ins.
Spanker boom	63 ft. 0 ins.	diameter 16 ins.

This represents an increase in size as compared with when she was originally ship rigged. Then it was 55 ft. 6 ins. in length with a diameter of 13 ins.

Bowsprit		
Length outboard	21 ft. 0 ins.	diameter 28 ins.
Length including housing	39 ft. 3 ins.	
Jibboom	47 ft. 0 ins.	diameter 15½ ins.
Flying jibboom	44 ft. 0 ins.	diameter 8 ins.

AUDACIOUS, INVINCIBLE, IRON DUKE, VANGUARD

Ship rigged, on first class scale

Lower masts, deck to cap		
Foremast	79 ft. 0 ins.	diameter 36 ins.
Mainmast	84 ft. 6 ins.	diameter 37 ins.
Mizen	67 ft. 6 ins.	diameter 24 ins.

In later years the mast and spar records of these vessels show slight increases in the diameter of the lower masts of amounts from ⅛ inch to ½ inch.

Topmasts		
Fore and main	65 ft. 0 ins.	diameter 22 ins.
Mizen	46 ft. 6 ins.	diameter 14½ ins.
Topgallant masts		
Fore and main	49 ft. 0 ins.	diameter 12 ins.
Mizen	36 ft. 0 ins.	diameter 8½ ins.

The topmasts lengths given above represent a reduction of about 6 ft. 0 ins. as compared with when they were first completed. The *Vanguard* was further lightened aloft by receiving topmasts of only 55 ft. 6 ins. in length, with a diameter of 17½ ins. on her fore and mainmasts, and topgallants masts of 42 ft. 0 ins. with a diameter of 10 ins. Like her sisters when reduced to a barque she received a mizen topmast, with its pole of 66 ft. 0 ins. in length, and a diameter of 14 ins. Late in their careers the mizen poles were lengthened by 5 ft. 6 ins. in some of the surviving vessels.

Dimensions of yards		
Fore and main masts		
Lower yards	96 ft. 0 ins.	diameter 23 ins.
Topsail yards	71 ft. 0 ins.	diameter 15½ ins.
Topgallant yards	46 ft. 0 ins.	diameter 11 ins.
Royal yards	34 ft. 0 ins.	diameter 7 ins.
Mizen		
Crossjack	64 ft. 0 ins.	diameter 15½ ins.
Topsail yard	49 ft. 0 ins.	diameter 11 ins.
Topgallant yard	33 ft. 6 ins.	diameter 8 ins.
Royal yard	25 ft. 6 ins.	diameter 5 ins.

The fore and main upper yards of the *Vanguard* were further reduced in size to: –

Topsail yards	68 ft. 0 ins.	diameter 15 ins.
Topgallant yards	42 ft. 0 ins.	diameter 9 ins.
Royal yards	30 ft. 6 ins.	diameter 6 ins.

Towards the end of her active career, the *Audacious* also received upper yards of these dimensions.

Gaffs		
Fore and main	41 ft. 0 ins.	diameter 11 ins.
Spanker gaff	45 ft. 0 ins.	diameter 10½ ins.

About 1874 both the *Audacious* and *Invincible* were given spanker booms, 57 ft. 6 ins. and a diameter of 13½ ins. which could double up for use as a derrick.

Bowsprit		
Length outboard	24 ft. 6 ins.	diameter 25 ins.
Length with housing	43 ft. 6 ins.	

In later years the length of the housing was reduced by 2 to 3 feet.

Jibboom	52 ft. 9 ins.	diameter 13½ ins.
Flying jibboom	40 ft. 6 ins.	diameter 7 ins.

BELLEROPHON, LORD CLYDE, LORD WARDEN

Ship rigged, on second class scale

Lower masts, deck to cap		
Foremast	78 ft. 0 ins.	diameter 33 ins.
Mainmast	84 ft. 0 ins.	diameter 35 ins.
Mizen		
Bellerophon	67 ft. 0 ins.	diameter 24 ins.
Lord Clyde, Lord Warden	69 ft. 6 ins.	diameter 22 ins.
Topmasts		
Fore and main	55 ft. 0 ins.	diameter 18 ins.
Mizen	41 ft. 0 ins.	diameter 13 ins.
Topgallant masts		
Fore and main	42 ft. 0 ins.	diameter 10 ins.
Mizen	36 ft. 0 ins.	diameter 9 ins.
Dimensions of yards		
Fore and main masts		
Lower yards	91 ft. 0 ins.	diameter 22 ins.
Topsail yards	65 ft. 0 ins.	diameter 14½ ins.
Topgallant yards	41 ft. 6 ins.	diameter 10 ins.
Royal yard	30 ft. 6 ins.	diameter 6 ins.
Mizen		
Crossjack		
Bellerophon	67 ft. 6 ins.	diameter 16 ins.
Lord Clyde, Lord Warden	64 ft. 0 ins.	diameter 15½ ins.
Topsail yard		
Bellerophon	51 ft. 6 ins.	diameter 11½ ins.
Lord Clyde, Lord Warden	49 ft. 0 ins.	diameter 11 ins.
Topgallant yard	33 ft. 6 ins.	diameter 8 ins.
Royal yard	25 ft. 6 ins.	diameter 5 ins.
Gaffs		
Foremasts	41 ft. 0 ins.	diameter 10½ ins.
Mainmast		
Bellerophon	45 ft. 0 ins.	diameter 11½ ins.
Lord Clyde, Lord Warden	46 ft. 0 ins.	diameter 11½ ins.
Spanker	37 ft. 0 ins.	diameter 8½ ins.
Spanker boom	56 ft. 0 ins.	diameter 12½ ins.
Bowsprit		
Length outboard		
Bellerophon	20 ft. 0 ins.	diameter 28 ins.
Lord Clyde	20 ft. 0 ins.	diameter 28 ins.
Lord Warden	28 ft. 11 ins.	diameter 16 ins.
Length with housing		
Bellerophon	38 ft. 0 ins.	
Lord Clyde	41 ft. 0 ins.	
Lord Warden	52 ft. 7 ins.	
Jibboom		
Bellerophon	52 ft. 0 ins.	diameter 14½ ins.
Lord Clyde	61 ft. 10 ins.	diameter 14½ ins.

The *Lord Warden* did not have a jibboom, only a 4 ft. 7 ins. pole to the end of her bowsprit.

Flying jibboom		
Bellerophon	51 ft. 0 ins.	diameter 8½ ins.

CALEDONIA, OCEAN, PRINCE CONSORT, ROYAL ALFRED, ROYAL OAK, ZEALOUS.

Ship rigged, on second class scale

Although originally authorised to be rigged in a different manner, by about the year 1870 they were all rigged as follows.

Lower masts, deck to cap		
Foremast	77 ft. 6 ins.	diameter 36 ins.
Mainmast	83 ft. 0 ins.	diameter 37 ins.
Mizen	67 ft. 8 ins.	diameter 24 ins.
Topmasts		
Fore and main	55 ft. 0 ins.	diameter 18 ins.
Mizen	41 ft. 0 ins.	diameter 13 ins.
Topgallant masts		
Fore and main	42 ft. 0 ins.	diameter 10 ins.
Mizen	36 ft. 0 ins.	diameter 9 ins.
Dimensions of Yards		
Fore and main masts		
Lower yards	91 ft. 0 ins.	diameter 22 ins.
Topsail yards	68 ft. 0 ins.	diameter 15 ins.
Topgallant yards	44 ft. 0 ins.	diameter 10 ins.
Royal yards	30 ft. 6 ins.	diameter 6 ins.
Mizen		
Crossjack	64 ft. 0 ins.	diameter 15½ ins.
Topsail yard	49 ft. 0 ins.	diameter 11 ins.
Topgallant yard	33 ft. 6 ins.	diameter 8 ins.
Royal yard	25 ft. 6 ins.	diameter 5 ins.
Gaffs		
Foremast	36 ft. 10 ins.	diameter 10 ins.
Mainmast	40 ft. 6 ins.	diameter 10½ ins.
Spanker	37 ft. 0 ins.	diameter 8½ ins.
Spanker boom	56 ft. 0 ins.	diameter 12½ ins.
Bowsprit		
Length outboard	19 ft. 0 ins.	diameter 28 ins.
Jibboom	52 ft. 0 ins.	diameter 14½ ins.

Only with *Zealous* do we know all of her bowsprit measurements, but these can no doubt serve as a guide for those of her sisters.

Length outboard	18 ft. 4 ins.	diameter 28 ins.
Length with housing	34 ft. 10 ins.	
Jibboom	50 ft. 4 ins.	diameter 14½ ins.

Those vessels of this class which were rigged as barques with double topsails and roller reefing for a brief period of their careers, the proposed dimensions as set out for the *Royal Alfred* still survive at the National Maritime Museum, Greenwich, and although countermanded before they could be put into effect, give an insight to her sisters which were rigged in this manner.

Lower masts, deck to cap		
Foremast	77 ft. 6 ins.	diameter 36 ins.
Mainmast	83 ft. 0 ins.	diameter 37 ins.
Mizen	67 ft. 8 ins.	diameter 24 ins.

Topmasts, including poles		
Fore and main	80 ft. 0 ins.	diameter 21½ ins.
Mizen	62 ft. 0 ins.	diameter 14½ ins.
Dimensions of yards		
Fore and main masts		
Lower yards	96 ft. 0 ins.	diameter 23 ins.
Topsail yards	71 ft. 0 ins.	diameter 18 ins.
Upper topsail yards	50 ft. 0 ins.	diameter 12½ ins.
Gaffs		
Fore and main	33 ft. 6 ins.	diameter 9½ ins.
Spanker gaff	43 ft. 0 ins.	diameter 10½ ins.
Spanker boom	60 ft. 0 ins.	diameter 15 ins.
Gaff topsail yard	24 ft. 0 ins.	diameter 6 ins.
Bowsprit		
Length outboard	25 ft. 0 ins.	diameter 36 ins.
Length with housing	Not given	
Jibboom	70 ft. 0 ins.	diameter 16½ ins.

REPULSE

Ship rig, on second class scale

Lower Masts, deck to cap		
Foremast	74 ft. 6 ins.	diameter 36 ins.
Mainmast	80 ft. 0 ins.	diameter 36 ins.
Mizen	67 ft. 8 ins.	diameter 24 ins.
Topmasts		
Fore and main	55 ft. 0 ins.	diameter 18 ins.
Mizen	41 ft. 0 ins.	diameter 13 ins.
Topgallant masts		
Fore and main	42 ft. 0 ins.	diameter 10 ins.
Mizen	36 ft. 0 ins.	diameter 9 ins.

About 1872 the length of her mizen topgallant mast was reduced to 30 ft. 11 ins.

Dimensions of yards		
Fore and main masts		
Lower yards	91 ft. 0 ins.	diameter 22 ins.
Topsail yards	68 ft. 0 ins.	diameter 15 ins.
Topgallant yards	44 ft. 0 ins.	diameter 10 ins.
Royal yards	30 ft. 6 ins.	diameter 6 ins.

From about 1881 the lower yard on her main was increased in size to 96 ft. 0 ins. with a diameter of 23 ins.

Mizen		
Crossjack	64 ft. 0 ins.	diameter 15½ ins.
Topsail yard	49 ft. 0 ins.	diameter 11 ins.
Topgallant yard	33 ft. 6 ins.	diameter 8 ins.
Royal yard	25 ft. 6 ins.	diameter 5 ins.
Gaffs		
Foremast	36 ft. 10 ins.	diameter 10 ins.
Mainmast	40 ft. 6 ins.	diameter 10½ ins.
Spanker	37 ft. 0 ins.	diameter 8½ ins.
Spanker boom	56 ft. 0 ins.	diameter 12½ ins.
Trysail masts		
Foremast	54 ft. 6 ins.	diameter 12 ins.
Mainmast	60 ft. 0 ins.	diameter 12 ins.
Mizen	58 ft. 0 ins.	diameter 12 ins.
Bowsprit		
Length outboard	22 ft. 0 ins.	diameter 28 ins.
Length including housing	37 ft. 8 ins.	
Jibboom	52 ft. 0 ins.	diameter 14½ ins.

HECTOR, VALIANT

Barque rig, on second class scale

Lower masts, deck to cap		
Foremast	77 ft. 6 ins.	diameter 36 ins.
Mainmast	83 ft. 0 ins.	diameter 37 ins.
Mizen	67 ft. 8 ins.	diameter 24 ins.
Topmasts		
Fore and main	55 ft. 0 ins.	diameter 18 ins.
Mizen, including pole	57 ft. 0 ins.	diameter 11 ins.
Topgallant masts		
Fore and main	42 ft. 0 ins.	diameter 10 ins.
Dimensions of yards		
Fore and main masts		
Lower yards	91 ft. 0 ins.	diameter 22 ins.
Topsail yards		
Hector	65 ft. 0 ins.	diameter 14½ ins.
Valiant	68 ft. 0 ins.	diameter 15 ins.
Topgallant yards	44 ft. 0 ins.	diameter 10 ins.
Royal yards	30 ft. 6 ins.	diameter 6 ins.
Gaffs		
Foremast	36 ft. 0 ins.	diameter 10 ins.
Mainmast	40 ft. 6 ins.	diameter 10½ ins.
Spanker	37 ft. 0 ins.	diameter 8½ ins.
Topsail gaff	20 ft. 0 ins.	diameter 6 ins.
Bowsprit		
Length outboard	20 ft. 0 ins.	diameter 28 ins.
Length including housing	37 ft. 3 ins.	
Jibboom	58 ft. 4 ins.	diameter 16½ ins.

DEFENCE, RESISTANCE

Ship rig. on second class scale

Lower masts, deck to cap		
Foremast	77 ft. 6 ins.	diameter 31 ins.
Mainmast	83 ft. 0 ins.	diameter 32 ins.
Mizen		
Defence	67 ft. 8 ins.	diameter 22 ins.
Resistance	67 ft. 3 ins.	diameter 22 ins.
Topmasts		
Fore and main		
Defence	55 ft. 0 ins.	diameter 18 ins.
Resistance	55 ft. 6 ins.	diameter 17½ ins.
Mizen	41 ft. 0 ins.	diameter 12½ ins.
Topgallant masts		
Fore and main	42 ft. 0 ins.	diameter 10 ins.
Mizen	36 ft. 0 ins.	diameter 8½ ins.
Dimensions of yards		
Fore and main masts		
Lower yards	91 ft. 0 ins.	diameter 22 ins.
Topsail yards	68 ft. 0 ins.	diameter 15 ins.
Topgallant yards		
Defence	44 ft. 0 ins.	diameter 10 ins.
Resistance	42 ft. 0 ins.	diameter 9 ins.
Royal yards	30 ft. 6 ins.	diameter 6 ins.
Mizen		
Crossjack	64 ft. 0 ins.	diameter 15½ ins.
Topsail yard	49 ft. 0 ins.	diameter 11 ins.
Topgallant yard	33 ft. 6 ins.	diameter 8 ins.
Royal yard	25 ft. 6 ins.	diameter 5 ins.

Gaffs		
Fore and main	40 ft. 6 ins.	diameter 10½ ins.
Spanker	37 ft. 0 ins.	diameter 8½ ins.
Spanker boom	56 ft. 0 ins.	diameter 12½ ins.
Bowsprit		
Length outboard		
Defence	21 ft. 0 ins.	diameter 28 ins.
Resistance	20 ft. 0 ins.	diameter 26 ins.
Length including housing		
Defence	40 ft. 0 ins.	
Resistance	41 ft. 0 ins.	
Jibboom		
Defence	59 ft. 6 ins.	diameter 14½ ins.
Resistance	45 ft. 6 ins.	diameter 14½ ins.
Flying jibboom		
Resistance	48 ft. 0 ins.	diameter 9 ins.

When the *Resistance* was finally reduced to a barque, the alterations were made to her head gear as well as her mizen. Her flying jibboom was discarded, and her new bowsprit was: –

Length outboard	21 ft. 0 ins.	diameter 28 ins.
Length including housing	40 ft. 0 ins.	
Jibboom	55 ft. 6 ins.	diameter 14 ins.
Mizen		
New topmast	57 ft. 0 ins.	diameter 12½ ins.
Topsail gaff	20 ft. 0 ins.	diameter 6 ins.

Both her spanker gaff and spanker boom remained unaltered.

FAVORITE, PALLAS, PENELOPE

Ship rig, on third class scale

Lower masts deck to cap		
Foremast		
Favorite	60 ft. 0 ins.	diameter 27 ins.
Pallas	59 ft. 0 ins.	diameter 28 ins.
Penelope	61 ft. 0 ins.	diameter 28 ins.
Mainmast		
Favorite	66 ft. 0 ins.	diameter 28 ins.
Pallas	64 ft. 0 ins.	diameter 30 ins.
Penelope	66 ft. 0 ins.	diameter 30 ins.
Mizen		
Favorite	53 ft. 5 ins.	diameter 20 ins.
Pallas, Penelope	52 ft. 0 ins.	diameter 21 ins.
Topmasts		
Fore and main	48 ft. 6 ins.	diameter 18 ins.
Mizen		
Favorite, Pallas	37 ft. 6 ins.	diameter 13 ins.
Penelope	38 ft. 6 ins.	diameter 13 ins.

During her 1878 refit at Chatham, the *Penelope* was given topmasts on her fore and mainmasts of 50 ft. 0 ins. in length.

Topgallant masts		
Fore and main	42 ft. 0 ins.	diameter 10 ins.
Mizen		
Pallas, Penelope	32 ft. 6 ins.	diameter 8 ins.
Favorite	33 ft. 2 ins.	diameter 8 ins.
Dimensions of yards		
Fore and main masts		
Lower yards	78 ft. 6 ins.	diameter 19 ins.
Topsail yards	59 ft. 0 ins.	diameter 13 ins.
Topgallant yards	40 ft. 0 ins.	diameter 9½ ins.
Royal yards	29 ft. 6 ins.	diameter 6 ins.
Mizen		
Crossjack	55 ft. 0 ins.	diameter 13 ins.
Topsail yard	43 ft. 0 ins.	diameter 9½ ins.
Topgallant yard	29 ft. 6 ins.	diameter 7½ ins.
Royal yard	23 ft. 0 ins.	diameter 4½ ins.

During her 1888 refit at Portsmouth, the *Penelope* lost all her royal yards.

Gaffs		
Fore and main		
Favorite	36 ft. 4 ins.	diameter 9½ ins.
Pallas, Penelope	35 ft. 0 ins.	diameter 9½ ins.
Spanker		
Favorite	36 ft. 3 ins.	diameter 8 ins.
Pallas, Penelope	35 ft. 6 ins.	diameter 9½ ins.
Spanker boom		
Favorite	48 ft. 6 ins.	diameter 11½ ins.
Pallas, Penelope	53 ft. 0 ins.	diameter 12½ ins.
Trysail gaff		
Favorite	15 ft. 3 ins.	diameter 6 ins.
Penelope	24 ft. 6 ins.	diameter 8 ins.

During her 1878 refit, the *Penelope* is shown as having trysail masts,

Foremast	43 ft. 0 ins.	diameter 10¼ ins.
Mainmast	46 ft. 6 ins.	diameter 10¼ ins.
Nizen	44 ft. 0 ins.	diameter 10¼ ins.
Bowsprit		
Length outboard		
Favorite	16 ft. 0 ins.	diameter 24 ins.
Pallas	16 ft. 6 ins.	diameter 24 ins.
Penelope	25 ft. 0 ins.	diameter 24 ins.
Length including housing		
Favorite	33 ft. 3 ins.	
Pallas	31 ft. 6 ins.	
Penelope	38 ft. 7 ins.	
Jibboom		
Favorite	48 ft. 6 ins.	diameter 12 ins.
Pallas	40 ft. 6 ins.	diameter 11½ ins.
Penelope	30 ft. 8 ins.	diameter 11½ ins.
Flying jibboom		
Pallas	40 ft. 0 ins.	diameter 8 ins.
Penelope	33 ft. 9 ins.	diameter 6¾ ins.

During her 1878 refit, the jibboom of the *Penelope* was lengthened to 32 ft. 1½ ins., and at the same time her flying jibboom to 34 ft. 6 ins.

TEMERAIRE

Brig, rig, on first class scale

Lower masts, deck to cap		
Foremast	83 ft. 6 ins.	diameter 40 ins.
Mainmast	89 ft. 6 ins.	diameter 40 ins.
Topmasts		
Fore and main, as designed	55 ft. 0 ins.	diameter 23½ ins.
as completed 1877	56 ft. 0 ins.	diameter 23½ ins.
Topgallant masts		
Fore and main, as designed	55 ft. 0 ins.	diameter 12½ ins.
as completed 1877	49 ft. 0 ins.	diameter 12½ ins.

Dimensions of yards
As originally designed

Lower yards	111 ft. 0 ins.	diameter 26½ ins.
Topsail yards	78 ft. 0 ins.	diameter 16½ ins.
Topgallant yards	56 ft. 6 ins.	diameter 12½ ins.
Royal yards	43 ft. 0 ins.	diameter 10 ins.

As first completed in 1877 her yards were modified as follows.

Lower yards	111 ft. 0 ins.	diameter 26½ ins.
Topsail yards	78 ft. 6 ins.	diameter 16½ ins.
Topgallant yards	49 ft. 0 ins.	diameter 11 ins.
Royal yards	34 ft. 0 ins.	diameter 7½ ins.

During the early 1890s when she had been reduced to harbour service further changes were made to the dimensions of her upper masts and her yards.

Gaffs		
Fore and main	49 ft. 0 ins.	diameter 11 ins.

In later years the diameters were reduced to 10 ins, while the gaff on her main was lengthened to 50 ft. 0 ins.

Bowsprit		
Length outboard	20 ft. 0 ins.	diameter 28 ins.
Length including housing	39 ft. 0 ins.	

Later in her career the total length of her bowsprit was increased by about 1 ft. 10 ins.

Jibboom	51 ft. 6 ins.	diameter 15¾ ins.

Originally it was proposed that she should have both a jibboom and flying jibboom, but these were discarded for the one given above.

Jibboom, as proposed	52 ft. 0 ins.	diameter 15 ins.
Flying jibboom, as proposed.	44 ft. 0 ins.	diameter 8 ins.

ALEXANDRA

Barque rig, on first class scale

Lower masts, deck to cap		
Foremast	77 ft. 6 ins.	diameter 35 ins.
Mainmast	86 ft. 6 ins.	diameter 37 ins.
Mizen	75 ft. 6 ins.	diameter 25 ins.
Topmasts		
Fore and main	53 ft. 0 ins.	diameter 21 ins.
Mizen, including pole	78 ft. 0 ins.	diameter 17 ins.
Topgallant masts		
Fore and main	42 ft. 8 ins.	diameter 12 ins.
Dimensions of yards		
Fore and main masts		
Lower yards	105 ft. 0 ins.	diameter 25 ins.
Topsail yards	78 ft. 0 ins.	diameter 16½ ins.
Topgallant yards	52 ft. 0 ins.	diameter 12 ins.
Royal yards	38 ft. 0 ins.	diameter 7½ ins.
Gaffs		
Fore and mainmast	39 ft. 0 ins.	diameter 9½ ins.
Spanker	45 ft. 0 ins.	diameter 10½ ins.
Gaff topsail yard	24 ft. 0 ins.	diameter 9½ ins.
Bowsprit		
Length outboard	23 ft. 0 ins.	diameter 29 ins.
Length including housing	42 ft. 0 ins.	
Jibboom	64 ft. 0 ins.	diameter 16 ins.
Flying jibboom	62 ft. 9 ins.	diameter 8½ ins.

When modernised in 1890, her bowsprit and other head gear was removed, and her lower masts shortened to,

Foremast	63 ft. 3 ins.
Mainmast	69 ft. 0 ins.
Mizen	60 ft. 3 ins.

Existing topmasts and topgallant masts removed, and replaced with pole topmasts.

Foremast	50 ft. 0 ins.	diameter 12 ins.
Mainmast	63 ft. 0 ins.	diameter 15 ins.
Mizen	45 ft. 6 ins.	diameter 12 ins.

All her yards were replaced by lighter signal yards. Initially, two yards on her mainmast, and a single yard on both her foremast and mizen. By the end of the century, a second yard had been added to both these masts as well.

Signal yards		
Foremast	50 ft. 0 ins.	diameter 12 ins.
Mainmast, lower yard	50 ft. 0 ins.	diameter 12 ins.
upper yard	31 ft. 0 ins.	diameter 8 ins.
Mizen	45 ft. 0 ins.	diameter 10 ins.
Gaffs		
Mainmast, and mizen	25 ft. 0 ins.	diameter 9 ins.

Other additions included military tops to both foremast and mizen 16 ft. 8 ins. diameter, with a depth of 2 ft. 8 ins. A steel derrick 41 ft. 4 ins. with a diameter of 15 ins. was added abaft her foremast, while a similar one was added to her main towards the end of the century, and at the same time the gaff was removed from this mast. By 1894 she received an early form of mast head semaphore, carried low on her main topmast, but before the end of the century these had been substituted for a set of the larger kind with 12 feet long arms at the truck of her main topmast. In 1899, she had the distinction of having a wireless gaff added to her mizen topmast, the first ship in the Royal Navy to be so fitted.

CERBERUS

Jury rig, as barque for voyage to Australia.

In the normal course of events she would not justify mention as a rigged vessel, but as she was given temporary masting, and a modest spread of canvas for her voyage down under, a record of their dimensions may not be out of place.

Lower masts, deck to cap		
Foremast	64 ft. 0 ins.	diameter 23 ins.
Mainmast	71 ft. 0 ins.	diameter 24 ins.
Mizen	53 ft. 6 ins.	diameter 15 ins.
Topmasts		
Fore and main	46 ft. 0 ins.	diameter 14 ins.
Mizen, including pole	47 ft. 0 ins.	diameter 9½ ins.
Topgallant masts		
Fore and main	19 ft. 6 ins.	diameter 9 ins.
Dimensions of yards		
Fore and main masts		
Lower yards	67 ft. 0 ins.	diameter 14 ins.
Topsail yards	49 ft. 6 ins.	diameter 10 ins.
Topgallant yards	32 ft. 0 ins.	diameter 6½ ins.
Gaffs		
Fore and mainmasts	32 ft. 0 ins.	diameter 8½ ins.
Spanker	34 ft. 0 ins.	diameter 8½ ins.
Topsail gaff yard on mizen	17 ft. 0 ins.	diameter 4¾ ins.
Bowsprit		
Length outboard	12 ft. 0 ins.	diameter 14 ins.
Length, including housing	24 ft. 0 ins.	
Jibboom	33 ft. 0 ins.	diameter 10 ins.

APPENDIX F

A SUMMARY OF THE MORE IMPORTANT ALTERATIONS MADE TO THE VESSELS BY THE YEAR 1870

MINOTAUR	Before completion, poop and forecastle added. Total change of armament, extra fittings for fire service, and fitted for a flagship.
AGINCOURT	Before completion, poop and forecastle added. Total change of armament, extra fittings for fire service.
NORTH-UMBERLAND	While building important alteration in armour plating, change of armament, and various fittings.
ACHILLES	While building important addition to armour plating and greatly increased defensive power. Armament entirely changed. Modification of rig.
WARRIOR	Small alterations, additional fittings, reduction of bowsprit, partial change of armament.
BLACK PRINCE	Small alterations, additional fittings, reduction of bowsprit, partial change of armament.
BELLEROPHON	Light deck added while building. Alterations for accommodation of officers and men. Armament fittings altered.
HERCULES	
LORD CLYDE	Alterations to armour plating, equipment and armament while building. Fitted for a flagship.
LORD WARDEN	Alterations to armour plating, equipment and armament while building.
MONARCH	
CAPTAIN	
PRINCE CONSORT	Alterations in fittings, modification in rig. Scuttles and scuppers added, change of armament, forecastle added.
CALEDONIA	Fitted for a flagship. Modification in rig. Scuppers, scuttles, poop and forecastle added, change in armament.
OCEAN	Alterations in fittings, modifications in rig, poop added, change of armament.
ROYAL OAK	Alterations in fittings, modification in rig, forecastle added, scuttles and scuppers added.
ROYAL ALFRED	Alterations in armour plating, armament etc. Poop and forecastle added. Fitted for a flagship.
ZEALOUS	Fitted for a flagship. Armament changed, poop and forecastle added.
REPULSE	
HECTOR	Alteration in fittings, change of armament, alteration to rig.
VALIANT	Alteration in fittings, change of armament, alteration to rig.
DEFENCE	Small alterations, additional fittings, modification to rig, forecastle added, armament changed.
RESISTANCE	Small alterations, additional fittings, modification to rig.
AUDACIOUS	Transverse armour-plated bulkheads to upper deck battery added, bottom sheathed with wood.
INVINCIBLE	Transverse armour-plated bulkheads to upper deck battery.
IRON DUKE	Transverse armour-plated bulkheads to upper deck battery.
VANGUARD	Transverse armour-plated bulkheads to upper deck battery.
FAVORITE	Alteration of armament, scuttles added.
PALLAS	Light deck added while building.
PENELOPE	
TEMERAIRE	
ROYAL SOVEREIGN	Several changes made since first designed, turret gear altered.
PRINCE ALBERT	Several changes made since first designed, turret gear altered.

APPENDIX G

SOURCES FOR 'THE BLACK BATTLE FLEET'

The Editors hope that Admiral Ballard's accounts of these vessels will encourage readers to explore even more widely the subject of the early and mid Victorian navy. They are well aware of the difficulties which beset those living out of easy reach to London with its museums and libraries in obtaining suitable reference material, but provincial libraries do have means of borrowing from a central source specialised books for relatively short periods, although some times there is a long waiting period for a particular publication to become available.

For those able to visit London are almost endless possibilities for research, and perhaps pride of place must go to the National Maritime Museum at Greenwich. Here they have one of the largest libraries of maritime books in the country, to which access can be obtained by advance application in writing for a Reader's Ticket. Its Admiralty Collection of ship plans not only includes all the vessels dealt with in this book, but many more beside. The Museum's Draught Room staff would be pleased to provide information as to what is available concerning individual vessels, and to provide details of the Museum's plan copying service, and of which plans are available through it. Also held by the Draught Room are the so called 'Ships Covers', which are in effect the Director of Naval Construction's bound files of documents relating to the design, building and subsequent alterations to a class of ships. The earliest surviving cover, and dealing with one of the ships mentioned in this book, is that on the *Glatton* of 1871, but it should be said that there is very little material in many of the early covers. The Museum also owns a number of models of these vessels, chiefly official half block models built to a scale of 1:48. Finally there are the Museum's collections of oil paintings, prints, drawings and photographs totalling many thousands of pictures. Provided a negative exists within the Museum, there is usually no difficulty to supplying further prints to order.

Because the volume of material now owned by the Museum has grown so considerably over recent years, much of it has to be kept in store well away from the Museum, so, if the intending visitor is not to be disappointed, it is essential that an appointment be arranged with the department concerned well in advance of the date of the intended visit.

Next in importance to the researcher is the Science Museum at South Kensington, where they have a very fine collection of ship models, which extends to include marine engines, boilers, propellers, and auxiliary machinery. They too have a very fine library, but in their case, maritime matters form only a very small section of it.

Then there is the Imperial War Museum in Lambeth. While they are primarily concerned with the events of the two World Wars, during the 1920s, the late Doctor Oscar Parkes discovered two very old collections of photographs, the work of Symonds of Portsmouth, and Long of Plymouth, hidden away in a loft, and persuaded that Museum to purchase them, with the result they now own the finest collection of Victorian warship photographs in the country. Visitors to their Photograph Library will only be seen by appointment, which can be made either by letter or telephone.

Although London has so much to offer, the provincial museums, especially those cities and towns with strong maritime connections, be it ship owning, or ship building, can produce some pleasant surprises, and for all those to whom London is too remote, we would like to offer the suggestion that you might find what you want right on your own door-step. Nor should it be forgotten that almost all Museums have more material than they can hope to exhibit, so there is the slender chance that the item you are seeking could be tucked away in a vault.

G.A.O. N.A.M.R.

FOOTNOTES TO THE CHAPTERS

CHAPTER ONE

1 page 17 In practice they nearly always made an Atlantic passage under sail.
2 page 18 A minor difference also existed in the proportions of the lowermast heads.
3 page 21 But see page 164 for an exception.

CHAPTER TWO

1 page 29 It was conventional for the flagshig to anchor at the head of the lee line, therefore lying nearest to the shore.

CHAPTER THREE

1 page 45 The corvette *Bacchante*, with the future King George V on board as a naval cadet, also took part in this cruise, through not included in the divisional organisation.
2 page 46 During her commission in the Mediterranean the *Achilles* had 12 fatal accidents, involving the deaths of a petty officer, a leading seaman, 6 A.B.'s and 4 o.s.

CHAPTER FOUR

1 page 50 The loss of the *Birkenhead*, one of these converted frigates, became one of the most celebrated Victorian disasters.
2 page 51 i.e. line-of-battleship.
3 page 56 Although officially known as *C77*, to those at the oil depot she is still the *Warrior*. The government announced in the Commons on 12 June 1979 that they would hand her over to the Maritime Trust for preservation.

CHAPTER SIX

1 page 78 An alloy of 40 per cent. zinc and 60 per cent. copper.

CHAPTER EIGHT

1 page 106 She ran steaming trials as a three-decker in 1858, although not rigged or armed.

CHAPTER ELEVEN

1 page 145 This was changed in 1866 to a two-bladed one before she sailed to the Pacific.
2 page 147 Better known today by the Spanish form of the name, Punta Arenas.
3 page 153 Or three times if one includes the nuclear submarine *Repulse* as a capital ship.

CHAPTER FIFTEEN

1 page 188 Shadwell was not only a well known eccentric, but a leading practical astronomer and mathematician, whence his *Star Tables* and other aids to navigation.

CHAPTER SIXTEEN

1 page 192 The other members of it were the armoured sloops *Research* and *Enterprise* and the armoured gunboats *Viper*, *Vixem* and *Waterwitch*.
2 page 195 A class of ten 22-gun corvettes, all to have been built in Royal Dockyards, of 1857 tons burthen, 225 feet long.

INDEX OF SHIPS' NAMES

(Principal references only)
Note: Page numbers in **bold type** refer to the general history of that ship.

GENERAL INDEX

Note: Having regard to Admiral Ballard's very detailed recital of the design, armour, armament, machinery, masts and spars and sails of each ship, coupled with the very detailed appendices, this Index is not intended to reflect items so dealt with.

This book is published under the direction of
AMI GUICHARD

Editorial responsibility and supervision by Tim Chilvers

Produced under the direction of Charles Riesen

Designed by Max Thommen

Printed by GEA, Milan
and bound by Maurice Busenhart, Lausanne

Printed in Italy